AF679130

Developing Eco-Cities Through Policy, Planning, and Innovation:

Can It Really Work?

Information Resources Management Association
USA

A volume in the Trending Topics
Book Series (TTBS) Book Series

IGI Global
DISSEMINATOR OF KNOWLEDGE

Published in the United States of America by
IGI Global
Engineering Science Reference (an imprint of IGI Global)
701 E. Chocolate Avenue
Hershey PA, USA 17033
Tel: 717-533-8845
Fax: 717-533-8661
E-mail: cust@igi-global.com
Web site: http://www.igi-global.com

Copyright © 2020 by IGI Global. All rights reserved. No part of this publication may be reproduced, stored or distributed in any form or by any means, electronic or mechanical, including photocopying, without written permission from the publisher.
Product or company names used in this set are for identification purposes only. Inclusion of the names of the products or companies does not indicate a claim of ownership by IGI Global of the trademark or registered trademark.

Library of Congress Cataloging-in-Publication Data

Names: Information Resources Management Association, editor.
Title: Developing eco-cities through policy, planning, and innovation : can it really work? / Information Resources Management Association, editor.
Description: Hershey, PA : Engineering Science Reference, [2020] | Includes bibliographical references and index. | Summary: "This book examines the economic, political, social, and environmental objectives essential to the planning and support of future communities"--Provided by publisher"-- Provided by publisher.
Identifiers: LCCN 2019023048 (print) | LCCN 2019023049 (ebook) | ISBN 9781799804413 (hardcover) | ISBN 9781799804420 (paperback) | ISBN 9781799804437 (ebook)
Subjects: LCSH: Municipal engineering. | City planning. | Urban ecology (Sociology)
Classification: LCC TD159 .D48 2020 (print) | LCC TD159 (ebook) | DDC 628--dc23
LC record available at https://lccn.loc.gov/2019023048
LC ebook record available at https://lccn.loc.gov/2019023049

This book is published in the IGI Global book series Trending Topics (TTBS) (ISSN: pending; eISSN: pending)

British Cataloguing in Publication Data
A Cataloguing in Publication record for this book is available from the British Library.

The views expressed in this book are those of the authors, but not necessarily of the publisher.

For electronic access to this publication, please contact: eresources@igi-global.com.

Trending Topics Book Series (TTBS)

Mehdi Khosrow-Pour, D.B.A.
Information Resources Management Association, USA

ISSN:pending
EISSN:pending

MISSION

Every day all over the world, researchers are making groundbreaking discoveries that address current challenges and solve complex problems across a variety of fields. Due to advancements in technology, changes in demographics, adjustments to societal norms, the development of political and economic structures, and more, new theories, methods, strategies, techniques, and applications are constantly being established in response to both existing and evolving issues.

Based on the success of IGI Global's latest Trending Topics Campaign, which provides news snippets and trending research breakthroughs relevant to IGI Global content, IGI Global has launched this new book series, titled the **Trending Topics Book Series**, which offers small anthologies of reprinted IGI Global book chapters and journal articles hand-selected by IGI Global's Executive Editorial Board on topics currently trending and stirring up discussion and/or controversy.

Spanning across topics such as the multigenerational workforce, legislation for sustainability, emotional branding, privacy on mobile devices, robotics in healthcare, and bio-based product development, this book series aims to provide researchers, academicians, professionals, and students with the most up-to-date studies in order to advance knowledge across all industries.

COVERAGE

- Library Budget Challenges
- Privacy on Mobile Devices
- Environmental Legislation and Sustainability
- Multi-Generational Workforce
- Bio-Based Product Development
- Aging Workforce
- Emotional Branding
- Inclusive Learning
- Eco-Cities
- Plan S and Open Access Movements

IGI Global is currently accepting manuscripts for publication within this series. To submit a proposal for a volume in this series, please contact our Acquisition Editors at Acquisitions@igi-global.com or visit: http://www.igi-global.com/publish/.

The Trending Topics Book Series (TTBS) Book Series (ISSN pending) is published by IGI Global, 701 E. Chocolate Avenue, Hershey, PA 17033-1240, USA, www.igi-global.com. This series is composed of titles available for purchase individually; each title is edited to be contextually exclusive from any other title within the series. For pricing and ordering information please visit http://www.igi-global.com/book-series/trending-topics-book-series/228608. Postmaster: Send all address changes to above address. Copyright © 2020 IGI Global. All rights, including translation in other languages reserved by the publisher. No part of this series may be reproduced or used in any form or by any means – graphics, electronic, or mechanical, including photocopying, recording, taping, or information and retrieval systems – without written permission from the publisher, except for non commercial, educational use, including classroom teaching purposes. The views expressed in this series are those of the authors, but not necessarily of IGI Global.

Table of Contents

Preface

The rampant use of artificial resources and chemically engineered matter in all aspects of daily life, from mass transit to construction, is causing lasting environmental damage to the atmosphere and the world at large. Not only are they threatening the sustainability of natural resources, but they are also creating an abundance of public health concerns.

Seeking a solution for these growing ecological problems can be quite costly, hindering economic growth and prosperity. The public sector has a responsibility to stimulate the marketplace in favor of more resource-efficient and less-polluting goods, services, and works in order to support environmental and wider sustainable development objectives. Eco-cities, human-made communities built on self-sustainable models using the natural ecosystems around the area, are emerging around the globe in response to environmental crises and are creating ecologically-healthy cities to combat the expanding environmental issues.

Urban planners and city designers focus their design efforts by connecting urban planning with ecology, public participation, and green legislative policies to create environmentally-healthier cities through means such as planting trees along sidewalks, building solar greenhouses, and encouraging public transportation. By living in harmony with the nature around society, humanity greatly reduces their ecological footprint left behind. While numerous organizations have different definitions of what makes an eco-city, some standard criteria include but are not limited to utilizing local resources in a self-contained economy, promoting renewable energy, supporting walkability and bikeable paths to reduce carbon emission, maximizing water efficiency, managing recycling, restoring environmentally-damaged areas within the community, and ensuring affordable housing for all income levels.

While the global goal should be to convert existing cities into eco-cities while also designing new areas, budgetary issues and existing infrastructure make this task daunting and highly difficult. However, by tackling some smaller aspects, like planting more trees and promoting recycling in public areas, local governments can still enhance these locations into more sustainable areas. Major eco-cities in the world including Reykjavik, Iceland; Portland, USA; Vancouver, Canada; and Copenhagen,

Denmark remain sustainable and green by focusing on public transportation, green spaces, treating soil contaminations, and other small changes that yield large results.

In response to the need for advanced research that can address these current issues and offer innovative solutions, this compilation, *Developing Eco-Cities Through Policy, Planning, and Innovation: Can It Really Work?* intends to adequately address the issue of eco-city planning including new city layouts, logistical analyses to improve existing spaces, and often overlooked areas such as green waste management and sustainable transportation.

Serving as an anthology of reprinted book chapters and journal articles in the areas of environmental science and urban land use planning, this reference source represents content that has been hand-selected by IGI Global's executive editorial board and is sure to appeal to urban planners, city designers, legislators, architects, researchers, policymakers, and government officials as these research findings will equip them with the necessary knowledge to adequately design and plan new eco-cities while also functionally updating existing cities to become more sustainable. Cities can thrive and the inhabitants will become more empowered with a better understanding of modern green issues taking place in urban areas.

This book is the second release within IGI Global's Trending Topics Book Series, which aligns with IGI Global's weekly trending topics campaign that showcases the latest research breakthroughs that are stirring up discussion and/or controversy in a variety of fields. It centers on conceptual, methodological, and technical aspects and provides insight into emerging topics including but not limited to environmental sustainability, modernizing public spaces, smart cities, waste management, ecosystem services, sustainable energy consumption, and urban green spaces.

Developing Eco-Cities Through Policy, Planning, and Innovation: Can It Really Work? is organized into three sections that provides coverage of several important topics. The sections include:

1. Environmental Policies and Legislation
2. Green Technologies and Innovations
3. Sustainable Urban Development and Planning

The following paragraphs prove a summary of what to expect from this invaluable reference source:

Section 1, "Environmental Policies and Legislation," opens this detailed reference source by addressing the latest challenges city planners and local governments must overcome when designing eco-cities and creating policies that support a green public space. In the first chapter of this section, "Conceptual Foundations of Creating Sustainable Development Strategy of Smart Cities: Environmental Aspect,"

Prof. Olga Burmatova from the Institute of Economics and Industrial Engineering of the Siberian Branch of the RAS, Russia suggests cities build environmental strategies, such as the environmental mission, vision of the future, goals and priorities, implementation, and result assessment, as a long-term landmark of their sustainable development. The second chapter of this section, "The Development of the Environmental Impact Assessment Process in Botswana," by Prof. Eagilwe M. Segosebe from the University of Botswana, Botswana investigates the steps Botswana has taken since the 1970s to become an environmentally-sustainable country, like prioritizing environmental protection efforts. In the following chapter, "Modeling the Role of Government, Firm, and Civil Society for Environmental Sustainability," the authors, Prof. Humaira Yasmeen, Prof. Ying Wang, and Prof. Hashim Zameer from Nanjing University of Aeronautics and Astronautics, China and Prof. Hina Ismail from NUML Multan Campus, Pakistan, describe a game-theory model design to outline the role of government, firm, and civil society for environmental sustainability using cost analysis as a factor for governments making sustainability decisions. In a concluding chapter, "Reverse Logistics and Solid Waste: Challenges for the National Waste Policy (PNRS) in Brazil," Prof. Hermes de Andrade Júnior from Catholic University, Portugal promotes a selection of work collected over 19 years that analyzes the need and evolution of reverse logistics into the context of the national policy on solid waste in Brazil. In the final chapter of this section, "Environmental Reporting and Accounting," the authors, Prof. Radiah Othman, Prof. Nirmala Nath, and Prof. Fawzi Laswad from Massey University, New Zealand, examine the context in which "accounting-sustainability hybrids" were constructed in anticipation of the Environmental Reporting Act (ERA) passed in 2015 in New Zealand. The authors highlight how legislation facilitates action in sustainability policies.

Section 2, "Green Technologies and Innovations," offers chapters that discuss emerging technologies that support smart cities and renewable energy applications. In the first chapter of this section, "Leveraging Internet of Things to Revolutionize Waste Management," Prof. Mirjana Maksimovic from the University of East Sarajevo, Bosnia and Herzegovina analyzes the role of a smart and safe IoT-powered waste management system, highlights its benefits, and possibilities of implementation and evaluation. In the following chapter, "Analysis of the Nexus Between Smart Grid, Sustainable Energy Consumption, and the Smart City," the authors, Prof. Luke A. Amadi and Prof. Prince I. Igwe from the University of Port Harcourt, Nigeria, investigate the nexus between smart grid, sustainable energy consumption, and the smart city and look to connect electricity with sustainable energy consumption. In another noteworthy chapter, "Sustainability Evaluation of Green Urban Logistics Systems: Literature Overview and Proposed Framework," Prof. Jesus Gonzalez-Feliu from Ecole des Mines de Saint-Etienne, France proposes a methodological framework to use and combine existing methods of urban logistics assessment and evaluation to assess scenarios, and not a "black-box" model of software ready to use. In the final chapter of this section, "Operationalising Sustainability Within Smart Cities: Towards an Online Sustainability Indicator Tool," the authors, Prof. Conor Mark Dowling, Prof. Stephen Diarmuid Walsh, Prof. Stephen M. Purcell, and Prof. William M. Hynes from Future Analytics, Ireland and Prof. Mary Lee Rhodes from Trinity College, Ireland, devise sustainable energy indicators designed to be flexible enough to address the vast extent of urban sustainability processes and examine how these indicators need to be designed to reflect the broad range of data sources that reflect the three pillars of sustainability, namely environmental, economic, and social elements.

Section 3, "Sustainable Urban Development and Planning," presents coverage on strategies that companies and city planners can utilize to promote effective city layouts and urban development for both existing spaces and newly designed areas. In the first chapter, "Developing á Sustainable Eco-City in Pre-Olympic Tokyo: Potential of New Methods and Their Limits in an Urban Era," Prof. Mariko Ikeda of the University of Tsukuba, Japan elucidates the emergence of sustainable urban development in Tokyo in light of the upcoming 2020 Olympics by exploring various administrative and commercial practices, such as an urban development plan with rooftop and wall greening or river projects in the inner city. In the second chapter, "The Role of Urban Green Spaces in the Transformation of Community Ecosystem in Developing Countries," the authors, Prof. José G. Vargas-Hernández from the University of Guadalajara, Mexico; Prof. Justyna Anna Zdunek-Wielgołaska from Warsaw University of Technology, Poland; and Prof. Karina Pallaggst from Technische Universität Kaiserslautern, Germany, analyze some important features of urban green spaces such as the components, functions, services, community involvement, initiatives, and actions from an ecosystem perspective. The analysis begins from the assumption that urban green spaces are ecosystems of vital importance in enhancing the quality of life in an urban environment and supplying ecosystem services such as biodiversity and climate regulation. Meeting the needs of users is related with the functions and services that urban green spaces provide to communities. The methods employed in this analysis are the empirical literature and documents review, analysis of existing data on uses and users, interviews with authorities, and more detailed examination of a case specific data. Also, as some concluding remarks, some wider environmental, economic, and social initiatives for local authorities and communities are suggested that can justify initiatives by all represented and involved stakeholders. In the final chapter of this section, "Assessing Urban Ecosystem Services: Different Methodological Approaches Applied in Brazil, Germany, and Portugal," the authors, Prof. Marise Barreiros Horta from the Federal University of Minas Gerais, Brazil; Prof. Maria Inês Cabral from

Martin Luther University, Germany & iDiv, Germany; Prof. Iva Pires from Nova University of Lisbon, Portugal; Prof. Laura Salles Bachi from Federal University of Minas Gerais, Brazil; Prof. Ana Luz from Nova University of Lisbon, Portugal; Prof. Gerardo Wilson Fernandes from Federal University of Minas Gerais, Brazil; Prof. Maria Auxiliadora Drumond from Federal University of Minas Gerais, Brazil; and Prof. Sónia Carvalho-Ribeiro from Federal University of Minas Gerais, Brazil, summarize different participatory approaches for assessing ecosystem services in urban areas of three countries: Brazil, Germany, and Portugal. Each country and study demonstrate the potential benefits of utilizing the ES assessment approaches on urban landscapes, especially for better understanding the interactions between people and nature in urban sites.

Although the primary organization of the contents in this work is based on its three sections, offering a progression of coverage on important concepts, methodologies, technologies, applications, social issues, and emerging trends, the reader can also identify specific contents by utilizing the extensive indexing system listed at the end of this volume.

Section 1
Environmental Policies and Legislation

Chapter 1
Conceptual Foundations of Creating Sustainable Development Strategy of Smart Cities:
Environmental Aspect

Olga Burmatova
Institute of Economics and Industrial Engineering of the Siberian Branch of the RAS, Russia

ABSTRACT

This chapter is devoted to the study of the role of ecological subsystem in the structure of the sustainable development program of smart city. The author suggests the logic of building the environmental strategy of the city as a long-term landmark of its sustainable development including the environmental mission, vision of the future, goals and priorities, programs and their implementation, target indicators for assessing results, and consequences of realization programs. Certain attention is paid to the city as an object of research with a focus on environmental problems. The characteristics of the factors affecting the development of the ecological situation in the city are shown. A system of criteria and indicators that can be used to assess the impact of the planned environmental activities is proposed.

DOI: 10.4018/978-1-7998-0441-3.ch001

Copyright © 2020, IGI Global. Copying or distributing in print or electronic forms without written permission of IGI Global is prohibited.

INTRODUCTION

Smart City Concept combines various factors of development of the city into one system, including economics, management, energy, transport, environment and population. At the same time it relies on the increased role of human capital and the strengthening of the importance of information technology in the urban environment. The obligatory feature of any Smart City is its long-term sustainable development that can be achieved, if the balance of the three spheres - economic, social and environmental – is enforced complied. That means that any administrative decision shall be without prejudice to any of the components of the city subsystems - society, the economy, environmental conditions, etc. Therefore, for a Smart City is extremely important to have effective management and analytical tools to promptly predict potential negative externalities, as much as possible to internalize them and manage them. One of these tools is able to serve, in particular, strategic planning and management. More and more cities are on the path of development strategy, which is based on the concept of a better future and available opportunities and threats.

Obviously, the easiest way to create a Smart City from the ground up on a single project. It is much more difficult to transform an existing city, especially large one, in the Smart City. Most often within existing cities projects for creating "smart" blocks are being developed and being implemented, and these or other elements of the "smart" cities are gradually introduced. In world practice of the functioning and development of cities such attempts are quite numerous (mart Cities, 2014; Hollands, 2008; Songdo, n.d.; In Japan, 2014). For example, Vienna, Barcelona and Copenhagen are included to the category of Smart Cities.

Usually in existing cities the idea of "smart" cities are being introduced gradually and consistently, from "smart" solutions to improve the urban environment and the mobility of the population with further coverage of the various spheres of the city, including economics, management and ecology, to create ultimately in the long term Smart Model of life in the city.

Attempts to create a Smart Cities are undertaken in Russia as well (The Smart Cities, 2014), in particular, in Novosibirsk. For example, under the Program re-industrialization of the Novosibirsk Region's economy until 2025 (Program, 2009) the concept of Smart Region is proposed. It is aimed primarily at improving the quality of the urban environment and mobility through the use of information and communication technologies. The concept is aimed at improving the urban environment, and assumes management of the city and its economy, social sphere, transport system, the environment and life through smart technology (Program, 2009, p. 44-48). The aim of the concept is to make the people living in the city as comfortable and safe. One of the important elements of the Smart Cities is ecological. The author focuses attention on the integration of environmental aspects into the strategy of the city (with an emphasis on the big city).

The aim of this study is to develop a methodological and methodical fundamentals of formation of environmental strategies within the framework of ensuring sustainable development of Smart City.

Realization of this goal involves the following major tasks:

- Consider the city as an object of research;
- Show the place of ecological subsystems in the structure of socio-economic development of the city and to analyze its relationship with other urban areas;
- Show the possible directions of predicting the environmental impacts of economic activity in the city;
- Formulate the most important directions of formation of city environmental strategies and analyze the main factors of its formation, including spatial, industrial and institutional aspects of the selection variant conservation strategy;
- Propose a system of criteria and indicators that can be used for assessing the impact of the planned environmental activities;
- Identify conditions for the implementation of environmental policies;
- Carry out the practical application of the proposed approach in the development of socio-economic development of a particular city (on the example of novosibirsk).

The City as an Object of Research

The range of problems covering various aspects of interactions in the system "big city – the environment", is very broad and includes, in particular, such problems as the formation of the favorable environmental conditions as an important factor of the human environment; ensuring stability and the maintenance of a sustainable equilibrium of the state of ecological systems of the city; management of environmental processes in the big city; identification of the role of public authorities and local self-government in maintaining a healthy ecological environment of the metropolis; the development of the green economy as a mechanism modernization and innovative development of the territory; creation of the effective operational environmental monitoring in the conditions of a large city; increasing the level of ecological culture of the population and the extent of civil society participation in the formulation and implementation of environmental policy and other.

All of the above allows you to look at the city as a complex functional system in which an important place belongs to environmental problems. The spectrum of relevant issues covers the different aspects in the system "city - environment", which, in turn, allows fully and comprehensively review the formation of the environmental situation in the cities, to identify the arising problems and suggest possible ways of their solutions.

Every city is unique not only in its architecture, historical conditions of development, location and climate, but also on the emerging transport and economic relations and features of the production, including combinations of various industries, their composition, their branch specificity, the scale and nature of location, which is extremely important for the formation of the ecological situation in the territory.

The study of the ecological specificity of each city is the task extremely important because it determines the conditions of people's lives, their health, longevity and the comfort of the environment. This is particularly the relevant problem for large cities (Figure 1), because a large city - it, respectively, and the high concentration of population, industry, energy, transport, construction, etc. that leads to increased pressure on the environment and the formation of anthropogenic landscapes over large areas.

Place and Role of the Ecological Subsystem in the Structure of the Programs for Sustainable Development of Smart City

The concept of Smart City is based on the provisions of the sustainable development concept, the essence of which can be briefly reduced to the requirement of preserving and increasing the three potential - natural, technical and human, that is possible under the condition of simultaneous increase of the technical level of production, human health and the preservation of the proper quality of the natural environment, required to meet the needs of present and future generations. Thus, sustainable development is understood as the systematic unity of nature, economy and human, in the context of sustainable development. At the Summit in Rio de Janeiro ("Rio + 20") in 2012 the increasing significance of sustainable development has been demonstrated and acknowledged that ensuring the long-term successful development is possible only on the basis of the transition to "green" economy (Towards to a "green" economy, 2011). One way to achieve the goals of "green" economy is to ensure the environmental sustainability of cities (Resolution, 2015).

Ensuring sustainable environmental and economic development of cities requires joint consideration of the environment, production and population as elements of a single territorial-production system. One of the ways to reflect the environmental consequences of economic activity in the city is the development of ecological subsystem (or the environmental block) as part of its socio-economic development

Figure 1. Possible consequences of impact of the large city on the environment
Source: composed by the author using (Burmatova, 2015a; Lvovich, 2006; Urbanization, 2011; Chubik, 2006; Matugina et al,, 2015) All this accompanied by a multifaceted impact on the environment, affecting all of its components - the atmosphere, hydrosphere, flora and fauna, soil, relief, climate, etc. As a result, in large cities, a new and largely artificial environment is created.

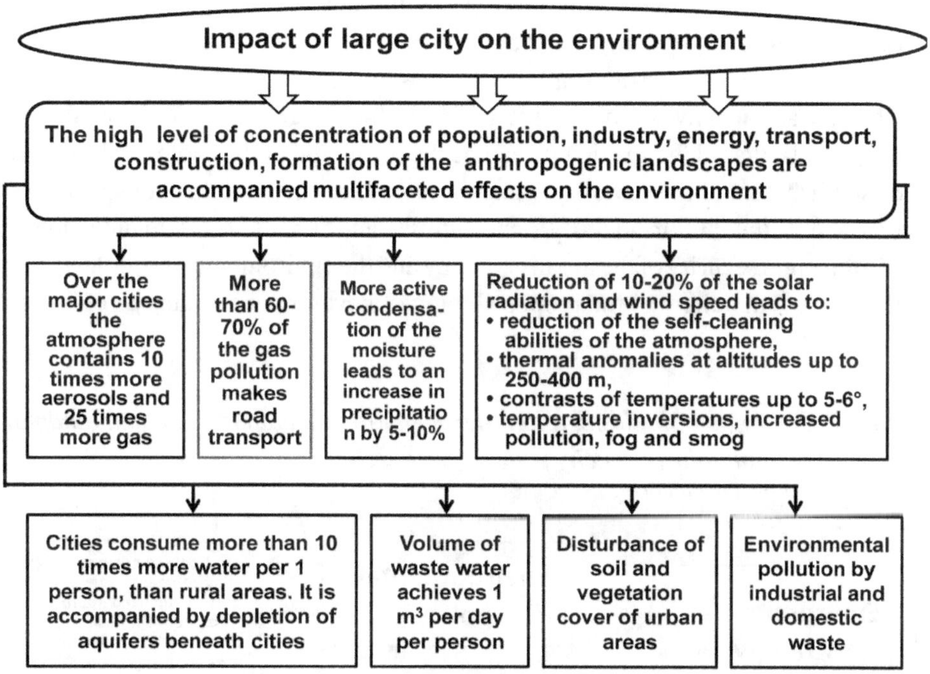

programs. Such subsystems ultimately encompass the main directions of the relationship of environmental, economic and social policies in the city. The result of the development of the ecological subsystem of socio-economic development of the city is to develop environmental management strategies, which can be regarded as a synthesis of its environmental problems and complex of nature protection activities aimed at their solution.

Strategic planning is an essential management function at various levels (including regional and local, including urban). It covers the process of selecting the city's development goals and ways to achieve them together into space (by performer), in time and finance. One of the elements strategic management tools are the environmental strategies and, accordingly, one of the subprograms as part of targeted programs of regions and cities are the subprograms for environmental protection. Environmental problems are usually long-term and require a strategic approach to their solution.

The logic of building a long-term environmental strategy as the landmark environmental policy is shown in Figure 2. As you can see, the development of environmental strategy involves a definition of the environmental mission of the city and its ecological image of the future, the choice of priority environmental goals and objectives, their detailing in particular for environmental programs, the mechanism of their implementation of the production, analysis and evaluation of the results and consequences of programs. At the same time the environmental mission of the city and its environmental image of the future are based on the analysis of internal and external factors that influence the formation of the ecological situation, the establishment of the advantages and disadvantages in the field of environmental protection in the city, to identify opportunities and threats to the maintenance of environmental security of the urban economy and the environment of people. In what follows, we will adhere to the proposed logic of developing a strategy in Figure 2.

Figure 2. The logic of building a long-term environmental strategy as a landmark of environmental policy in a city
Source: composition of author

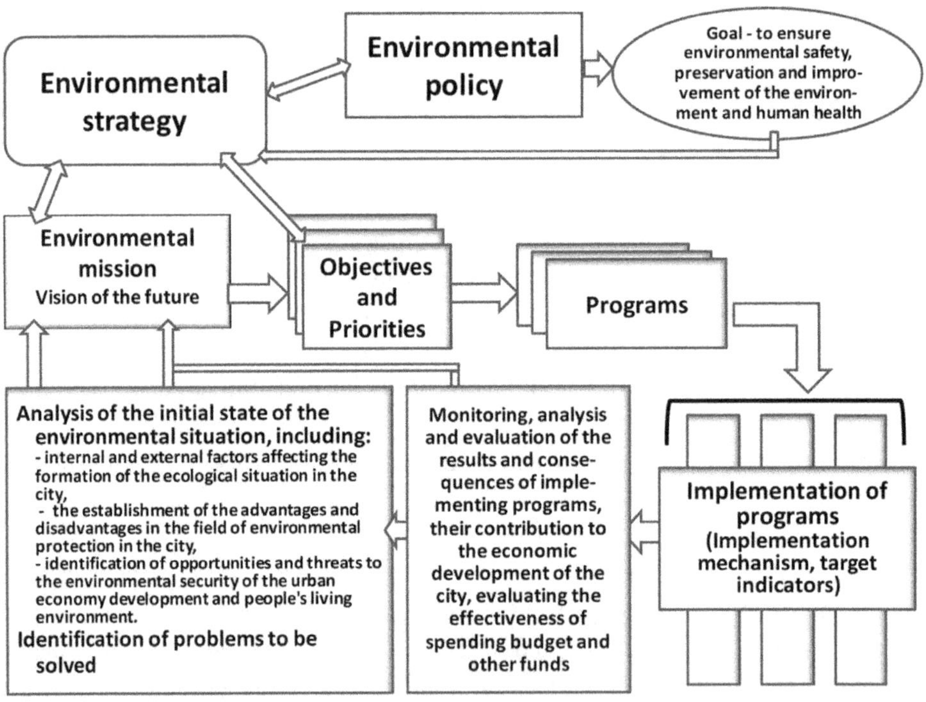

Consider the environmental subprograms in terms of their place in the structure of the regional programs of socio-economic development and to identify the relationship arising in the general system of the city, especially between the environmental, economic and social spheres.

The essence of the regional (municipal) programs as a form of implementation of the program-target method of planning, management and organization of social production, is to provide the solutions scientific, technical, economic and social problems of development of the territory, taking into account the specifics of its climatic, environmental and other conditions. In accordance with this, the regional program includes a set of measures to ensure a comprehensive and balanced (proportionate) solution of the problems of economic growth in a given period in a particular area. Regional programs reflect, as a rule, the preferred embodiments of the socio-economic development of cities and strategic concepts of social and economic policy.

Sustainable Development Program for the city can be conventionally represented as a synthesis of the three main territorial subsystems (or blocks) - the economic, social and environmental (Figure 3). Each of them has its own goals and objectives, a set of measures, the composition of which depends on the resource and ecological potential of the territory, the current and future production and the spatial structure of the economy of the city, as well as character of the relationship between the respective subsystems. Let us dwell on the analysis of the place of the ecological subsystem among these elements of municipal programs.

Place of the environment subsystem in the structure of environmental strategy for sustainable development of the city is determined by the necessity of linking of the prospects of economic and social development of the city in compliance with certain environmental requirements. Accordingly, between the ecological subsystem and other key subsystems of the city - economic and social - are set close forward and backward linkages (Figure 3).

Relationships between the ecological and economic subsystems provide the exchange of the following information. From the economic subsystem it's possible to receive the information about a perspective production and the spatial structure of the city economy, on the basis of use of which in the environmental subsystem the compliance of the proposed economic development of the city to adopted the environmental requirements is verified. At the same time the environmental compatibility of various industries analyzed, the ability to Output above permitted standard pollution is determined, the potential violation of state of green plantations are assessed and etc.

The results of the analysis are the recommendations on the composition, arrangement, production scale, cost, etc., which allows to specify the possible production and the spatial structure of the economy taking into account environmental requirements.

From a social subsystem information on existing and forecast population, existing and future systems of settlement and level of development of social infrastructure goes. In the social subsystem requirements for the quality of the environment in terms of human health, the organization of recreational zones and tourism, the formation of the environment, etc. set (Figure 3).

Feedback from environmental to social subsystem provides for the establishment of the correspondence between the emerging environmental situation and adopted sanitary and medical requirements and on the basis of this - the formulation of proposals for the selection of possible modes of nature use in order to ensure favorable conditions for people's lives.

In Figure 3, in addition to links with the economic (especially industrial) and social subsystems, also logic of the developing the environmental subsystem, its main objectives and some of the most important conditions for the implementation are shown. The content of the environmental subsystem is heavily dependent on the intended directions and scales of development and distribution of productive forces within the study area, the formation of the territorial organization of the economy, the choice of settlement systems, etc. That is, there is a close relationship between the generation of environmental management strategies and mode of economic and social activity in the region.

There are three main elements in the environmental section, corresponding to the following research areas (Figures 3 and 4):

1. Analysis of the current state of the environment in the city;
2. Forecast possible changes in the state of the environment and the risk of ecological trouble;
3. Assessment of consequences of the impacts of human activities on the environment and developing proposals on the composition of environmental measures.

To assess the state of the environment in the city is needed, first of all carefully analyze the various factors that influence on the formation of the ecological situation. One of the important results of the analysis of the initial state of the environment in the city is the allocation of problematic situations (most environmentally hazardous sites and their combinations), and problem areas (parts of the territory with complicated environmental conditions). This is particularly relevant for the

Figure 3. Place of ecological subsystem in the program of socio-economic development of the Smart City (relationships of economic, social and ecological subsystems city)
Notes: The blocks marked by a dotted line city describe environmental subsystem. The blocks are located below the dotted line characterize the basic conditions of implementation of the environmental strategy of the city
Source: composition of author

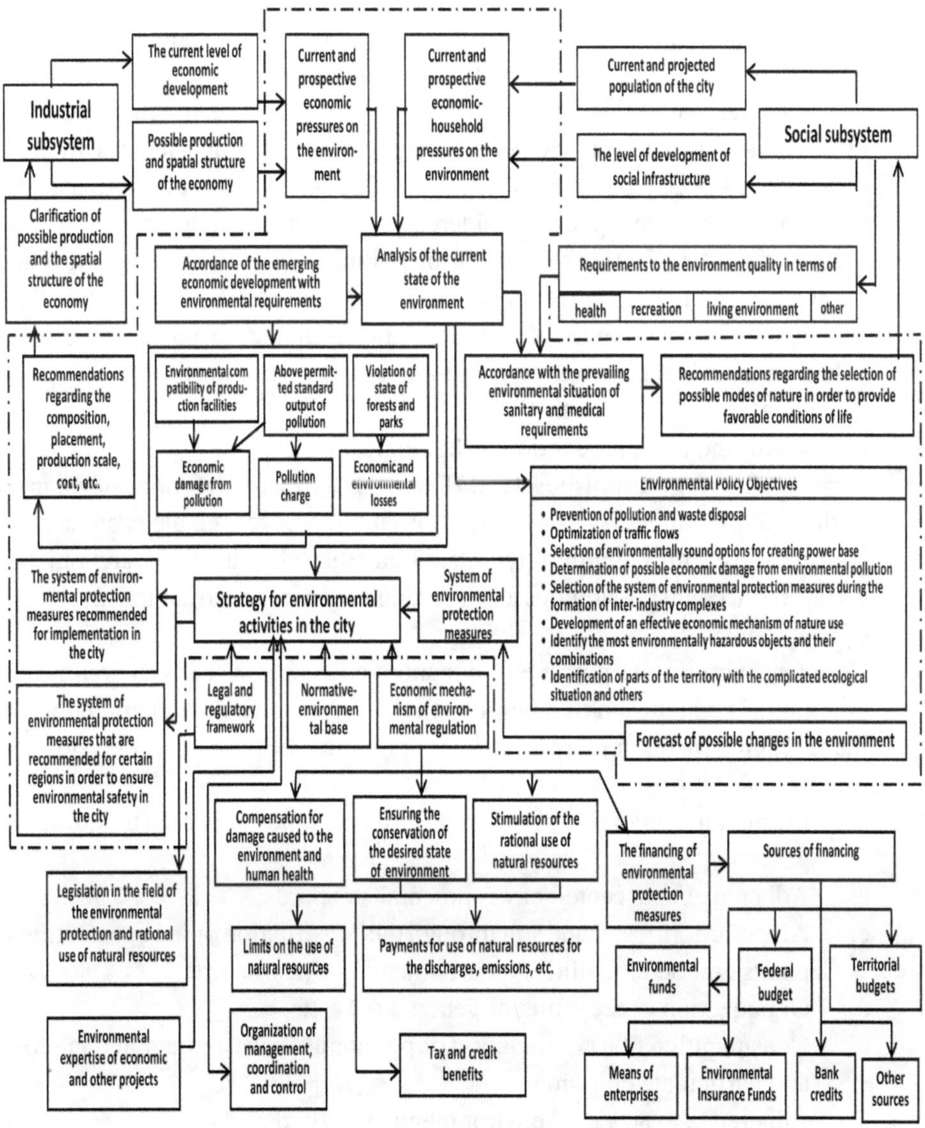

environmentally disadvantaged cities, because it allows you to identify "bottlenecks" with environmental positions and direct the forces and means in the first place on their unravel.

The following elements can be distinguished in the composition of the environmental subsistem:

1. Environmental facilities (wastewater treatment plants, equipment for the protection of the atmosphere, waste processing plants, and so on), for which in the process of solving problems is used to select treatment technology options, capacity, type, placement, taking into account their links with other elements of the economic complex of the city, as well as taking into account the environmental indicators considered production technologies;
2. Waste streams, reflecting the output of pollution in the environment; while the formation of the level of air pollution and water bodies should take into account all possible sources of emissions and discharges on the different types of hazardous substances and their combinations on condition the neutralization of pollutants emissions within acceptable limits;
3. Options for the main production technologies;
4. Arrangements for waste disposal of different types with the definition of volumes of their output, directions for use and distribution of recyclable products;
5. Measures to ensure conditions the wastewater discharge in open water objects, taking into account possibilities for water dilution, storage and natural movement of contaminants;
6. Costs for environmental purposes and economic damage caused by environmental pollution. The inclusion of environmental subsystem (environmental unit) in the measures of the regional programs provides, in particular, decision the following questions:
    a. Formation and the distribution of different types of waste (liquid, solid, gaseous);
    b. Adherence to accepted environmental standards;
    c. Air pollution, water and soil through the construction of sewage treatment plants and other facilities environmental infrastructure;
    d. Organization of recycling of generated waste;
    e. Consideration of a combination of pollutants in wastewater and gaseous and particulate emissions;
    f. Adherence to accepted environmental standards;
    g. Air pollution, water and soil control through the construction of sewage treatment plants and other facilities environmental infrastructure;
    h. Organization of recycling of generated waste;

i. Consideration of a combination of pollutants in wastewater and gaseous and particulate emissions;
j. Some specific conditions for the reproduction of water and land resources (including the conditions of formation and distribution of waste water, limitations on the possible volumes of their discharge, the conditions of the remediation work in the field of disturbed land, and others.).

In general, the environmental subsystem acts as a specialized tool for analyzing economic activity relationships and the environment in the city. Its specific composition is determined by specificity each city - its natural, economic and social conditions, the peculiarities of the territorial organization of the production and development prospects, as well as put forward by environmental objectives and nature of the problems to be solved.

The development and use of environmental subsystem of the urban programs allows not only analyze the interaction of elements of the economy, population and environment in the region, identify the necessary conservation measures, make a

Figure 4. Scheme of formation of ecological subsystems in the system the environmental strategy of the city
Source: composition of author

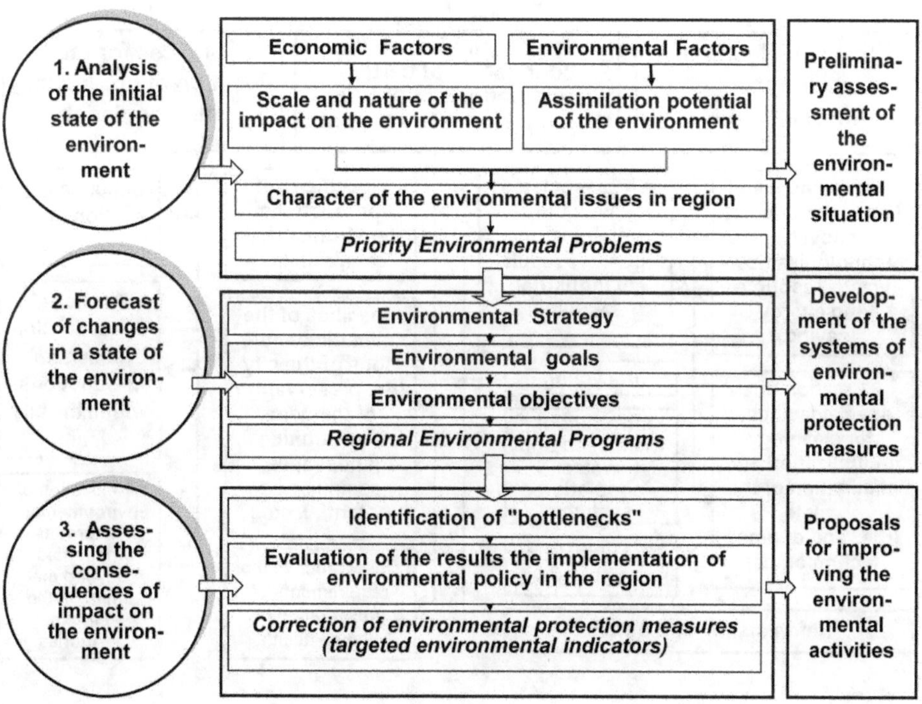

choice of options of developing and the distributing the production and population in its territory, under condition of fulfillment of the established environmental regulations, but also to determine the amount of costs required for environmental purposes, as well as the number of labor resources, the demand for local natural resources, infrastructure services needed to prevent negative environmental impacts of economic activities in the city.

Possible Directions of Predicting the Environmental Impacts of Economic Activity in the City

At forecasting the environmental consequences of economic activity in the city, there are the following methods of accounting for the impact of environmental factors on the elements of territorial and production systems (Figure 5).

Figure 5. The main directions of integration of environmental requirements in predicting the formation of the city's economy
Source: composition of author

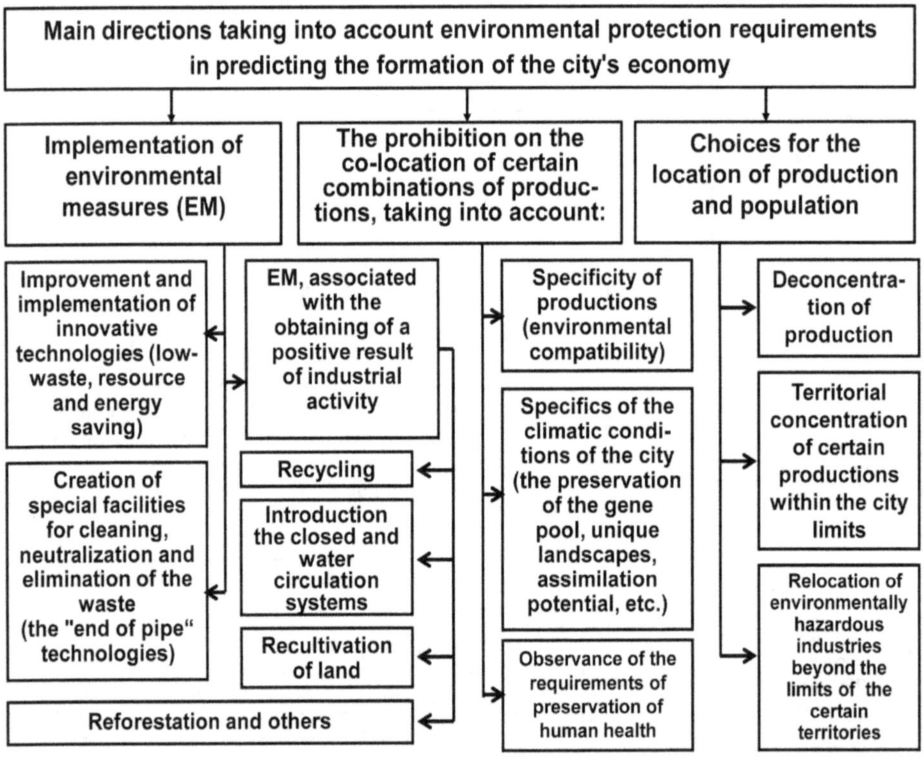

The first way envisages carrying out the complex of nature protection measures aimed at the prevention, reduction or elimination of the negative impact of various objects in the environment.

Depending on the nature of the impact of an object on the environment, the possible negative effects of such an impact, the technical possibilities of their prevention, etc. implementation of environmental measures can go in three main directions.

1. Improvement and introduction of innovative processes, which necessarily accompanied by improvement of production and environmental indicators (i.e. indicators characterizing the impact of technology on the environment and complete use of natural resources). This direction includes the expansion of the share of energy-saving and low-waste technologies. Reflecting such measures is possible primarily in the variants of functioning production facilities. At the same time improvement of technologies aimed at reducing their negative impact on the environment and applies both to the main production processes and on methods of waste treatment and environmental protection equipment.
2. Create special installations, equipment and facilities, other activities intended mainly for reducing pollution (the "end of pipe" technologies). Due mainly unidirectional action of this kind objects and events (to reduce pollution), they are called special-purpose. There are the sewage treatment plant for the treatment of industrial and domestic waste water, gas cleaning and dust removal equipment, Water-cooled plant and equipment, measures for dust suppression and dispersal of dust and gases, providing dilution of contaminated wastewater with fresh water, the establishment of buffer zones, gardening of territory and others. Accounting for such measures when setting targets to minimize the negative impact on the environment is possible by selecting the appropriate option power structures, methods and techniques of neutralization of contaminants, cleaning efficiency indicators, the required expenses, etc.
3. Implementation of the various activities aimed not only at reducing certain types of human impact on the environment, but also connected with additional obtaining of a positive result of industrial activity (for example, issue of additional products, saving of resources, etc.). This is the so-called multi-purpose activities. They include waste management; implementation of sequential systems, recycling and closed water supply; land reclamation, the disturbed surface mining of minerals; reforestation and others.

A special place in the overall system of environmental protection measures belongs, in our opinion, to the organization of waste disposal, which is a necessary element in the overall chain of a system of non-waste productions. Greening of production is impossible without the addition of production facilities in special facilities designed for the processing of all kinds produced industrial and domestic waste. Disposal of waste, on the one hand, contributes to the reduction of allocation to the environment of various harmful substances, reduction of the scale of the negative impact of production on the state of the landscape, flora and fauna, etc. All of this eventually accompanied by a reduction of air pollution and water pollution, preservation of the landscape, the release of the territory as a result of the elimination of waste dumps, etc. On the other hand, waste disposal reflects the possibility of obtaining economic benefits - additional sources of release of a useful product, the possibility of expanding the raw material base of production through the use of waste and therefore saving natural resources, the ability to reduce the cost of the resulting utilization of waste products (through the use of cheaper raw materials, which is the waste), etc.

The second way of the integration of environmental factors when predicting the economic development of the area is the introduction of restrictions on co-location and functioning in a particular place different combinations of production, unfavorable from the viewpoint of sanitary conditions (Figure 5). The need for such a way is due primarily to the fact that, on the one hand, in the present available technical possibilities do not provide one hundred percent neutralization of waste and, on the other hand, focus on adherence to accepted environmental standards is not yet a reliable guarantee for the preservation of all the components of the environment (including human). Therefore, in the process of the implementation of environmental activities within the different areas additional measures of prohibitive character may be required.

A ban on the environmentally incompatible co-location facilities can be based on the following requirements.

Firstly, taking into account the peculiarities of enterprises considered in the limits of the city. Among these features should be called the peculiarity of the individual ingredients of emissions and discharges of different companies to interact with each other, leading to a deterioration of the quality of the environment or to the complication of the problem of waste disposal (effect of synergism). Harmful substances that have the effect of synergism, in conditions of joint presence may lead to the formation new compounds of heightened toxicity or enhancing the harmful effects of each other.

It is noted that the environmental compatibility of the plants largely depends on environmental danger of production technology, which is defined as by the physicochemical properties of the substances released into the environment, and their degree of the conversion and accumulation in the environment. A synergistic effect in the atmosphere possess, in particular, sulfur dioxide and chlorine, nitrogen oxides and hydrocarbons, sulfur dioxide and moisture, hydrogen fluoride and methyl mercaptan and others (Lystsov & Skotnikova, 1991; Petin & Synzynys, 1998).

Synergistic effect of the influence in an aqueous environment have, in particular, ions of copper and lead (the presence of which combination of in sewage causes more damage than either of them separately), pesticides and oils, etc. (Combined (complex) action,n.d.).

Furthermore, there are persistent (non-degradable) contamination - substances and poisons, such as aluminum cans, mercury salts, phenolic compounds with long chain

DDT, which in the natural environment or are not destroyed at all, or break down very slowly. For these substances, there is not natural processes, which could decompose them at the same rate as they are introduced into the ecosystem. Such non-destructive pollutants not only are accumulated, but often "biologically intensified" as far as passing in biogeochemical cycles and food chains. In addition, they can form other toxic substances, connecting with other substances surrounding environment.

Unfortunately, the account of the environmental compatibility of various industries belongs to the least studied and are usually ignored when taking economic decisions problems. At the same time, the interaction of pollutants entering the environment can lead to undesirable consequences for human health, the state of natural complexes and other recipients.

Secondly, when a ban on the co-location of certain industries need to consider not only the mentioned features of these industries, but particularly the territory of accommodation and its climatic conditions. This can be important when the target area has a unique landscape, valuable and rare species of plants and animals, etc. In this case, the decision to ban the co-operation of various industries in an area may be caused by requirements of the conservation of its natural complexes and a whole gene pool. Accounting for features of the area is of great importance and when the area is characterized by low assimilation capacity, i.e. low ability of ecosystems to neutralize harmful waste of human origin. This may be due to the unfavorable potential of climatic factors (associated with low air environment diffusing capacity), extreme climatic conditions (leading to a slowdown contamination decomposition processes), high already established load in the city on the soil, air and water (resulting in self-cleaning capacity different elements of the environment is already largely exhausted), etc. The presence of these conditions can lead to rapid and irreversible changes in the environment that requires particularly careful attention to such areas.

Thirdly, the establishment of restrictions on co-location and functioning of certain plants may be dictated by the requirements of preserving people's health (Characteristics of the main pollutants of the environment, n.d.). Different substances contained in wastewater and gaseous emissions, can have different impacts on public health. For example, nitrogen oxides cause a sharp irritation of the respiratory system; damaging effect on the respiratory system has and sulfur dioxide, which also causes an increase in susceptibility to infections, metabolic disorders, and others. In addition, under certain conditions the interaction of certain harmful substances may lead to extremely dangerous for human health compounds, for example, reacting benz(a) pyrene with nitrogen oxides leading to formation nitrabenzapirena which is highly active mutagen etc. (Burmatova, 2009). It is undesirable also the connection in one place for such activities which emissions in the air can lead to the formation of toxic or photochemical smog, which is also dangerous for human health.

The third way to integrate the protection of the environmental factor in predicting the formation of an economic complex of the city is associated with use of opportunities of the location of production and population to solve environmental problems (Figure 1.7). Such opportunities consist in primarily in the fact that considering a number of different variants of production operation (each of which is characterized by its environmental friendliness) it is possible to provide a choice of scale territorial concentration of production and the composition of such enterprises in each industrial site, that meet environmental requirements. The character of the placement and operation of production, the level of its territorial concentration and sectoral structure, existing settlement system largely determine the environmental situation in each place.

Thus, the environmental role of location and operation of production and population is mainly in the fact that with its help it is possible to regulate the concentration or dispersion within a single area of production and population in order to preserve the natural environment. At the same time the territorial dispersion of objects-pollutants makes better use of assimilation potential of the territory. Solving problems of protection of the environment by controlling the distribution of productive forces seems justified especially in cases where there are no sufficient technical and economic opportunities for the prevention, reduction or elimination of various adverse environmental impacts of economic activities. Therefore, this way of contributing to a partial solution of environmental problems, should be considered as one of the possible, and in some cases, necessary additions to other environmental practices, aimed primarily at the absolute reduction of harmful anthropogenic impact on the environment.

In general the forecast of possible changes in the state of the environment in the city is based on the account of economic development prospects within it, the current and expected production structure of the economy, the characteristics of sources of impact on the environment. In the case if production in the investigated territory is concentrated in certain parts of the city, it should be take into account that this entails, and increase of load on the environment in the relevant parts of the city and possible complications in their environmental situation. If the forecast the focus should be not so much to assess the impact of the individual objects in the environment, as the assessment of the total impact of territorial-production combinations on various elements of the environment, taking into account of background contamination, as well as other, already taking place, violations of the environment.

According to the results of evaluation of possible environmental impacts of the creation and functioning of the various objects the comprehensive forecast of the influence of the intended impact on the environment in the city are carried out. This forecast is directly related to the choice of the system of environmental protection measures, which should provide an approximation formed indicators of possible impacts on the environment (especially pollution) to the normative.

Prerequisites of the Development of Environmental Strategy

As already was mentioned, one of the most important tools of management the environmental sphere of the region and of forecasting its condition is strategic planning. The strategic planning process provides the basis for managing the city as a whole and its individual spheres, including, in particular, environmental. Development of a regional environmental strategy involves determining the environmental mission of the region and its ecological image of the future, the choice priority objectives and goals, detailing their in specific projects and programs, developing mechanisms for implementation, analysis and evaluation of results and consequences of implementing projects and programs. Consider on the example of Novosibirsk and Novosibirsk region[1] the main methodological aspects of the development of environmental strategy.

Environmental policies in the Novosibirsk region is aimed at maintaining the integrity of natural systems and providing a favorable environment for people. At the same time it's necessary to take into account that not only is a lot of accumulated in the past and so far unresolved environmental problems, but also the fact that the prospects of socio-economic development of the region associated with a possible further increase of the load on the environment. This, in turn, determines the relevance of development and implementation in the medium and long term, an adequate system of environmental measures that could form the basis for the strategy of environmental activities in the region and, consequently, to determine the main directions of improving the environmental regulatory system in this region that minimize the negative impacts of human activities on the environment and in general of ecological safety of the economic development of the Novosibirsk region.

The basis of the system of prediction of the environmental situation in the Novosibirsk region is the development and implementation of a number of interconnected documents, including the following elements:

1. The section "Ensuring environmental safety and environmental protection" as part of the Strategy for Socio-Economic Development of the Novosibirsk region for the period up to 2025 (Strategy, 2009);

2. Long-term target program "Environment of the Novosibirsk Region" within the Program of the Socio-Economic Development of the Novosibirsk region up to 2015 (The Program, 2009 The Concept, 2009);
3. Strategic Plan for Sustainable Development of the city of Novosibirsk and complex target programs. - Novosibirsk, 2004. - 245 p. (Strategic Plan, 2004).
4. Subsection "Solving the environmental problems" of the Strategy of socio-economic development of Siberia up 2020 (Section IV. Priority Interbranch development directions of Siberia) (Strategy of socio-economic development of Siberia, 2007; Economy, 2009).

It is important to note that forward-looking strategic developments in the field of environmental protection are part of the general system of strategic planning documents of socio-economic development in the region (Economy, 2009; Kuleshov& Seliverstov, 2016; Seliverstov, 2012; Seliverstov & Melnikova, 2013) and have a close connections with economic and social issues in the region, as oriented on a comprehensive solution economic, social and environmental problems. In accordance with this, the formation of long-term target program "Environmental Protection Novosibirsk Region" (hereinafter - the Program) suggests as a first phase the elaboration of the project concept of this program, which are built with taking into account the connection with the Program of socio-economic development of the Novosibirsk region in 2025 (Strategy, 2007; The program, 2009; Seliverstov, 2012). The concept of long-term program "Environmental Protection the Novosibirsk region" is intended to further the formation of the long-term program, as well as definitions of the main contours of the internal structure of the section dedicated to environmental protection, its place in the program of socio-economic development of the Novosibirsk region and the total system of strategic planning area.

Development and implementation of the Program should begin with an analysis of the initial state of the environment in the region and should end with the elaboration of defined measures for the desired adjustments of the nature management and forming a healthy environment within the territory under conditions of the quality control of the entire process of development and implementation of the Program, the necessary coordination in the relevant activities and evaluation of the results (Figure 6)[2].

Factors Determining the Formation of Environmental Situation in the City

Environmental policy in the region or city is largely dependent on how adequately in its implementation taken into account factors affecting the development of the environmental situation. Analysis of these factors in relation to the possibility of the

Figure 6. Structure and scheme of the process of the development and implementation of the Program "Environmental Protection the Novosibirsk region"
Source: composition of author

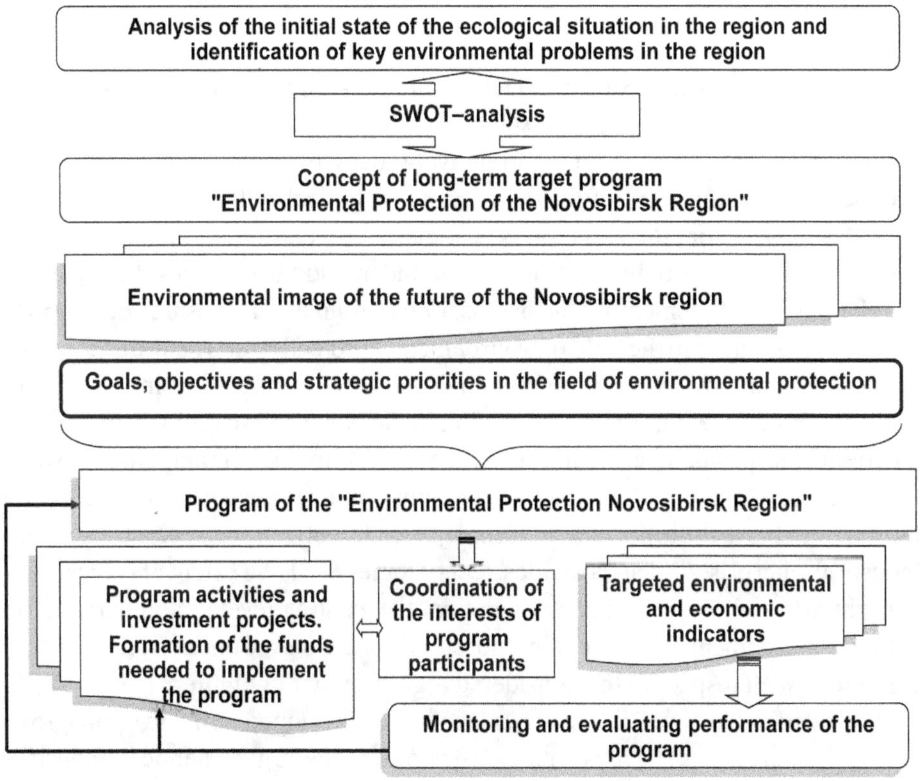

environmental governance allows you to identify the most significant for a specific period of time environmental problems and offer the ways for their solutions. It is important to be clear about whether the critical factors amenable to control by the environmental agencies, whether they are internal or external, within the scope of regulation of environmental field or external conditions, on which directly affect the regional authorities are not able.

In every city the environmental situation, priority environmental issues and finding approaches to their solution are determined, as a rule, by three groups of aggregated factors:

1. The specific conditions of the region or city (spatial aspect);
2. Industry features productions that are presented in the region or in the city (production aspect);
3. Approaches to solving environmental problems in the framework:

a. Separate economic entities (organizational and technological aspects),
b. The city as a whole, including the environmental policy at the level of the city and its districts (management aspects).

In more detail, these factors are presented in Figure 7.

Results of imposition of these factors on the territory of the Novosibirsk city (Figure 8) testify that from the standpoint of the formation conditions of the environmental situation, the region relatively lucky only with branch structure of production (meaning first of all the absence of typical polluting industries). For the remaining factors the situation is sufficiently urgent that in the perspective of socio-economic development of the region under study determines the urgency of developing and implementing the necessary environmental measures that form the basis of the program in the field of environmental protection and determine the main directions of improving the system of environmental management in the region.

Features of the environmental situation in the Novosibirsk region and emerging environmental problems are mainly caused by local climatic conditions and character of influence on them the region's economy (industry, energy, transportation, utilities and agriculture), which in turn depends largely on the specific location of industrial enterprises, their capacities, technologies used, the extent of the territorial concentration of production and population, the existing level of a violation of the natural environment in the region and other conditions. Concretization of these conditions with respect to the considered region shown in Figure 8.

Projected in the Novosibirsk region economic development involves the growth of the fuel industry, ferrous and non-ferrous metallurgy, chemical industries, construction materials industry (cement) and freight turnover transport (Strategy, 2007). This can lead to increased pressure on all parts of the environment that will require conducting adequate environmental measures.

Analysis of the Initial State of the Environment and the Major Problems

A high concentration of industrial production and population In Novosibirsk causes an increased flow of pollutants into the environment of the city, posing a threat to human health. Novosibirsk is the third city in Russia by the number of residents in the population (1.6 million. Pers.). Environmental situation in Novosibirsk, as well as other big cities, is dependent not only on the released into the air of harmful substances, but also on a variety of adverse weather factors, such as calm, temperature inversions and fog (which are conducive to the accumulation of harmful substances in the surface layer of the atmosphere). According to studies of Roshydromet[3],

Figure 7. Factors influencing the choice of environmental situation on a territory
Source: composition of author

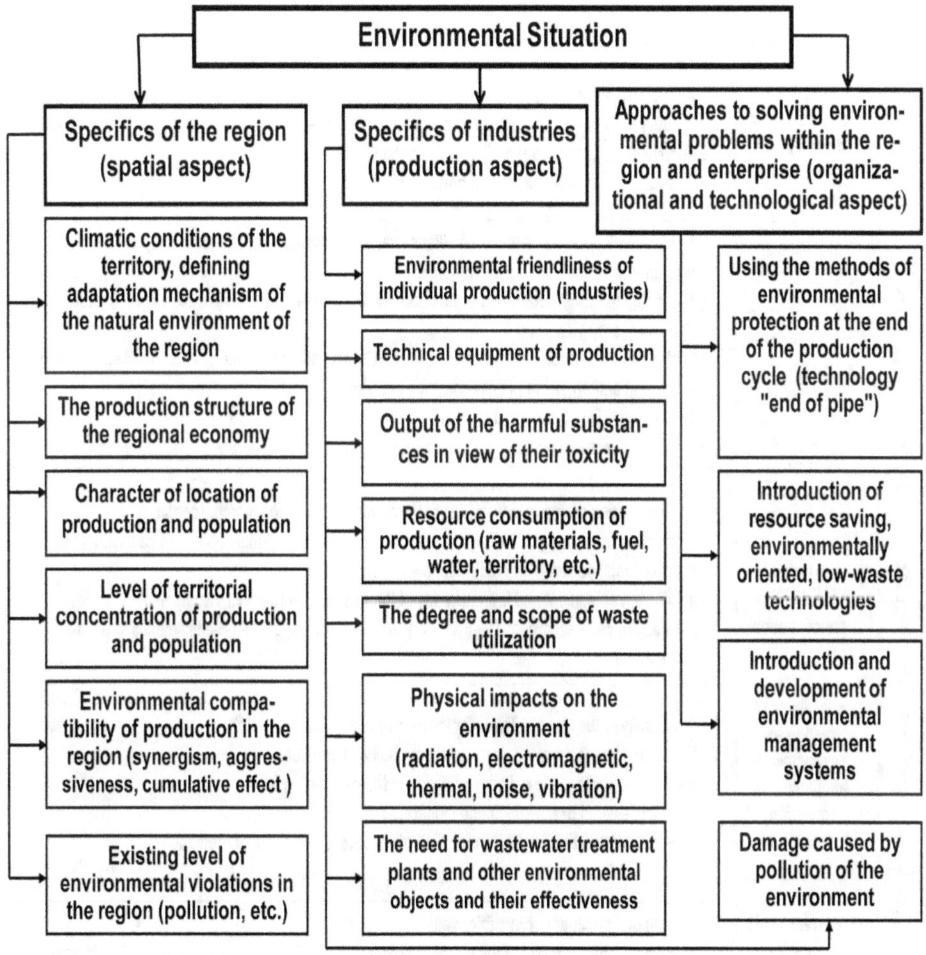

Novosibirsk is located in the area of high potential of air pollution[4] (PAP), i.e. in the zone of unfavorable meteorological conditions for the dispersion of contaminants, so that in some periods of intense accumulation of harmful substances can occur in the atmosphere and the formation of high levels of pollution. In general, the scattering ability of the atmosphere in Novosibirsk much better than, for example, in Eastern Siberia, or Kuzbass (the Kemerovo region), but they still do not reach the appropriate level, which is observed in the European part of Russia, for this reason, the city has increased the meteorological pollution potential[5].

Figure 8. Features of the Novosibirsk from positions forming the ecological situation (in italics are marked favorable factors)
Source: composition of author

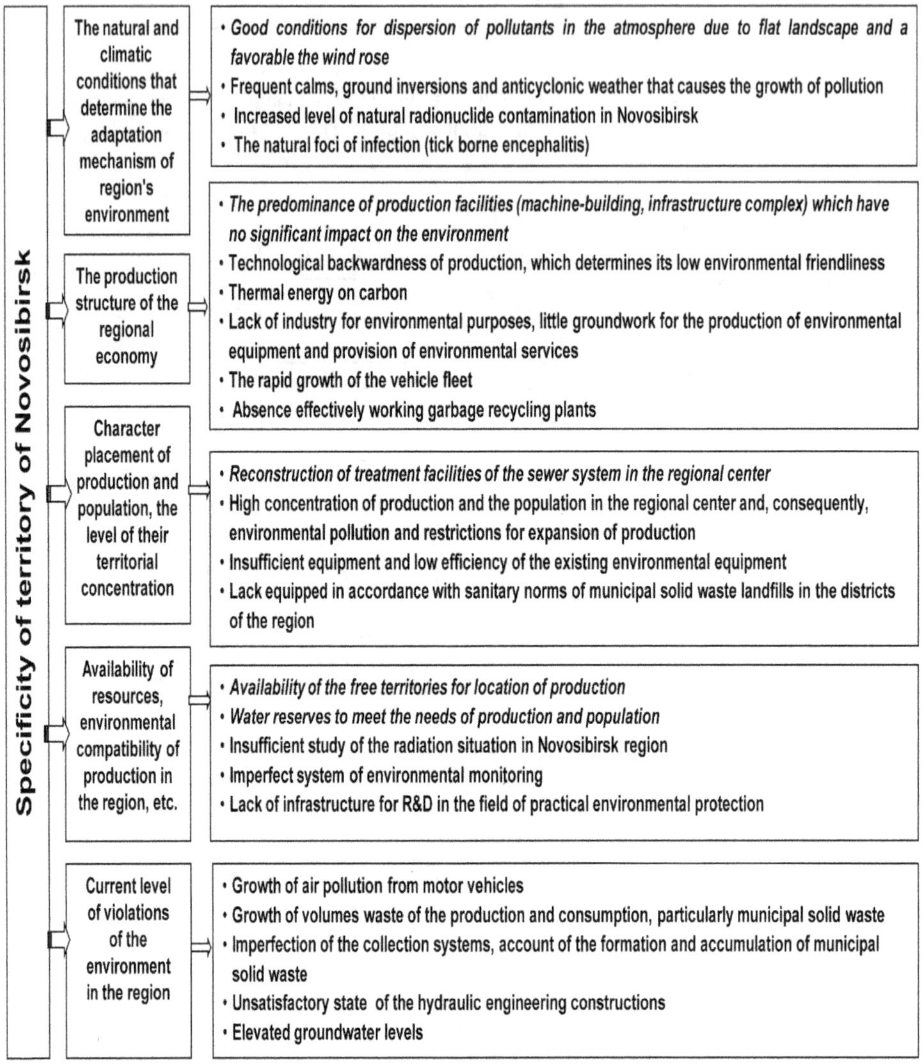

Main indicators characterizing the state of the environment in the territory of the Novosibirsk region from positions of atmospheric air, water basins and waste in the dynamics are shown in Table 1 (Burmatova, 2015a; State reports, 2015; State report, 2016a).

Despite the growth in industrial production and an increase in the number of vehicles, air quality in large settlements of the Novosibirsk region (including

Novosibirsk) in recent years has remained relatively stable (Burmatova, 2015a; State reports, 2015; State report, 2016a; The State Report, 2016b) (Table 1 and Figures 9 and 10).

Notes:

1. The table is compiled using data of the Novosibirskstat and the Department of Natural Resources and Environmental Protection of the Novosibirsk region[6], as well as sources (Burmatova, 2015a; State reports, 2015; State report, 2016a; The State Report, 2016b).
2. Calculations are made by the Department of Natural Resources and Environmental Protection of the Novosibirsk region under the simplified procedure developed by "SRI Atmosphere", using specific emission factors.
3. In the denominator is indicated polluted waste water without treatment (in total volume of wastewater discharges).

For the past 7 years, emissions of air pollutants from stationary sources per capita decreased from 88.1 to 66.9 kg (Figure 10).

Table 1. Main indicators characterizing the impact of economic activity on the environment of the Novosibirsk region[1)]

Indicators	2009	2010	2011	2012	2013	2014	2015
Extraction of water from natural water bodies - total, million m³, including:	755,2	763,6	676,0	703,9	646,6	634,8	639,73
- from surface water bodies	659,5	696,5	613,4	642,8	586,19	574.83	579,59
- from underground sources	95,7	67,1	62,6	61,1	60,4	59,97	60,14
Wastewater discharges into surface water bodies total, million m³, including:	588,2	604,2	527,1	544,2	512,16	500,6	491,7
- discharge of polluted wastewater, million м³	98,4	106/ 48,3[3)]	92/ 34,9[3)]	112,5/ 40,1[3)]	114,7/ 33,96	109,35/ 33,93	106,94/ 30,99
- discharge normatively treated wastewater, million m³	254,1	280,18	231,77	248,23	222,0	220,28	216,19
Air emissions, thousand tons: - from stationary sources	233,5	228,4	234,0	224,5	195,7	207,8	184,7
- from motor vehicles	358,4	319,9	287,4	286,2	310,2	276,5	275,2
Pollutant emissions from stationary sources, per 1 inhabitant, kg	88,1	85,7	87,1	83	71,65	75,7	66,9
Waste generation of production and consumption, million tons	1,91	2,07	2,5	1,8	1,863	1,95	3,89

Figure 9. Dynamics of emissions of harmful substances into the atmosphere of the Novosibirsk region in 2009-2015 (thous. tonnes / year)

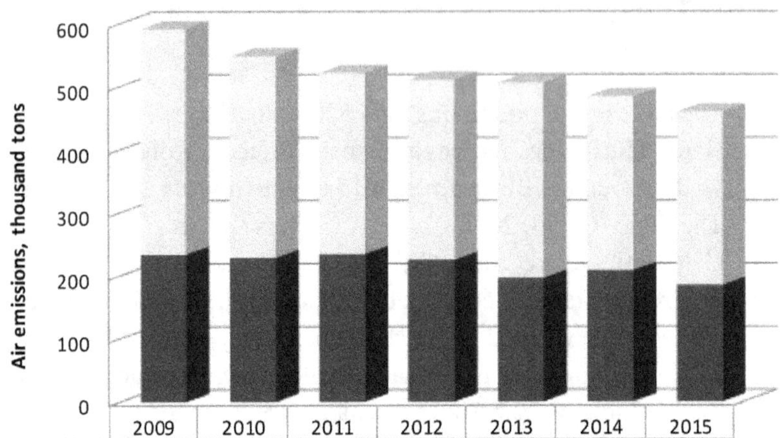

	2009	2010	2011	2012	2013	2014	2015
from motor vehicles	358.4	319.9	287.4	286.2	310.2	276.5	275.2
from stationary sources	233.5	228.4	234	224.5	195.7	207.8	184.7

Table 2 lists the most important ecological problems of Novosibirsk that require focus and solution in the mid- and long-term perspective, as well as possible ways of solving them.

Figure 10. Change of air emissions of pollutants from stationary sources (kg / per 1 person) on years

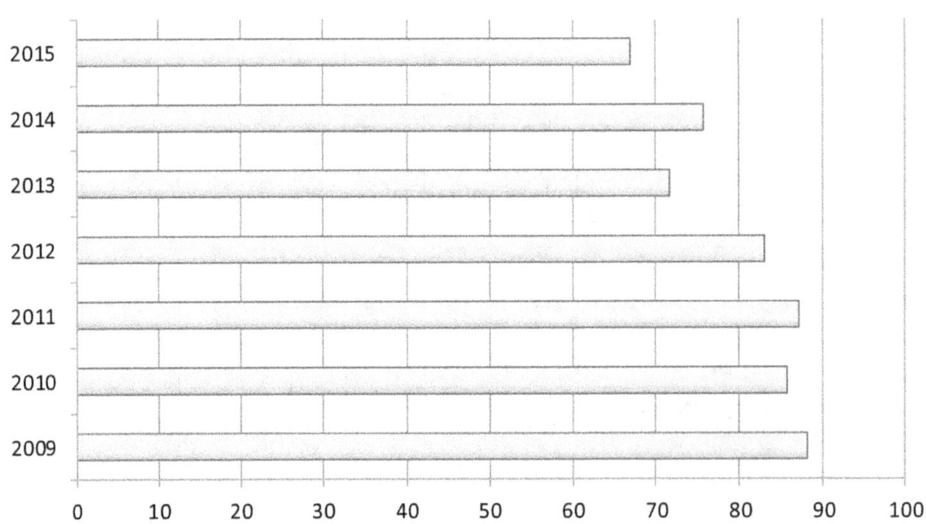

Table 2. Most important ecological problems in Novosibirsk and possible ways of solving them

Problem	Possible Ways to Solve
1. Air pollution by stationary emission sources	• Implementation of ecologically oriented technologies at businesses of the city, both active and planned; modernization of equipment at a number of active businesses. • Continued operations to supply gas to communal and government-office furnaces; closing of ecologically unfavorable furnaces (the number of which in Novosibirsk exceeds more than 100 objects). • Improving control of atmospheric emissions, primarily in the districts of the city where concentrations of harmful admixtures in the air regularly exceed the maximum permissible concentration. • Greening of urban territories (tree planting, increasing urban forest area). • Greening of urban territories (tree planting, increasing urban forest area). • Improving the fining system for negative anthropogenic impact on air quality. • Development for city objects of regulations on maximum permissible emissions of harmful substances into the air. • Revealing of problem areas of air pollution in the city
2. Air pollution from mobile emissions sources (auto transport)	• Road repairs and construction of new transport interchanges and bypasses in Novosibirsk. • Ensuring conditions for rapid transfer of part of the auto fleet (including municipal transport) to motor fuel. • Organization of industry and implementation of auto exhaust neutralizers. • Strengthening of state control over compliance with norms for permissible smog and harmful substance emissions by auto transport, which accounts for around 70% of the total harmful emissions in the city. • Optimization of road traffic, ordering of transport flows in city, determination of places where auto transport is concentrated in populated areas. • Improving the fining system for environmental pollution from mobile sources.
3. Organization of rational use and conservation of water resources	• Creation of combined water supply systems (using surface and underground water). • Reconstruction of sewer and water main networks. • Modernization of treatment facilities at active businesses of the city. • Measures on reducing volumes of dumping polluted sewage into surface waterbodies. • Implementation of water recycling systems. • Organization of repeat (sequential) use of treated waste water between businesses of city. • Improving drainage sewers in city. • Accountability of ecocompatibility of different industries from the standpoint of water pollution and organization of where to locate water discharge points. • Placement of new industries with allowance for requirements of waterbody conservation. • Improving the payment system for pollution of waterbodies. • Improving the system for monitoring the water basin, expansion of a continuously acting observation network for the state of surface waterbodies. • Revealing of "narrow" areas in the city from the standpoint of pollution of waterbodies. • Continuation of operations on restoring, cleaning, and conserving small rivers bordering Novosibirsk and other cities of the region.

continues on following page

Table 2. Continued

Problem	Possible Ways to Solve
4. Improving the situation in dealing with industrial and consumer waste	• Construction of areas of solid household waste in Novosibirsk, their development using modern technology. • Construction of new waste treatment plants. • Organization of solid waste sorting. • Elimination of unsanctioned dumps. • Utilization of thermal energy from ash and slag created at factories by using it as raw materials at construction industry entities. • Collection and utilization of household waste from the private sector with full coverage of private homes. • Creation of industrial businesses for secondary waste recycling (including paper).
5. Ensuring soil protection	• Measures on combating wind and water erosion of soils. • Compensation for losses from confiscation of agricultural holdings for construction. • Recultivation of damaged lands.
6. Improving the situation with radiation	• Improving and developing a system for observing and controlling radiation in the region. • Monitoring of the level of impact of natural radon on the inhabitants of Novosibirsk. • Testing of the territory of Novosibirsk for radon hazard.
7. Ensuring reproduction of fish stocks	• Measures on conserving and raising commercial fish species (the Ob River and the Ob reservoir). • Measures on preventing and compensating damage to the fishing industry in connection with the destruction of spawning grounds and the death of young fish and egg. • Intensification of combat with poaching in rivers and lakes of the city and region.
8. Reduction in the water level of the Ob River and Ob reservoir, progress-ive underflooding the territory of city	• Measures on ensuring safe functioning of hydroengineering installations • Fortification of the shoreline of the Novosibirsk reservoir, improvement of water conservation zones in the Ob River basin. • Improvement of urban drainage networks
9. Development of an environmental monitoring system	• Creating a network of operational monitoring and control of the environmental situation
10. Creating a unified system of financing environmental protection measures	• Attraction of investments for nature conservation. • Use of budgetary means of different levels. • Strengthening of the role of the oblast budget in financing ecological programs and measures. • Introduction of mandatory ecological insurance of a number of potentially hazardous industries and technologies.
11. Education in ecological culture	• Involvement of society in preparing and implementing ecologically important solutions. • Creation of centers for ecological education of the population

Source: The table is compiled using data of the Department of Natural Resources and Environmental Protection of the Novosibirsk region[7], as well as sources Burmatova, 2015a; State reports, 2015; State report, 2016a; The State Report, 2016b).

Environmental Vision of the Future of the Region, the Challenges in Creating a Favorable Ecological Situation, Strategic Environmental Goals, Objectives and Priorities

By analyzing the ecological situation in Novosibirsk and Novosibirsk region, it is possible to form an ecological vision of this region's future, delineate challenges in creating a favorable ecological situation, and determine the strategic ecological priorities and directions of nature conservation activity.

The main elements of the vision of this region's future are characteristics that presuppose improvement in the quality of the environment and ecological conditions for human vitality, including the creation of a healthy environment, ecologized industry, organization of an efficient ecological sector of the economy, and nature conservation and protection (Figure 11). The ecological vision of the future (which is, however, in the territorial aspect sufficiently universal, since it is based on common social, economic, and ecological requirements) envisages an aligned foundation on which the necessary conditions should be created to make the harmonic interaction of the regional economy and ecology to become a reality.

In modern conditions of economic development the maintenance of the ecological safety in Russia faces serious challenges and threats which require for their solution the participation of the authorities at all levels of management.

The main challenges in creating a favorable environment in the regions of the country, including the Novosibirsk region are the following circumstances.

1. The need for transition to a new life and environmental safety standards, the urgency of the introduction of resource-saving and environmentally friendly technologies which are conditioned by the needs of modernization and increase of competitiveness of the economy, including tightening of conditions for access to the international markets for goods and services in accordance with accepted international environmental standards and increasing demands for environmental quality and safety of products, including in environmental technology parameters.
2. The possibility of modernizing the economy during the economic crisis, which involves the need for the introduction of energy-saving and environmentally friendly technologies and to increase public funding of environmental measures. This, in turn, requires a strengthening of environmental priorities in the state economic and social policies that will eventually lead to an increase in the competitiveness of Russian companies on the world markets.

Figure 11. Environmental vision of future for Novosibirsk and Novosibirsk region
Source: comosition of author

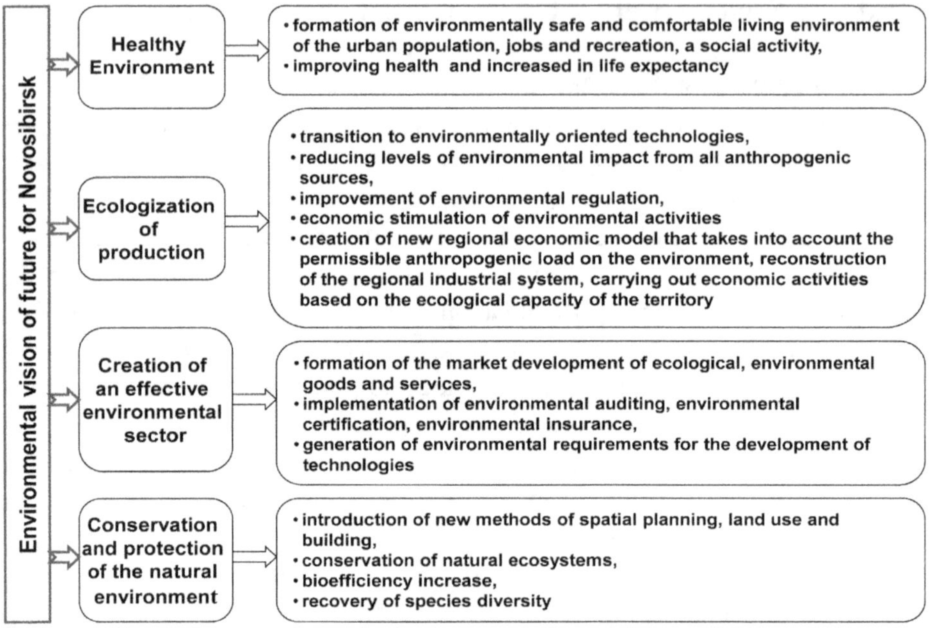

3. The weakness of the current state of environmental policy in the Russian Federation, which is characterized in particular by the following main features:
    a. the imperfection of assessment of the taxable base of the using natural resources (low interest rate payments for the use and reproduction of natural resources; the practical absence of the payments for recycling and re-side use of natural resources), etc.;
    b. low specific weight of resource taxes and environmental payments in the payment system, forming the budgets of all levels;
    c. insufficient financing environmental activities from the budgets of all levels, specialized funds, funds of enterprises and etc. One result of inefficient state environmental policy is unsatisfactory state of the environment in many regions of the country;
    d. existing practice of levying payments for negative impact on the environment has a number of weaknesses:
        i. are extremely low base rates, which is why their role is purely symbolic;
        ii. the legal status of these payments is not adjusted, the relevant law is not adopted so far, although it stems from the need of the law "On Environmental Protection (as amended December 29, 2015)";

    iii. refusal of targeted use of the relevant funds in the budgets of the Russian Federation and the regions;
    iv. a set of substances for which payments have been established, is far from complete;
    v. insufficient consideration of inflation factor - adjustment of payments occurs with a delay, and the value of the correction factor to the regulatory charge is not comparable with the actual rate of inflation.

4. The need to develop and implement new and effective tools in the field of environmental regulation, allowing to stimulate, on the one hand, the ecological modernization of production, development and use of environmental technologies, the formation of the market environment-friendly products and environmental services; and, on the other hand - an environmentally responsible behavior of business. Such tools should be supported by both legislation and provided with appropriate implementation mechanisms.
5. Technological backlog of industrial enterprises (for example, most of the technology engineering companies refer to the 4th technological way), physical obsolescence of technological equipment park that is directly related to the environmental risk and economic inefficiency of the technology used, and requires a transition to new environmentally safe technologies.
6. The need for improvement and strict compliance with environmental legislation, which, despite all the positive intentions, remains almost unchanged. Emasculation of the essence of environmental legislation as a result of numerous reforms has led, in particular, to the fact that the environmental impact assessment of projects (including the environmentally dangerous) turned out to be unnecessary, resulting in the growth of the environmental risks and is fraught with unpredictable consequences for the safety of the environment and human health. The main drawback of the existing environmental legislation is the absence of it instruments to promote the use of environmentally friendly technologies. As a result, objects-pollutants is more profitable to pay fines for environmental pollution (because their level is still quite low), than to carry out the actual environmental activities.
7. Imperfect methods to determine the economic damage caused by economy and people's health by pollution of the environment; the need to eliminate accumulated environmental damage from past economic activity.
8. The imperfection of statistical reporting on the use of natural resources and environmental protection, weak control financial discipline of resource and environmental charges; the absence of a modern system of monitoring, evaluation and prediction of the environment, emergency situations of natural and man-made disasters and adverse climate changes.

9. The tightening of conditions for access to international markets from the standpoint of environmental standards and regulation; increased international competition due to increasing demands for environmental quality and safety of products, the transition to the integration of environmental parameters, not only products, but also the technology used for its production.
10. Low environmental responsibility of the business, which is manifested especially in the weak economic incentives for usingf natural resources to comply with environmental requirements, and as a result - low investment activity in environmental protection measures. Ignoring the ethical aspects in the field of ecology and generally low ecological culture of the population. Ecological awareness in Russia is at the initial stage of formation, so it is extremely important not only the formation of environmental ethics, a respectful attitude to natural environment, the strengthening of the principles of eco-efficiency and environmental justice, but also timely ethical evaluation by the country's leadership the facts of gross violation of the environment.
11. The need for a purposeful state economic policy focused on the greening of production, provided a systematic approach to solving problems of structural and technological transformation of the economy in favor of resource-saving and environmentally friendly industries that would not only lay the foundations and to strengthen the innovation economy, but also to provide both economic and environmental benefits.
12. The need for the use of strategic planning and management in the formation of the state environmental policy, since environmental problems are usually long-term and require a strategic approach to their solution.
13. Nature conservation activity in Novosibirsk consists of measures aimed at solving revealed and formulated ecological problems that have both occurred on its territory and may occur when implementing planned investment projects, as well as of a system of nature conservation measures determined by the need to alleviate or warn about possible ecological problems (see Table 2).

In accordance with this logic, let us consider the main steps in developing an environmental protection strategy in the region (Fig. 12).

The strategic role of the regional ecological program can be formulated as providing ecological safety in Novosibirsk region by stabilizing and improving the ecological situation, and conserving and restoring the integrity of natural ecosystems. This aim can be achieved by solving the following problems:

- Conservation and improvement of the quality of the natural environment, reduction of the negative impact on it during scientific and technological developments of the economy;

Figure 12. Scheme of the process of development and implementation of the Program "Environmental Protection Novosibirsk Region"
Source: composition of author

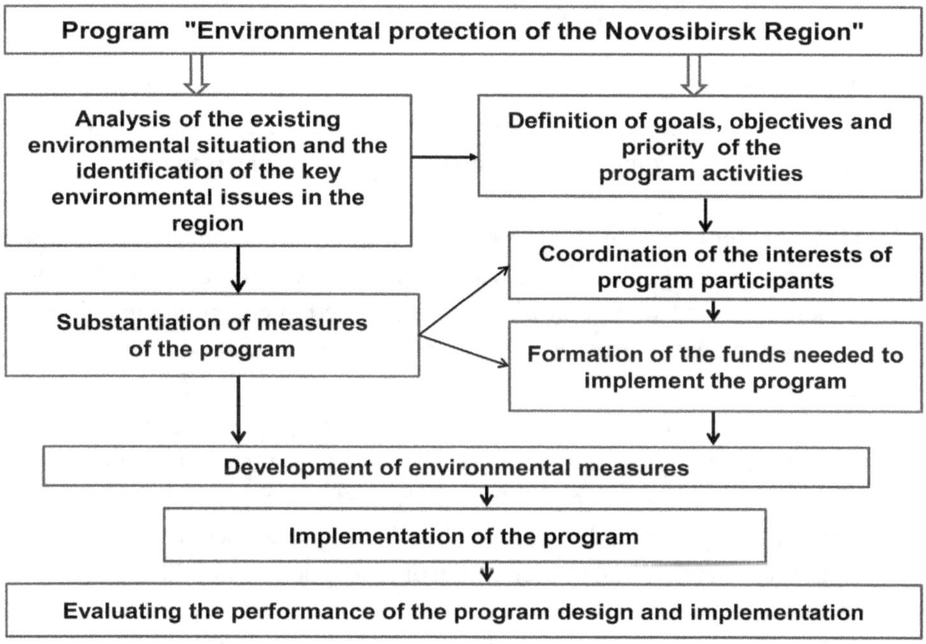

- Ensuring protection of the population, objects of the economy and territory of the city and region from harmful water impact;
- The creation of an effective system for dealing with industrial and consumer waste;
- Protection of the city from radiation, as well as reduction to a socially acceptable level of the risk of radiation impact on people and the environment they inhabit;
- Implementation of measures for preserving and reproducing (restoring) renewable natural resources as components of the natural environment;
- Observance of ecological regulations on economic activity envisaging correspondence of its scales and risk of impact to the ecocapacity of natural systems;
- Development of a network of specially protected natural territories and conversation of biodiversity;
- Improvement in the system for regulating natural resource use, environmental conservation, and the corresponding regulation mechanisms;
- Increasing the level of ecology education and advocating care for nature.

In accordance with the stated aims and problems of the program "Environmental Conservation of Novosibirsk region" and the main positions of the Socio-economic Development Strategy of Novosibirsk region for the Period up to 2025, the overall strategic priorities for environmental conservation are as follows:

1. Wide use of ecologically clean technologies (at both active and planned plants) to provide a technological basis for ecologically safe economic activity with transfer of the regional and state economy on the whole to an innovative path of development;
2. The equipping of businesses with nature conservation equipment, technological reequipping and gradual retirement of businesses with obsolete equipment, reduction in total water consumption in production and communal housing, and development of a system for using secondary resources, including waste recycling.
3. Ecological regulation of economic activity that bring into correspondence with established ecological standards the scales of the latter and the impact risk of individual industrial objects (primarily the fuel energy, metallurgical, and chemical complexes as the ecologically most important sectors of the oblast's economy) and territorial concentration of production in individual parts of the region;
4. Elaboration and implementation of measures to strengthen territorial administrative structures active in environmental conservation, the development of an ecological monitoring system, expansion of ecological control of industry and types of activity, including potentially hazardous types, independent of administrative jurisdiction and forms of property;
5. Prediction of the level of ecological expenditures for the oblast as a whole and for individual elements of its economic complex with allowance for planned production growth rates;
6. An ecological accountability mechanism for the economic activity of subjects for negative impact on the environment and increased interest in implementing nature conservation activity, including improvement in the order of fines for negative impact on the environment.

In accordance with these ecological priorities, nature conservation activity in Novosibirsk should be directed at the following: sequential reduction in negative technogenic impact on the environment; prevention of degradation of natural complexes when implementing new investment projects; use of a nature factor for restoring and improving people's health (primarily by development of tourism and creation of recreation zones); increased quality of the drinking water supply; combating underflooding of the territory and feral herd infection (tick-borne encephalitis).

Orientation toward these priorities will make it possible to gradually improve environmental quality in Novosibirsk and, on this basis, implement principles of stable development of the oblast in the mid- and long-term perspective with allowance for solving problems on environmental conservation.

Targeted Environmental Indicators

To assess the effectiveness of the ecological policy in the region from the standpoint of achieving the stated aims and tasks, a system of indices can be used that characterize the ecological processes occurring here and include the entire set of parameters that adequately describe the state of the natural environment. The system of indices were substantiated previously and published by us in Burmatova (2015b).

The tasks of assessing the ecological and economic effectiveness of the measures of environmental conservation of Novosibirsk region are as follows: to obtain quantitative criteria on the admissibility or inadmissibility of implementing a particular measure; to ensure a choice of planned economic activity with the least losses; to obtain quantitative criteria on assessing the effectiveness of planned nature conservation measures.

The main generalizing indices of the efficiency of carrying out these measures can be as follows:

1. The total amount of emissions of pollutants into the environment (atmosphere and water basin, in tons and cubic meters, respectively, per person per year);
2. A reduction in the volume of air pollution from stationary sources per unit of gross regional product (GRP);
3. A reduction in the volume of air pollution from mobile sources per unit of GRP;
4. The mean annual growth (reduction) of the volume of pollutants dumped into water bodies per unit of GRP;
5. A decrease in the volume of unrecycled industrial and consumer waste.

A reduction in pollutants into the environment per unit of GRP means a strengthening in the ecocompatibility of the applied technologies, growth in the efficient operation of gas-purification equipment, a reduction in the energy intensity of production, improvement in the quality of the environment, and a decrease in the negative impact of the economy on health. Increasing the level of recycling and rendering wastes safe will entail a reduction in the ecological hazard of waste accumulation and will serve as a characteristic of the effectiveness of the waste regulation system. The achievement of planned targeted environmental targets would indicate the possibility of not only preserving the achieved quality of the environment, but also its progressive improvement.

In addition, to assess the results of achieving the aims and solving the tasks of the program, specific particular indices can be used that characterize different aspects of the impact on the region's environment taking into account its economic, social, and ecological specifics.

To solve the stated problems, it is necessary to have the targeted ecological guidelines—indices for the effectiveness of environmental quality control, which are listed in Table 3 (prognostic estimates have been made using a mobilization development scenario for Novosibirsk region (Strategy, 2007)).

The air pollution forecast for Novosibirsk in terms of the most widespread pollutants coming from stationary sources for the period of 2005–2025, taking into account the aforementioned measures and parameters of the mobilization scenario (Strategy, 2007), testify the real possibility of a trend toward gradual improvement in air quality in the region (Figure 13).

The target ecological indices reflecting the strategic ecological priorities and key ecological problems are designed to be used by regional authorities as a tool for monitoring activity in the sphere of ensuring stable territorial growth. This makes it possible to simultaneously improve and adjust the measures of the program along with ensuring that the ecological goals in the region's development are met in case this activity is ineffective.

Meanwhile, these indices should be accessible and clear to other interested stakeholders, including social organizations and the regional population. This is a very important condition, because ensuring the ecological welfare depends equally on the actions of individual people and on the work carried out by regional authorities.

Table 3.Forecasted target environmental indicators for the Novosibirsk region

Indicators	2010-2013	2014-20016	2017-2020
Average annual increase in the volume of emissions into the atmospheric air from stationary sources per unit of the GRP	8-10	7-9	5-7
Average annual increase in the volume of emissions into the air from mobile sources per unit of the GRP	13-15	12-14	10-12
Average annual increase in the volume of waste water discharged into water bodies per unit the GRP	16-18	12-14	9-11
Volume of reduction of the unprocessed waste production and consumption, thousand tons	120–150	160–200	210–250

Source: calculations of author

Figure 13. Forecast of air pollution by the major pollutants by 2025
Source: calculations of author

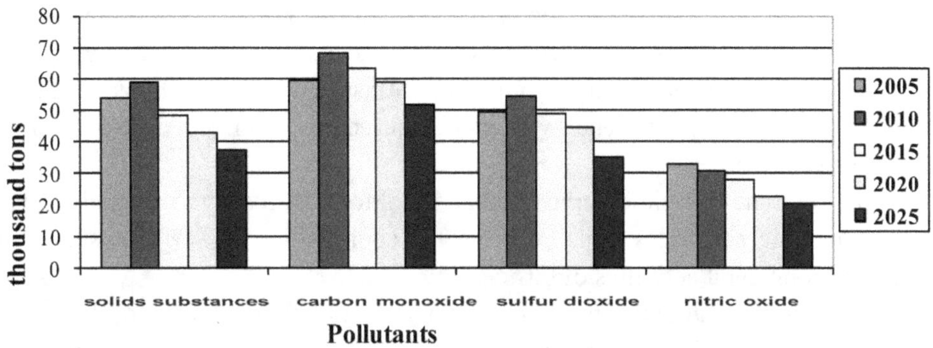

The corresponding structural administrative subdivisions of a federal subject collect data and prepare reports on the actual values of the indices for stable development so that all interested parties can see how effectively the program has been implemented.

One of the most important conditions for effective implementation of strategic ecological developments in the region is strengthening of state regulation in environmental conservation. The particular urgency in relation to this assumes the creation of an economic mechanism of ecological regulation aimed at stimulating rational natural resource use and environmental conservation; spurring and supporting ecologically responsible businesses; and improving the administrative regulatory structure and legal bases of environmental conservation. In many aspects, the effectiveness of implementing mid- and long-term prognostic documents concerning ecology also depends on the quality of the corresponding mechanism.

Using a Program-Oriented Approach, the Possible Variants of Solutions of the Problems and Assessment of Risks

Integrated solution of the mentioned problems of natural resource management and environmental protection in Novosibirsk requires the use of program-oriented approach, which allows you:

1. To take into account the magnitude, complexity and diversity of environmental problems of the region, the solution of which requires the consolidation of efforts and funding sources for the development and implementation of complex interrelated specific tasks, resources, and implementing activities of different nature to achieve the goals;

2. To coordinate the goals and objectives of the Program with goals and objectives of the other long-term programs of the Novosibirsk region ("Development and distribution of productive forces the Novosibirsk region", "Development of innovative activity in the economy and the social sphere in the Novosibirsk Region", "Development of transport infrastructure Novosibirsk Region", "Energy development, energy efficiency and energy security of the Novosibirsk region", etc.);
3. To combine the administrative and control tools of management and market economic principles, thus ensuring the coordination of various aspects of environmental activities of businesses;
4. Consistently integrate environmental objectives into the process of the socio-economic development of the area in order to ensure sustainable development;
5. To ensure the harmonization of setting and achieving balanced current and long-term environmental objectives;
6. To establish a clear priority in meeting the investment needs in the field of environmental protection, given the limited resources.

Thus, the program "Environment of the Novosibirsk Region" is seen as a key tool for planning, forecasting and implementation of regional environmental policies, as well as coordination of environmental activities in the Novosibirsk region. At the same time, the program acts as a method of implementation of the Strategy of socio-economic development of the Novosibirsk region for the period until 2025 and the Socio-Economic Development of the Novosibirsk region for the period until 2020.

The task can be solved by the several options corresponding mobilization (baseline) scenario Strategy of socio- economic development of the Novosibirsk region until 2025 (Strategy, 2007). Under this scenario, we consider two possibilities, reflecting the minimum and maximum options of mobilization scenario. Minimum scenario simulates the development of the Novosibirsk region on inertial type (reproducing the conditions and restrictions of 2000-2005). Maximum scenario describes the most complete use of the basic potential conditions for the development of the Novosibirsk region. Accordingly, there are two possible options for solving environmental problems.

The first option largely reflects current trends of development and distribution of productive forces of the Novosibirsk region and provides for the implementation approach to solving environmental problems, including mainly the establishment and improvement of methods and means to protect the environment at the end of the production cycle. In this case, the completion (or addition) of existing fixed manufacturing technologies or individual objects by different systems disposal of waste is carried out in order to prevent certain scale negative impacts on the environment (including treatment facilities for treatment of contaminated wastewater, installations for dust and gas extraction, water recycling system, the organization of waste management, construction of waste treatment plants, etc.) as well as activities to eliminate negative already committed violations in the state of the environment.

However, the possibility of the first variant, first and foremost in terms of economic and environmental efficiency of the technologies "end of pipe" are rather limited mainly due to the difficulties in achieving sufficiently high degree of purification of emissions and discharges, as well as in connection with an exponential relationship between the degree of extraction of harmful substances contaminants and level of expenditure for the necessary environmental protection measures.

At the same time, the failure to comply with of measures of this option significantly increases the risk of harm to the environment and human health. In addition, the cost of rehabilitation of territory in the event of violations (especially pollution) greatly exceeds the amount of investment required to prevent such violations.

In the capacity of the ecological risk assessments can be used indicators characterizing:

1. An increase in air pollution and acid rain formation (due to huge amounts of emissions of sulfur dioxide and nitrogen oxides produced during combustion, acid rain reduces the crop, destroys vegetation, destroys life in fresh water, destroys buildings, increases corrosion of metals and etc.);
2. Changes in the qualitative and quantitative status of surface and underground water sources (under the influence of excess pollution, violations of the hydrological regime of rivers caused by different kinds of human impact, etc.);
3. Formation of hazardous waste (toxic and radioactive) above permissible limits;
4. The expected economic damage (calculated and prevented) from possible contamination of the environment (air, water, soil, mineral wealth, etc.);
5. The volume of greenhouse gases (carbon dioxide, nitrogen oxides, methane, chlorine, etc.) and their accumulation in the atmosphere above certain concentrations (established the environmental standards for relevant ingredients);
6. Emissions of ozone-depleting substances (CFCs, chlorine and its compounds with oxygen, greenhouse gases); known that reducing the ozone layer by 1% leads to an increase of ultraviolet radiation by 1.5% and a corresponding increase in skin cancer from 2 - 3 to 5-7%. In addition, fall harvest crops, reduced phytoplankton productivity, the loss of many species of fish and marine invertebrates, etc. take place.

The procedure for evaluating the likely environmental risk is shown in Figure 14.

In the case of the using the second variant of solution to the problem in addition to the measures of the first variant is provided modernization and technical re-equipment of production through the introduction of resource-saving and low-waste

Figure 14. Procedure for evaluating the environmental risk
Source: composition of author

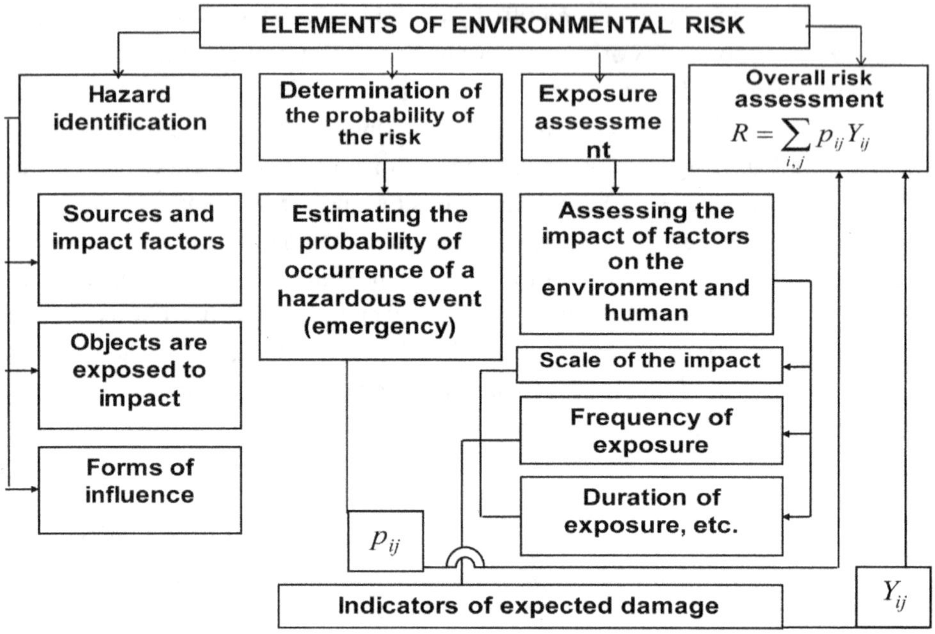

technologies. This variant is characterized by a high economic and environmental performance compared to the first, it will allow qualitatively change the ecological situation in the region through technical upgrading of existing facilities and the introduction of new facilities on the basis of high technologies for environmental safety operation of enterprises.

Currently, however, the second solution to the problem seems to be premature in view of the fact that in modern Russian conditions, exacerbated by the effects of a deep global economic and financial crisis the implementation of such large-scale, technically complex and expensive programs without preparatory activities and to develop mechanisms public-private partnership has a high degree of economic risk.

In this connection, preference is given to the first options of solution of the problem, according to which provides for the implementation of measures aimed at the progressive reduction to the lowest acceptable level of risk the negative impact of economic activities on the environment and the population.

Mechanism for Implementing Environmental Strategy

One of the most important conditions for effective implementation of the strategic developments in the region is to strengthen state regulation in the field of environmental protection. Of particular relevance in this connection acquires the solution of problems of the formation of the economic mechanism of environmental regulation aimed at promoting environmental management and environmental protection, promotion and support of environmentally responsible business, improving the organizational structure of management and legal foundations of environmental protection. From the quality of an appropriate mechanism depends largely the effectiveness of the implementation of medium-and long-term forecasting documents in the field of ecology.

Mechanism for the implementation of the regional program requires a definite complex of legal, economic, organizational, informational, and other measures, which are an integral part of the national environmental policy. Achieving favorable ecological situation as a prerequisite for a decent quality of life and health is possible only on condition of concerted action of regional authorities, business and the public in the field of environmental protection.

Conditions of implementation of environmental strategies should include, in our view, the following elements (Figure 15):

1. Economic mechanism of implementation of environmental policies;
2. Legal support of environmental policy;
3. Environmental financing;
4. Innovative aspects of environmental safety of economic development;
5. The organization of management, coordination and monitoring of the planned environmental protection measures.

Analysis of each of these directions of formation mechanism of the implementation of the environmental strategy requires an independent and sufficiently detailed exposition, therefore, restrict ourselves to brief remarks.

Main directions of the formation mechanism of the implementation of the environmental strategy depend on both the measures taken at the federal level and on the powers of the subject of the Russian Federation, in cooperation with local authorities and, in our opinion, should primarily include:

- Development and implementation of long-term and / or departmental target programs Novosibirsk region in cooperation with environmental problems;
- Development and implementation of projects and plans of the executive authorities of the Novosibirsk region, local governments;
- Implementation of investment projects and plans for development of enterprises that use natural resources;
- Improvement of the legal base in the field of environmental protection and environmental safety;

Figure 15. Possible structure of the mechanism of implementation of the environmental strategy
Source: composition of author

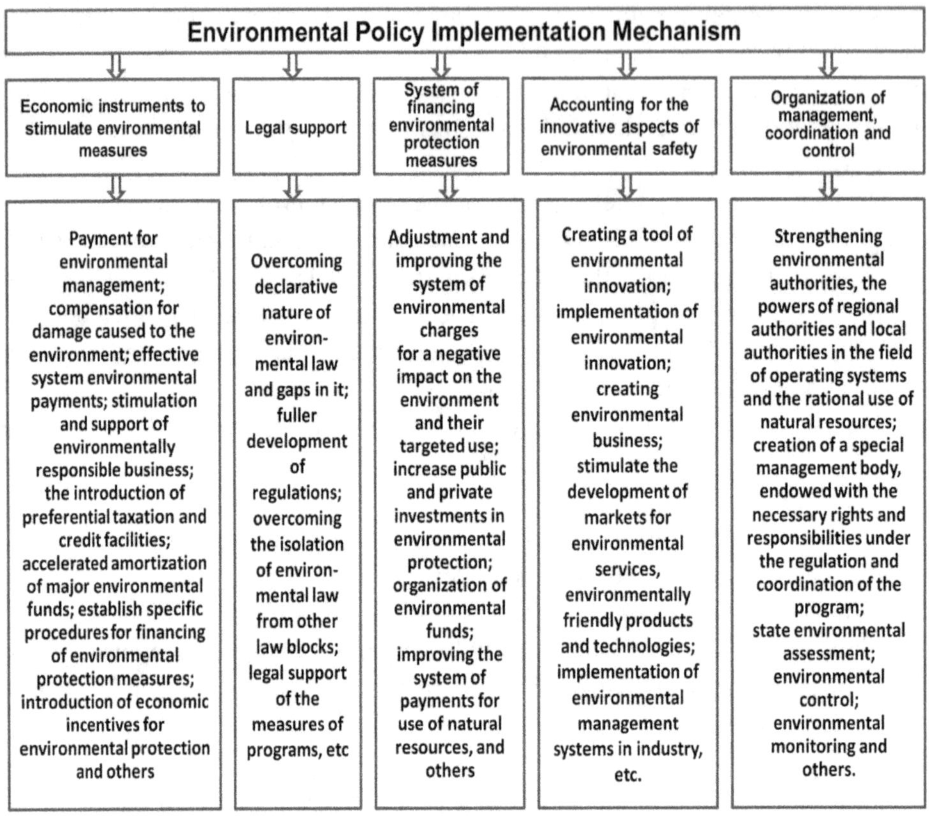

- Motivation of local governments, enterprises that use natural, scientific and expert community to actively participate in the implementation of planned activities;
- Attracting resources of the federal and local budgets, private sector and civil society organizations for the implementation of programs and projects;
- Maintaining the continuous monitoring and the analysis of the effectiveness of programs and projects.

Implementation of the strategic directions of environmental activities in the Novosibirsk region requires the creation of adequate financing environmental protection, which requires:

- Attracting investment in the environmental sector, mainly due to companies' own funds, increasing the share of equity in natural resources conservation activities;
- A clear delineation between the sources of funding for the protection of ambient between the company's own funds, extra-budgetary and budgetary sources;
- Improving the system of environmental charges and charges for using of natural resources;
- Use of the federal budget, increasing the financing of interregional environmental measures at the expense of the federal budget as co-financing;
- Strengthening the role of regional budgets in financing environmental programs and environmental protection measures, increase in funds for environmental measures as a part the budgets of the subjects of the Federation;
- Improvement of regional environmental funds;
- Increasing investment activity to the resource-saving technologies;
- Attraction of credits of the banks;
- Introduction of obligatory ecological insurance for series of potentially hazardous industries and technologies, etc.

For the purpose facilitating and supplement intraregional environmental financing as key directions are considered the mobilization of domestic resources which are the main source of funding for environmental activities, as well as more effective use of external resources. The main focus should be placed on expanding the budget and resource base and on improving the use of budgetary funds.

CONCLUSION

In general, a large city it is not only big problems (including environmental), but also a great opportunity for their decisions. Almost all environmental problems of the big cities not only have a tendency to be aggravated, but many of them more or less may be solved, as evidenced by the world practice, when the effective solutions for many ecological problems of big cities are being constantly seek and are being find. Formation of smart cities - one of them.

Ensuring sustainable development of the city as one of the ways to eliminate current environmental threats requires greening business activities in various fields, providing for a steady and consistent implementation of systems of technological, administrative, legal and other solutions to improve the efficiency of use of natural resources along with better (or at least preserving) quality of the environment. The success of the greening of the economy also largely depends on the greening of economic management tools and improve their institutional support.

Today's environmental challenges and threats put new challenges to society and require choice of new priorities, based on the principles of "green" economy, according to which economic growth and improving people's well-being must be at the same time reducing negative environmental impact. In other words, economic growth and protection of the environment are mutually reinforcing strategies.

Recommendations were made to improve the management mechanism of the regional eco-economic system. They include development and implementation of long-term and / or departmental target programs Novosibirsk region in cooperation with environmental problems; development and implementation of projects and plans of the executive authorities of Novosibirsk and the Novosibirsk region, local governments; implementation of investment projects and plans for development of enterprises that use natural resources; improvement of the legal base in the field of environmental protection and environmental safety; motivation of local governments, enterprises that use natural, scientific and expert community to actively participate in the implementation of planned activities; attracting resources of the federal and local budgets, private sector and civil society organizations for the implementation of programs and projects; maintaining the continuous monitoring and the analysis of the effectiveness of programs and projects.

Achieving the goals and objectives is aimed at addressing health improvement of the environment within the Novosibirsk region, reducing the anthropogenic load, maintaining acceptable levels of air pollution from stationary and mobile sources of emissions, organization of rational use and protection of water resources, improvement of the treatment of waste production and consumption, as well as identifying possible directions of improving control mechanisms in the field of environmental protection. Orientation to manufacture high-tech products, realization of the project on gasification of industrial and household sectors of the region and other projects provided for in Novosibirsk in the long term, will contribute (along with carrying out the conservation measures) improve environmental performance of the economy of the Novosibirsk region. Exit to the intended target environmental guidelines will testify about the possibility of not only preserving the achieved quality of the environment (primarily due to air quality), but also its progressive improvement.

In accordance with supposed ecological priorities, nature conservation activity in Novosibirsk should be directed at the following: sequential reduction in negative technogenic impact on the environment; prevention of degradation of natural complexes when implementing new investment projects; use of a nature factor for restoring and improving people's health (primarily by development of tourism and creation of recreation zones); increased quality of the drinking water supply; combating underflooding of the territory and feral herd infection (tick-borne encephalitis).

Orientation toward these priorities will make it possible to gradually improve environmental quality in Novosibirsk and, on this basis, implement principles of stable development of the oblast in the mid- and long-term perspective with allowance for solving problems on environmental conservation.

The results of fulfilled research allow to carry out a more informed choice of the main directions of environmental activities in the city under consideration, which avoid the possible risks and the best use of available resources. All of this is a prerequisite for the formulation of priorities in the environmental field and the development of strategies for the protection of the environment as an element of sustainable development of the city as a whole.

REFERENCES

Бурматова, О. П. (2009). Управление воздействием отраслей экономики на окружающую среду: Учебное пособие [Managing the impact of industries on the environment: Textbook]. Novosibirsk: NSU.

Чубик М.П. (2006). *Экология человека: Учебное пособие* [Human Ecology: Textbook]. Tomsk: Publishing House TSU.

Государственный доклад "О состоянии и об охране окружающей среды Новосибирской области в 2015 году." (2016a). [The State Report "On the state and Environmental Protection of the Novosibirsk region in 2015]. Novosibirsk. Retrieved from http://dproos.nso.ru/sites/dproos.nso.ru/wodby_files/files/wiki/2014/12/gosdoklad-2015_0.pdf

Государственный доклад «О состоянии санитарно-эпидемиологического благополучия населения в Новосибирской области в 2015 году». (2016b). [The State Report "On the state sanitary and epidemiological welfare of the population in the Novosibirsk region in 2015"]. Novosibirsk. Retrieved from http://54.rospotrebnadzor.ru/c/document_library/get_file?uuid=ea420274-bb3f-4a95-aef6-620cb759f4fc&groupId=117057

Государственные доклады «О состоянии и об охране окружающей среды Российской Федерации». (n.d.). [State reports "of the Russian Federation on the state and protection of the environment]. Retrieved from http://www.mnr.gov.ru/regulatory/list.php?part=1101

Характеристики основных загрязнителей окружающей среды. (n.d.). [Characteristics of the main pollutants of the environment]. Retrieved from http://www.projects.uniyar.ac.ru/publish/ecostudy/toxic2.html#0

Комбинированное (комплексное) действие ядов. (n.d.). [Combined (complex) action of poisons]. Retrieved from http://www.neonatology.narod.ru/toxicology/kompl_deistvie_jadov.html

Концепция охраны окружающей среды Новосибирской области на период до 2015 года. Утверждена распоряжением Губернатора Новосибирской области от 17.11.2009 N° 283-р. (n.d.). [The concept of environmental protection of the Novosibirsk region for the period up to 2015. Approved by order of the Governor of Novosibirsk region 17.11.2009 number 283-p.]. Retrieved from http://www.nso.ru/sites/test.new.nso.ru/wodby_files/files/migrate/activity/Socio-Economic_Policy/strat_plan /Documents/file895.pdf

Львович Н.К. (2006). *Жизнь в мегаполисе* [Life in the megalopolis]. Nauka.

Лысцов, В.Н., & Скотникова, О.Г. (1991). О возможности взаимного усиления вредных воздействий загрязняющих агентов окружающей среды [On the possibility of a mutually reinforcing harmful effects of polluting environmental agents]. *Журнал Всесоюзного Химического Общества им. Д.И. Менделеева, 1*, 61-65.

Навстречу «зеленой» экономике: пути к устойчивому развитию и искоренению бедности - обобщающий доклад для представителей властных структур. (2011). [Towards to a “green” economy: Pathways to sustainable development and poverty eradication - a synthesis report to representatives of authorities]. UNEP.

Петин, В.Г., & Сынзыныс, Б.И. (1998). *Комбинированное действие факторов окружающей среды на биологические системы* [The combined effect of environmental factors on biological systems]. Obninsk: IATE.

Программа реиндустриализации экономики Новосибирской области до 2025 года (утв. постановлением Правительства Новосибирской области от 01/04/2016 N° 89-п). (2016). [Program of reindustrialization of the Novosibirsk Region’s economy until 2025 (approved by the Government of the Novosibirsk region 04.01.2016 number 89-p)]. Retrieved from https://www.nso.ru/page/15755

Программа социально-экономического развития Новосибирской области до 2015 г. (n.d.). [The program of social and economic development of Novosibirsk region until 2015]. Retrieved from http://economy.newsib.ru/files/99713.pdf

Сибири, Э. (2009). стратегия и тактика модернизации [Economy of Siberia: The strategy and tactics of modernization]. Novosibirsk: Ankil.

Сонгдо — умный город будущего. (n.d.). [Songdo - smart city of the future]. Retrieved from http://green-agency.ru/songdo-umnyj-gorod-budushhego

Стратегический план устойчивого развития города Новосибирска и комплексные целевые программы. (2004). [Strategic Plan for Sustainable Development of the city of Novosibirsk and complex target programs]. Novosibirsk.

Стратегия социально-экономического развития Новосибирской области до 2025 года. (2007). [Strategy of Socioeconomic development of Novosibirsk oblast until 2025]. Novosibirsk. Retrieved from http://economnso.ru/files/1654.pdf

Умные города будущего: как строить полноценные интеллектуальные города в России. (n.d.). [The Smart Cities of the future: how to build a full-complete smart city in Russia]. Retrieved from http://sk.ru/utility/scripted-file.ashx?GroupKeys=net%2F1120292%2F&_cf=callback-standard-detail.vm&_fid=2609466&_ct=page&_cp=blogs-postlist&_ctt=c6108064af6511ddb074de1a56d89593&_ctc=1549&_ctn=7e987e474b714b01ba29b4336720c446&_cc=0&AppType=Weblog

Умные города: потенциал и перспективы развития в регионах России. Круглый стол (ГУ ВШЭ, Москва). (2014). [Smart Cities: Potential and prospects of development in the Russian regions. Round table. Retrieved from http://issuu.com/epliseckij/docs/bc9fac678b9405/5?e=7773934/7474790

Урбанизация и ее воздействие на состояние окружающей среды. (2011). [Urbanization and its impact on the environment]. Retrieved from http://phasad.ru/z1.php

В Японии официально открыт “умный город” Фудзисава. (2014). [In Japan, officially opened the “smart city” Fujisawa]. Retrieved from http://hitech.vesti.ru/news/view/id/6071

Burmatova O.P. (2015a). Nature conservation strategy for regional socioeconomic development. *Region Research of Russia, 5*(3), 286-297.

Burmatova, O. P. (2015b). Environmental and Economic Diagnostics of the Local Production Systems. In Functioning of the local production systems in Bulgaria, Poland and Russia theoretical and economic policy issues. Lodz: Łódź University Press.

Стратегия социально-экономического развития Сибири до 2020 года (Раздел IV. Приоритетные межотраслевые направления развития Сибири - Решение экологических проблем) – Утверждена распоряжением Правительства Российской Федерации от 5 июля 2010 г. N° 1120-р. (n.d.). [Strategy of socio-economic development of Siberia until 2020 (Section IV. Priority intersectoral directions of the development of Siberia -. Environmental issues) - Approved by the Federal Government on July 5, 2010 N° 1120-p]. Retrieved from http://www.sibfo.ru/strategia/strdoc.php

Hollands R. G. (2008). Will the Real Smart City Please Stand Up? *City, 12*(3), 303-320.

Kuleshov, V.V., & Seliverstov, V.E. (2016). Program for reindustrialization of the Novosibirsk Oblast Economy: Development ideology and main directions of implementation. *Regional Research of Russia, 6*(3), 214-226.

Matugina, E. G., Egorova, A.Y., & Palamarchuk, A. V. (2015, May). To the question of forming and development of ecological entrepreneurship. *SWorld Journal. Economy*, 66-68.

Resolution of the UN General Assembly adopted on 25 September 2015. (n.d.). Retrieved from https://documents-dds-ny.un.org/doc/UNDOC/GEN/N15/291/92/PDF/N1529192.pdf?OpenElement

Seliverstov, V.E. (2012). Regional monitoring: Information management framework for regional policy and strategic planning. *Regional Research of Russia, 2*(1), 60-73.

Seliverstov, V.E., & Melnikova, L.V. (2013). Analysis of Strategic Planning in Regions of the Siberian Federal District. *Regional Research of Russia, 3*(1), 96-102.

ENDNOTES

1 Novosibirsk is administrative center of the Siberian Federal District. It has the population of 1,6 million people (for 2016). Novosibirsk is the third city in Russia in terms of population and occupied territories. The leading industries are energy, gas supply, water supply, metallurgy, metalworking, mechanical engineering, they accounted for 94% of total industrial production of the city.

2 In this program, a key role belongs to Novosibirsk as a regional center and the city, concentrated basic economic potential of the region.

3 Roshydromet - Russian Federal Service for Hydrometeorology and Environmental Monitoring. Web site of Roshydromet: is http://www.meteorf.ru.

4 PAP - potential of air pollution - a combination of meteorological conditions that determine the dispersion (accumulation) of impurities coming in the form of emissions from industries and motor vehicles.

5 http://greenologia.ru/eko-problemy/goroda/novosibirsk.html - Экологическая ситуация в Новосибирске (The environmental situation in Novosibirsk).

6 https://dproos.nso.ru/ - website of the Department of Natural Resources and Environmental Protection of the Novosibirsk region.

7 https://dproos.nso.ru/ - website of the Department of Natural Resources and Environmental Protection of the Novosibirsk region.

This research was previously published in Smart Grid Analytics for Sustainability and Urbanization edited by Zbigniew H. Gontar, pages 27-72, copyright year 2018 by Engineering Science Reference (an imprint of IGI Global).

Chapter 2
The Development of the Environmental Impact Assessment Process in Botswana

Eagilwe M. Segosebe
University of Botswana, Botswana

ABSTRACT

The desire to protect the natural environment and the resources it hosts is at the core of every country's ambition to achieve long-term sustainable development. Yet, the history of development shows that environmental protection always came about as an after the fact activity. Nations around the world, including the industrial countries, most of the time reacted to adverse impacts of the environment rather than acting proactively to prevent undesirable outcomes of development. The 1970s decade is seen as a watermark in the development of legislation aimed at protecting the environment particularly in the developed world. This development and requirement to have projects subjected to environmental impact assessments spread to developing countries like Botswana in the 1980s. Botswana embraced the ideal to protect the environment to achieve sustainable development. Toward this end, the government of Botswana carried out Environmental Impact Assessment (EIA) on major projects voluntarily, even before the passing of the maiden EIA Act of 2005 and its supplement of 2011 and accompanying guidelines in 2012. After this and particularly in the 1990s, Botswana rigorously pursued its position on the international terrain where the country signed and ratified some international conventions and protocols. This chapter concludes by suggesting that the next area of focus for Botswana is on monitoring the outcomes of projects for which EIAs have been conducted.

DOI: 10.4018/978-1-7998-0441-3.ch002

Copyright © 2020, IGI Global. Copying or distributing in print or electronic forms without written permission of IGI Global is prohibited.

INTRODUCTION

This chapter begins by suggesting that environmental destruction is often inadvertent and that only in rare cases do people's activity seem to harm environmental resources deliberately as in the case of hunting. Then a global picture of the state of the environment is reviewed followed by a brief explanation of what Environmental Impact Assessment entails. Following on this a narration of how Africa was introduced to the new concept of environmental management is addressed. Then a brief on how the development of environmental impact assessment came about in Botswana is narrated, as well as the extent to which environmental concerns were considered in National Development Plans after independence. It is demonstrated that government concerns about environmental impact dates back to the period prior to the attainment of independence as shown by some of the Legislation that dates back as early as 1961. Notwithstanding the existence of legislation related to environmental impact, evidence on the ground showed growing concern by government about the ever-increasing pressure on natural resources due to increase in human population. Adverse environmental consequences drove government to motivate for the formulation and adoption of the National Conservation Strategy of 1992, as well as the enactment of the Environmental Impact Assessment (EIA) Act of 2005 and a revised Act of 2011 with associated regulations to ensure an effective application of the Environmental Impact Assessment procedure. What is of interest as suggested later is that government implemented EIA on some projects even before the EIA Act was enacted and adopted e.g. the Metsimotlhabe Water Transfer Scheme.

Perception on Environmental Problems

In order to appreciate the problems of the environment in Botswana, it is perhaps wise to begin by exploring fundamental human experiences and how people of the world perceive environmental issues. Thus knowledge of global experiences would better enable us to comprehend what is happening around us. As people we often take things occurring around us for granted and little do we realize that there is a purpose for everything that exists in the natural environment. We neither seem to always have foresight about the consequences of our activities nor comprehend the workings of nature. One must hasten to pronounce that the majority of ordinary people have no intention to destroy nature and its resources. Instead, their routine activities are aimed to eke out a living from natural resources. This is the case with livestock rearing, crop production, hunting, fishing, timber harvesting and so on. The purpose of these activities is not to destroy pasture or soil structure; deplete wildlife, fish, or timber, but to make a living out of them. But the reality is that with human population increase and multiple demands by individual users on the same resources, the natural capacity of natural resources (both fixed and renewable) are diminished. Even renewable resources are often overwhelmed, especially when the thresholds of their productivity is exceeded by increased exploitation. This applies to such resources as wildlife, fish, forests and water. Of course these would happen to varying extents depending on the nature of the resource in question. It is also worthy of note to point out that humanity is also concerned about imminent depletion of non-renewable resources such as minerals in the absence of readily available substitutes.

Because we do not always pay attention to what happens in our surroundings subsequent to our activities or otherwise, undesirable impacts often develop in our environment without being noticed, if humanity is not just showing indifference to such developments. In many cases impacts on the environment are often noticed after the fact or after we have implemented policies or projects. For example, it was only in the early 1970s, after damage on the environment had occurred that humanity began to show real concern about the state of the environment e.g. the introduction of National Environmental Policy Act in the United States America and Environmental legislation in Europe. It was only during the early 1970s that we saw the enactment of environmental laws by the forerunners of conservation in the developed world. Only then did these countries require Environmental Impact Assessments to be undertaken for major development projects. It is noteworthy to mention here that prior to these requirements environmental care has always been an after the fact initiative even in the industrial world. Pollution events such as those reported about the rivers Rhine and Thames in Germany and the United Kingdom, respectively and Rachel Carson's *Silent Spring* (1962) are cases that were reported way before the 1970s. Yet, it took nearly a decade after the publication of *Silent Spring,* which is credited with bringing the issues of pesticide pollution on the public arena. This was the past trend among the current industrial countries regarding the impact of natural resource exploitation i.e. that of complacency, indifference, or ignorance whatever the case may be.

It would seem trivial to suggest that all resource uses are exploited or destroyed for a reasonable course, particularly when it is common knowledge that some resources are destroyed out of fun or boredom. There are many people around the world who do hunting or fishing as a hobby, little realizing the destruction they inflict on natural resources. In some cases natural life is destroyed out of boredom as when someone kills a colony of ants on their busy trail when one happens to be near them or camping near their colonies. Of course this contrasts with the reality that more often than not environmental harm is inadvertent and largely results as side effect of good intentions of development to improve the wellbeing of the human population.

Global Picture of the State of the Environment

The destruction of the environment and subsequent social problems arising therein motivated concerned people to come up with the idea of a world revolution.

"The World Revolution is an idea for a new, global activist social movement for progressive social change..." http://www.worldrevolution.org/. According to this source there are four major issue areas of the World Revolution. These include peace, human rights, the environment, and world poverty. This chapter is concerned with the environment issue.

The World Revolution makes stunning revelation about the current state of the world. Among others it suggests that half of the world's forests have disappeared from the earth's land surface and that only one-fifth remains undisturbed. It also suggests that based on current trends the following developments are imminent:

- Between 10% and 20% of all species will be driven to extinction in the next 20 to 50 years.
- An estimated 34,000 plants; 5,200 animal species; one in eight bird species - face extinction.
- 25% mammal species will face extinction within 30 years.
- Up to 47% of the world's plant species are at risk of extinction.
- More than 20 percent of the world's known 10,000 freshwater fish species have become extinct, been threatened, or endangered in recent decades. Sixty percent of the world's important fish stocks are threatened from overfishing.
- Desertification and land degradation threaten nearly 25% of the land surface; over 250 million people are directly affected by desertification; and one billion are at risk.
- Global warming is expected to increase the Earth's temperature by 3C (5.4F) in the next 100 years; resulting in multiple adverse effects on the environment and human society, including widespread species loss, ecosystem damage, flooding of populated human settlements, and increased natural disasters.

The above claims cannot be far from the truth as science suggests that recent natural events around the world in the form of flooding, heat waves, and extended droughts are all manifestations of an unstable environment due partly to human activities. Therefore, it cannot be over emphasized that these events call for a concerted effort from the world community to minimize their effect on the environment through adaptation, technological development, and improved production processes. Environmental Impact Assessment has therefore been regarded as one of the few proactive processes that deal with impacts of human resource exploitation or development on the environment before serious damage is caused on the natural environment. The following section therefore provides a brief explanation of the EIA processes to provide context to the rest of the chapter.

Environmental Impact Assessment

Environmental Impact Assessment (EIA) is a process designed to gather information on components of the natural environment (such as soil, water, air, flora, landscape, fauna and drainage), and socio-economic environment (e.g. population, economic activities, health, policy and heritage) to identify pre-existing conditions in any given place prior to implementing development initiatives, be they policies or projects. The information gathered provides baseline data that informs decision-making to determine if a particular policy or project should be allowed to go ahead. Information is usually sourced by the developer from both primary sources and secondary sources based on experience from elsewhere. The components of the environment are then individually assessed using established methods of assessment (Morris and Therivel 2001) and observations to determine potential environmental effects of proposed development are made. The outcome of these assessments and observations will then be used to inform decision-making on development. Once the effects on the receptor are identified and known, alternative mitigation measures are proposed and subjected to a review together with the methods proposed therewith. The effectiveness of the mitigation measures proposed largely depends on the nature of the impact envisaged and the characteristics of the receptor. Therefore, ability to benefit from the EIA process depends on the traditions developed overtime toward the environment. Environmental impacts may be beneficial or adverse, short-term, medium, or long-term, reversible or irreversible, permanent or temporary (Morris and Therivel 2001), direct or indirect.

Africa

The 1970s was a turning point in the development of the African continent both from the socio-economic and environmental perspectives. While the general literature alludes to a period of agricultural self-sufficiency prior to the 1970s, the situation turned gloomy from this time onward. Since the 1970s, African countries experienced unprecedented extended periods of drought leading to widespread crop failures, food deficits, famines and loss of human life especially in sub-Saharan Africa. Consequently, according to UNEP (2002), during the decade of the 1970s environmental management was not considered paramount. At the time African governments and people concentrated on subsistence and survival means. They only awakened to the modern concept of environmental management following the adoption of the African Convention on the Conservation of Nature and Natural Resources of 1968 in Algiers, according to UNEP. Following this initial convention, Africa has since been a party to a number of international conventions, among them the Convention on Wetlands of International Importance especially as water fowl habitats (1971); the Stockholm Convention on the Human Environment (1972); the Convention on the International Trade on Endangered Species of wild fauna and flora that came into force in 1975. These conventions and many others introduced Africa to the modern environmental Agenda (UNEP 2002). Among these perhaps one that really set motion to the African continent toward the modern environmental agenda was the World Conservation Strategy by the IUCN (World Conservation Union) of 1980 which introduced the concept of sustainable development that was later popularized by the World Commission on Environment and Development (WCED 1987). The World Conservation Strategy encouraged African countries to develop their own national conservation strategies.

Botswana

Against the above background Botswana also jumped onto the bandwagon and started developing the Botswana National Conservation Strategy (NCS) in the late 1980s. In particular the government accepted the need to prepare the NCS in 1983 (Government of Botswana 1990 [GOB]). Interestingly and soon after this, the then Ministry of Local Government and Lands commissioned an Environmental Impact Assessment Statement in 1986 to oversee the development of dams and storages in the Kolobeng-Metsimotlhabe catchment area. Of particular interest here is the fact that the Government of Botswana started implementation of EIA on government projects nearly two decades (i.e. 19 years) prior to the enactment of the maiden EIA Act of 2005. This also occurred prior to the release of the Government Paper No. 1 of 1990 that constituted the National Policy on Natural Resources Conservation and Development for that matter. There must have existed, at the time, some major drivers to this initiative. The new environmental attitude and the sense of urgency to act by government emerged partly because the country experienced growing pressure on water resources, rangeland degradation, depletion of wood resources, rapid exploitation of some veld products and pollution of air, water, soil and vegetation resources (GoB 1990). The other reason that compelled government to implement EIA prior to enactment of EIA Legislation can be found in the quotation below.

In the absence of Environmental Impact Assessment legislation, environmental impact assessments on projects were carried out voluntarily. Most of the projects for which the environmental impact assessments were carried out were donor-funded. MFDP 1997:p.329 [14.26]

It would therefore not be unjustified to suggest that a major driver of the compulsion, on the part of government, to act voluntarily in carrying out Environmental Impact Assessments on government projects prior to the enactment of due Legislation was a result of pressure from donors (direct or indirect). There is no doubt that at this point in time it was common knowledge that rather than being localized environmental problems tended to be regional or global in nature and many donor agencies were uneasy about supporting projects for which EIAs had not been conducted. The difficulty of operating without an environmental Legislation in an economic-environment that is supported by donor funds is further supported by the following quotation,

This proved problematic in that each donor came up with a different set of environmental impact assessment procedures. 1997:p.329 [14.27]

The development of the NCS which later constituted the National Policy on Natural Resources Conservation and Development was the culmination of a long standing desire by government to balance economic growth with environmental protection. As illustration, Table 1 lists some of the legislation enacted since the early 1960s and prior to the EIA Legislation, which shows that there has been a long tradition of concern about environmental issues, albeit not acted upon. Despite the existence over a long period of time of environment-related legislation as illustrated in Table 1, there was no evidence on the ground that the legislation was implemented, let alone effective in easing out pressure on the environment. Even before the development of the NCS, government was already concerned with a number of environmental problems that besieged the country. These were later articulated in the NCS as follows: 1) rangeland degradation; 2) depletion of wood resources; 3) unsustainable harvesting of veld products, and 4) pollution of air, water, soil and vegetation or crop resources.

Table 1. Some environment related legislation in Botswana

Legislation	Date Enacted
Fauna Conservation Act	1961
National Parks Act	1968
The Forest Act	1968
Tribal Land Act	1970
Atmospheric Pollution (prevention) Act	1971
Agricultural Resources Conservation Act	1974
Health Act	1981

Source: MFDP 1991 (NDP 7)

This occurred despite the fact that NDP 7 acknowledged that development is not sustainable without effective conservation of resources. Perhaps this could partly be explained by the fact that emphasis was initially placed on resource pricing such as water tariffs to the down play of other equally important resource management instruments like implementation of legislation and incentive packages. In addition some government ministries did not express any concern about environmental problems in their development plans particularly over NDP 7 and 8.

Table 2 lists government ministries that had anything to do with natural resources to determine if they had expressed any concerns about environmental problems affecting their development over the two planning periods. Surprisingly the sectoral chapters of four out of the seven selected sectors (57%) did not consider or mention anything on environmental conservation in their sections of the National Development Plan. In particular Industry, Transport, Agriculture and Health were silent on environmental conservation issues over both National Development Plan 7 and National Development Plan 8 planning periods. The indifference toward environmental issues displayed by these sectors of government at the time came as a surprise because the four sectors are normally associated and should be concerned with pollution in their development plans. For example, to mention one problem associated with each of these sectors among many: carbon emissions from transport, hazardous chemicals from industry, agro-chemicals from agriculture and medical waste from health all pose real danger to the environment and people. It could also be added that although the energy sector rightly expressed concerns about environmental issues, there was an element of laxity as succinctly articulated in the quotation that follows from the energy section of the National Development Plan 8 (1991\98 – 2002\03).

Fortunately, there are some in-built security mechanisms against over-exploitation of wood fuel resources, which to a degree counteract this trend:

- *Increasing formal employment and associated incomes will allow people to switch to other fuels which are more convenient;*
- *Formal employment reduces the time available for wood collection; and*
- *The use of wood for fencing and shelter may decline, as the availability of other building materials increases. MFDP 1991:p.233 [10.60].*

The reasoning behind the above statement was that employment would allow people to adopt use of other fuels, which are convenient, perhaps because they can afford it. While this statement may be true for middle income people it may not apply to the low income people who constitute the majority of the population. The reality on the ground contradicts these assumptions. The absolute numbers of the unemployed far exceed that of people in formal employment. The majority of these people continue to use firewood for energy production and timber for

Table 2. Environmental concerns by selected ministerial sectors over NDP 7 and 8

Sector	NDP 7 1991 Concerns	NDP 8 1997 Concerns
Minerals	Air pollution; Landscape; health hazard; air quality	Air pollution; Landscape; health hazard; air quality
Industry	Not considered	Not considered
Transport	Not considered	Not considered
Energy	Wood depletion; Pollution	Considered not serious to be monitored in the future
Water	Pollution	Pollution
Health	Not considered	Not considered
Housing & Settlement	Conservation needs considered	Conservation needs considered
Agriculture	Not considered	Not considered
Tourism	Sustainability	Sustainability

Source: MFDP 1991; 1997 (NDP 7 & 8)

construction material. The assumptions of the Department of Energy came as a surprise because at this point in time the NCS had been adopted and the necessary institutional structures set up and running to promote environmental awareness and conservation. In addition, the National Development Plan 8 (NDP) was the first in a series of plans to include a chapter on Environmental Conservation and Land Use. Therefore, it is without doubt that by assuming as above the Department of Energy underestimated the importance of the energy problem or the general consequences of energy consumption in the long-term.

The goals of the Botswana National Conservation Strategy aimed at addressing distortions such as those expressed in the assumptions of the Department of Energy above, by developing a conservation strategy. Achievement of sustainable development could therefore only be realized through comprehensive evaluations of the impacts of major development initiatives prior to implementation (GoB 1990). Such evaluations would be delivered best in a conservation strategy. Following the establishment of the National Conservation Strategy, there was need to establish an authority to oversee the implementation of the strategy by the different sectors of government. Hence, institutional reforms were undertaken that resulted in the establishment of the NCS Advisory Board and the NCS Coordinating Agency. One of the major initiatives of these institutions was in ensuring the enactment of the NCS Act. These institutions initiated efforts that laid out the necessity for new developments projects (both public and private) to be preceded by professionally prepared and approved Environmental Impact Assessments (GoB 1990).

It would appear that the development of the national conservation strategy did not only help change government attitude toward development in terms of avoiding environmental impact, but it seems to have given impetus to government engagement with international concerns about environmental degradation. Since the development of the National Policy on Natural Resources Conservation and Development out of the National Conservation Strategy there has been a tremendous increase in the way government has taken interest and started participating in international conventions pertaining to the environment. Table 3 illustrates a chronological order of international conventions to which Botswana is a party. It is clear from the table that beside the top two, Botswana became a party to some international conventions/protocols at

the time when the Botswana National Conservation Strategy had just been developed and to many more thereafter. This is in stark contrast to the enactment of environment related legislation (Table 1) that began in the early 1960s. Table 3 also shows that within a short period of time (the 1990s decade) Botswana was able to join the rest of the world in becoming party to international conventions on environmental protection. At independence there was little, if any mention of environmental concerns in government cycles. Attainment of political independence following the long drought of the early sixties (one of the longest on record at the time) ensured that much emphasis of development focused on improving the economy of the country and the welfare of the people. There was little room for anything but economic growth. Notwithstanding government efforts pertaining to environmental conservation and sustainable development discussed above the first Legislation pertaining to EIA was enacted in 2005, thus for the first time paving a way for guided development.

EIA Legislation

The maiden EIA Act of Botswana was first published in 2005 and its supplement followed in 2011 with accompanying regulations published in 2012. The contributions of each one of these documents are summarised below.

EIA Act of 2005

As discussed earlier, the EIA Act came into operation rather late relative to the actual implementation in some selected government projects due to reasons provided above. Upon its inception the legislation was intended as;

An Act to provide for environmental impact assessment to be used to assess the potential effects of planned developmental activities; to determine and to provide mitigation measures for effects of such activities as may have a significant adverse

Table 3. Conventions and protocols to which Botswana is signatory

CONVENTION / PROTOCOL	DATE SIGNED / RATIFIED
The African Convention on Conservation of Nature and Natural Resources	Signed on 15th September 1968 Not ratified
The Convention on International Trade in Endangered Species of Wild Fauna and Flora (CITES)	4th November 1977
United Convention on the Law of the SEA	2nd may 1990
Vienna Convention on the protection of Ozone layer	4th December 1991
Montreal Protocol on Substances that deplete Ozone layer	4th December 1991
United Nations Framework Convention on Climate Change	27th April 1994
Permanent Okavango River Basin Commission	1994
SADC Protocol on Shared Water Courses	1995
Convention on Biological Diversity	12th October 1995
Convention to Combat Desertification (CCD)	26th January 1996
Helsinki Convention on the Law of Non-navigational Use of International Water Courses	1996
RAMSAR Convention	9th April 1997
London Amendment to Montreal Protocol	13th May 1997
Copenhagen Amendment to Montreal Protocol	13th May 1997
Basel Convention on the Control of Transboundary Movement of Hazardous Wastes and their Disposal	20th May1998
Convention Concerning the Protection of the World Cultural and Natural Heritage	23rd November 1998
SADC Protocol on Wildlife Management and Law	2001
Cartagena Protocol on Biosafety to the Convention on Biological Diversity	Signed on 1st June 2001 and ratified in June 2002

Source: Environmental Assessment Guidelines, 2005

impact on the environment; to put in place a monitoring process and evaluation of the environmental impacts of implemented activities; and to provide for matters incidental to the foregoing. EIA Act 2005. A.32

The Act stipulates critical requirements under the Legislation that have to be followed before any activity can be carried out. It is not intended to provide a full description to the EIA Act, suffice it to mention a few subsections. This is done cognizant to the fact that like any other legislation, the EIA Act provides for appeals where individuals feel violated. Recourse to the minister are also an option in case of international concerns. Among others, the Act lists the following requirements before an activity can be authorized:

1. States requirements that are essential to undertake an activity and accompanying penalties for contravening the various subsections of the Act.
2. Addresses issues of licensing and requirements that need to be satisfied before a proposed activity can be authorized
3. Indicates authorization requirements such as the need to provide preliminary EIA for projects and strategic environmental assessments for policies. A description of the proposed policy or project as the case may be is also required
4. Provides for public participation and input in the proposed activity together with the scoping exercise that gives an idea of the scale of the project and by extension the magnitude of potential impacts of the proposed activity.
5. Specifies qualifications and experience required of personnel to be engaged as consultants.
6. Emphasises the need for the public to be given the opportunity to review environmental impact statement.
7. States conditions of rejection or approval of the proposal.
8. Requires that a monitoring and evaluation report of implemented projects should be produced by the technical departments or local authority at times determined by them to check for compliance.

EIA Act of 2011

This Act just extends the previous one by including subsections that were left out in the previous one (2005) such as making provision for the establishment of the Board. It then suggests the composition of the Board and how the Board will be constituted such as who may be allowed into the Board and who may not. It also stipulates the powers of the Board. The 2011 Act also provides for the establishment of the position of an executive Board Secretary, whose responsibility it would be to run the day to day duties of the Board. It also spells out disciplinary procedures where a complaint has been lodged about the omission to act or bad conduct on the part of a practitioner or bad. The Act also has financial provisions regarding charges from the Board on the services they provide and how records on transactions are to be kept in compliance with standard practice. As these additions were not provided for in the 2005 Act they added value to the operations of the Board in executing its mandate.

Environmental Assessment Regulations

The Environmental Impact Assessment Act was followed by accompanying Environmental Assessment Regulations of 2012. They do so by identifying projects and policies for which EIA is mandatory. The regulations are meant to provide guidance as to the conduct of the EIA process in matter relating to project briefs, terms of references, registration and certification of practitioners. The regulations identify projects for which EIAs are mandatory e.g. projects that are located in environmental sensitive areas. Environmentally sensitive areas are defined as:

...any development in national parks, game reserves, wildlife management areas, wetlands, within 1kilometre of open surface water, flood plains, important breeding areas for fauna, within 2kilometres of important archaeological, historical, religious or cultural sites, areas protected underlegislation, hilly areas and areas containing rare or endangered flora and fauna.

There is also a wide range of development activities across a wide range of infrastructural, development, construction, recreational activities with given thresholds of development. In other words EIAs would be mandatory where given thresholds such as size of projects or hazard levels of materials dealt with would dictate that an EIA be conducted in such activities as agriculture, installations, manufacturing, energy and chemical industries to mention but a few. EIAs are also required for transboundary projects. Outside the mandated projects environment management plans are required or determination by the competent authority in the Department of Environmental Affairs.

CONCLUSION

After many years of infrastructural development, policy formulation and other developments that did not take due care for environmental protection since independence in 1966 Botswana finally began on the path toward environmental conservation. It started with the enactment of Legislation related to natural resources conservation. Albeit not implemented, it no doubt kept the conscience about environmental protection alive as witnessed in the developments towards the National Conservation Strategy, after which a lot was achieved towards the development of the EIA Legislation. But more needs to be done as the development of appropriate legislation has not been accompanied by equally robust implementation of monitoring. This the next critical stage for Botswana if environmental impact assessment efforts are to bring desired environmental benefits.

REFERENCES

Botswana Government Gazette. (2011). *Environmental Impact Assessment Act.* Author.

Carson, R. (1962). *Silent Spring*. Boston, MA: Houghton Mifflin Company.

Government of Botswana. (2005). Gaborone: Environmental Impact Assessment Act.

Government of Botswana. (2011). *Environmental Impact Assessment Act, 2011. Supplement A – Botswana Government Gazette. Gaborone. Government of Botswana. 2012. Environmental Assessment Regulations. Gaborone. Government of Botswana. 1990. Botswana National Conservation Strategy: National Policy on Natural Resources Conservation and Development.* Gaborone: Government Printer.

Ministry of Finance and Development Planning. (1991). *National Development Plan 7.* Botswana: Government Printer Gaborone.

Ministry of Finance and Development Planning. (1997). *National Development Plan 8.* Botswana: Government Printer Gaborone.

Morris, P., & Therivel, R. (2001). *Methods of Environmental Impact Assessment (2nd ed.).* Spon Press.

UNEP. (2002). *Africa Environment Outlook: past, present, and future perspectives.* Earth Print Limited.

Word Revolution. (n.d.). Retrieved from: http://www.worldrevolution.org/

This research was previously published in the Handbook of Research on Geospatial Science and Technologies edited by Joyce Gosata Maphanyane, Read Brown Mthanganyika Mapeo, and Modupe O. Akinola, pages 363-373, copyright year 2018 by Engineering Science Reference (an imprint of IGI Global).

Chapter 3
Modeling the Role of Government, Firm, and Civil Society for Environmental Sustainability

Humaira Yasmeen
Nanjing University of Aeronautics and Astronautics, China

Hashim Zameer
Nanjing University of Aeronautics and Astronautics, China

Ying Wang
Nanjing University of Aeronautics and Astronautics, China

Hina Ismail
NUML Multan Campus, Pakistan

ABSTRACT

The objective of this article is to design a game theory-based model to outline the role of the government, firm and civil society for environmental sustainability. The study used the dynamic game theory of complete information. Based upon the equilibrium analysis, the study highlights that when the punishment for non-compliance with environmental responsibility is smaller, the role of civil society would be higher for environmental sustainability. On the other hand, when the environmental responsibility cost is higher, then the role of a government is also higher for the implementation of environmental responsibility and to ensure the punishment. However, the authors found from model analysis that if the cost is low, the probability of firm to fulfill environmental responsibility is higher. In real life, the high cost of environmental responsibility is the main reason that the firm does not fulfill environmental responsibility. Under the high cost, the firm often has the phenomenon of bribery to the government and other means to avoid environmental responsibility. This article is a valuable policy guide for policy makers to cope with global environmental challenges.

DOI: 10.4018/978-1-7998-0441-3.ch003

Copyright © 2020, IGI Global. Copying or distributing in print or electronic forms without written permission of IGI Global is prohibited.

INTRODUCTION

Creating sustainable futures, countries need to emphasize on multiple initiatives including the reduction in greenhouse gas emissions, protecting a natural environment and focusing on human development (Sassi, 2016). Energy production and environmental issues have to be considered within the context of socio economic sustainability that enables and provides support for positive development of countries, communities and individuals (Epstein & Buhovac, 2014; Roseland, 2000). In the whole scenario of environmental sustainability, the communities have a main role. However, communities cannot be forced to adopt environmental friendly life style. But, they need to be offered an environmentally viable alternative to their current lifestyle. The goal of environmental sustainability cannot be achieved without simultaneously considering the quality of life.

Globally, in the context of developing countries, the living standards are very low as compare to developed countries (Poudel, 2016). However, in the context of developing country it is difficult to manage the environmental sustainability along with the quality of life. Currently, developing countries are facing environmental challenges due to an imbalance in the economic and social development of recent years (Qureshi, 2010). The major cities face haphazard, unplanned expansion due to the shift of the rural population toward urban areas that has worsen the situation (Waseem et al., 2014). The municipal authorities in developing countries also have limited resources; the haphazard urbanization is the core reason of deterioration of natural resources (Qureshi, 2010). Moreover, Pakistan a developing country is facing a severe energy shortage since last two decades similar to other developing countries of the world (Khalil & Zaidi, 2014; Valasai et al., 2017; Zameer & Wang, 2018). Presently, to overcome an energy shortage, the government has taken initiatives to install coal-based power plants (Massarrat Abid & Ashfaq, 2015; Edenhofer, Steckel, Jakob, & Bertram, 2018). On one end coal-based power plants will help to overcome the energy shortage in the country. On the other hand, it is harmful to the environment (Coutinho & Butt, 2014). At a time when developed nations are shifting their energy sources from coal to renewable and environmental friendly energy production, the initiative of Pakistani government to install coal power plants in the country is strange enough. No doubt, energy is an essential element for the quality of life and economic development. However, it's the responsibility of the government to ensure the availability of cheap, reliable and environmental friendly energy in the country.

Furthermore, the industrial sector is one of the growing sectors worldwide. As a result of industrial sector growth, environmental issues have evolved world-wide (Eliana Andréa Severo, de Guimarães, Dorion, & Nodari, 2015). However, in that perspective, the environmental degradation in any country can harm the overall ecosystem, and its negative impacts can also be felt in neighboring countries (Rosenbloom, 2001; Eliana Andrea Severo, de Guimarães, & Dorion, 2017). The study of (Eliana Andréa Severo & Guimarães, 2015) indicated that a large number of firms have consensus that environmental sustainability can be a strategic part of organizational development and are designing business models to incorporate environmental problems and embracing competitive advantages. Similarly, the firms operating in the developing countries can reduce their environmental effect by using an environmental friendly production system. Qudrat-Ullah and Karakul (2007) suggested that new policy incentives should be introduced to cope with environmental issues. Such as, to cope with environmental issues, industrialized nations have signed a second commitment (Doha Amendment to the Kyoto Protocol) in 2012 to reduce national CO_2 by 18% during year 2013-2020.

Even though, environmental policies are designed and negotiated in the parliament of democratic countries, but ultimately policy makers need to get support from the domestic public in order to implement policies in effective manners (Bernauer & Gampfer, 2013; Zürn, 2004). However, to make environmental policies effective and acceptable, scientific literature advocates that the government should engage civil society in environmental policy development (Bernauer, Gampfer, Meng, & Su, 2016; Betsill & Corell, 2008; Yamin, 2001). The involvement of civil society can improve the transparency of policy options and bargaining positions (Grigorescu, 2007). And it also, enables citizens to get more information about policy reforms that help them to assess whether their representatives are working in accordance with local interests, and to hold them accountable (Drews & Van den Bergh, 2016). Moreover, civil society can influence the government actors to ensure the transparency and publicize the relevant information (Van Rooy, 2004). Based upon subsequent discussion, we can argue that the government, firm and civil society have a crucial role for environmental sustainability. But, no study in the past has addressed the phenomena that when, how and to what extent the government, firm and civil society should play their role to ensure sustainable environment for the nations.

Therefore, to address theoretical shortcomings, the purpose of this study is to design a new game theory-based model to outline the role of a government, firm and civil society to cope with environmental issues worldwide. This study used the dynamic game theory of complete information to understand the equilibrium solutions for the government, firm and civil society. First time, this study provides the basis for policy makers to understand how and when the government, firm and civil society must intervene and play a vital role for environmental sustainability.

LITERATURE REVIEW

Environmental issues and pollution are considered as extremely dangerous hazards worldwide. Natural problems are approaching highly risky points, because people are unaware of their insecurity. Today, countries are emphasizing to cope with the elements those are creating environmental issues and making lives difficult for individuals. There are many ecological issues in developing countries, however, to address the ecological issues no serious effort has been made. Moreover, many environmental issues are inherent in developing countries, which are of great concern for its sustainable economic future. These environmental issues include water pollution (Azizullah, Khattak, Richter, & Häder, 2011), air pollution (Ali & Athar, 2008), noise pollution (Mehdi, Kim, Seong, & Arsalan, 2011), deforestation (Ahmed, Shahbaz, Qasim, & Long, 2015), pesticide misuse (Tariq, Afzal, Hussain, & Sultana, 2007), natural disasters (Finch et al., 2016), climate change (Khan, Khan, Ali, Ahmad, & Ahmad, 2016), urbanization (Azam & Khan, 2016), desertification (Zhang, Andam, & Shi, 2017) and energy production (Coutinho & Butt, 2014; Qudrat-Ullah, 2015).

The main contributor to the intensity of environmental issues is the industrial sector development along with population growth (Önder & Akay, 2015). According to World Factbook (CIA) Pakistan a developing country is 7th populous country in the world with a population more than 200 Million. The population in all other developing countries is growing at a fastest rate and according to estimates it will become double in 25 years. If the population in the developing continues to grow at this rate, it will create worse conditions for the environment. Unfortunately, the developing countries don't have enormous resources to sustain such a huge population. Moreover, to fulfill employment and other requirements of the people, the industrial sectors are continuously growing in these countries. Currently, the industrial sector contributes more than 20% in total GDP. The industrial sector in mostly developing countries is majorly consisting of food processing, chemical, agriculture and other industries. Due to these industries, climate change issues are extremely sensitive in the context of developing countries like Pakistan. According to the study of Muhammad Abid, Schilling, Scheffran, and Zulfiqar (2016), Pakistan is highly vulnerable to climate change. Climate change forecasts have anticipated severe and more frequent droughts which may lead to the major loss of trees and subsequent deforestation (March, 2016).

Furthermore, water is an essential element for human survival. The freshwater encompasses 3% of total water on the earth. However, a very small percentage (0.01%) of freshwater is available for human usage. Unfortunately, this small percentage of fresh water is under immense stress due to urbanization and unsustainable water consumption in industry and agriculture (Azizullah et al., 2011). Moreover, industrial disposal creates worse conditions for the availability of fresh water (Daud et al., 2017). Environmental pollution becomes more severe due to urbanization, population growth and industrialization (Önder & Akay, 2015). Similarly, population growth and urbanization increases the traffic flow in the big cities that creates noise pollution. The study of Fiedler and Zannin (2015) suggested that to control the noise pollution, the traffic flow should be minimized. However, it is difficult to control the traffic flow because most of the industries are situated in big cities. The smooth movements of labor, raw material and finished goods from industries need a high traffic flow that increases noise pollution in big cities. Moreover, the misuses of pesticides are also creating severe environmental problems in developing countries. For example, as the study has shown that the ground water in Punjab and Sindh (Pakistan) has been found contaminated due to the huge use of pesticides (Tariq et al., 2007). To cope with the environmental pollution coming from factories, the government needs to design a cooperative compliance with factories. Further, the government should also engage civil society to cooperate with the government to control the urbanization and misuse of pesticides in developing countries.

In addition, energy production is highly skewed towered thermal sources (Wakeel, Chen, & Jahangir, 2016; Zameer & Wang, 2018). According to scientific literature, the energy production from thermal sources create worse conditions for the environment (Anwar, 2016; Zaman, Abdullah, & Ali, 2017). However, it is the responsibility of the government to ensure environmental friendly energy production. The study of (Zaman et al., 2016) has suggested the for environmental sustainability, renewable energy is a better alternative of energy production. For instance, the current energy production mix of the China, India and Pakistan are creating severe conditions for the environment in South Asia. Past studies have indicated that more than 32% of CO_2 and 42% SO_2 emissions are emitting from the energy sector (Anwar, 2016; Harijan, Uqaili, Memon, & Mirza, 2011; Zaman et al., 2017). Therefore, it can be argued that the governments should emphasize on energy production from renewable energy sources to safeguard the environment in the country. As studies of (Farooqui, 2014; Sheikh, 2009; Zameer & Wang, 2018) has highlighted that Pakistan has 1100 kilometers (km) a coastal line having huge potential of wind energy production, but the country is producing a small part of energy from the wind. Wind energy considerably decreases dependency on fossil fuels, which strengthen global energy security. And also wind energy doesn't emit dangerous gases also it doesn't contribute to global climate change or global warming. Therefore, the governments in developing countries should play their role for environmental sustainability and shift energy production from current sources to renewable energy.

Summing up the aforesaid discussion, it can be argued that the government and factories both are responsible for continuous damage to the environment. However, the common man is being affected at large by environmental issues. Thus, authors believe it is highly significant that the government, firm and civil society should play their role to cope with environmental issues. But, when, how and to what extent each participant should intervene and play a vital role for environmental sustainability, need to be answered. Therefore, designing a model to outline the role or each participant is extremely important.

Designing a model to address environmental issues, game theory provides a good theoretical basis for modeling to understand and encounter environmental issues. As the study of (Li & Xu, 2018) used game theory for pollution governance in china and studied the role of local government, central government and firms. Shi, Wang, Fu, and Xue (2017) applied game theory to cope with environmental challenges and found that best way is to long-term cooperation which can help to sustain environment. Gao et al. (2018) applied game theory for environmental risk mitigation. Therefore, we believe game theory is best fit to be used to outline the role of the government, firm and civil society for environmental sustainability. In a real world, the strategy of one player can significantly influence the actions of other players. As the study of (McVea & Charalambu, 2014), argued that participants in any scenario are convinced to act strategically. Similarly, in the case of environmental sustainability, the decisions of the government can significantly influence the actions of firms operating in the country. Also, the role of civil society can significantly influence the decisions of the government and firms. The study of (Cheffins, 1997) stated that a game is a situation in which actions of players can significantly influence the utility of others. Similarly, green regulations introduced by the government can significantly influence the profitability of the firms and also the benefits of society. According to (Schelling, 2010) game theory helps explore and understand the situation in which players, make interdependent decisions which can be defined as strategic interactions. Colman (2013) argued that game theory is that there are two or more players (decision makers) involved, each of them have a choice to act according to best of his/her knowledge to maximize its utility. In this paper, authors analyzed the available options for the government, firm and civil society to cope with environmental challenges worldwide. Initially, game theory was developed by John von Neumann (Poundstone, 1993). Myerson (2013) stated that game theory can be divided into two distinguish types i.e. static and dynamic games. In a static game all players make decisions simultaneously, however in a dynamic game sequential moves are performed. The famous game of the prisoner dilemma is categorized as a static game. However, the chess game is an example of a dynamic game (Zameer, Wang, Yasmeen, Mofrad, & Saeed, 2018). It can be stated that information are known to all participants, the game is categorized as a game of complete information. Conversely, when all participants don't have equal information, the game is categorized as a game of incomplete information. In the context of environmental sustainability, the government and civil society knows about the firms operating in the country. Firms also know the rules and regulations of the country. So, it can be argued that all the players i.e. the government, firm and civil society have equal information. Therefore, this kind of situation can be categorized as the game of complete information. Further, the decisions of firm and civil society are dependent upon the rules and regulations introduced by the government. Therefore, this kind situation in which firm and civil society act based upon the initiatives of the government and all participants i.e. the government, firm and civil society have equal information, the case is considered as dynamic game of complete information. Thus, the study used dynamic game of complete information to classify the role of the government, firm and civil society to address environmental issues worldwide. Moreover, the solution concept used in game theory is called Nash-equilibrium. The concept was named as Nash-equilibrium after mathematician, John Forbes Nash Jr. who introduced this solution concept (Osborne & Rubinstein, 1994). Nash equilibrium is a fundamental concept in game theory, and it is widely used in predicting the outcomes of strategic interactions in the social sciences.

MODEL AND METHODOLOGY

The study used a dynamic game theory of complete information for the assessment of environmental responsibility. Based upon the proceeding discussion, the study designed a game theoretic model to address the environmental issues. The scientific literature and common knowledge suggests that the government and industries are liable for environmental issues worldwide. However, the common man is most affected by environmental problems around the globe. In the context of this study, it is presumed that the information's are fully known to the government, firm and civil society. Therefore, in the context of complete information, scientific literature suggests that a dynamic game of complete information provides good theoretical basis to reach at an equilibrium situation (Bajari, Hong, & Ryan, 2010; Rasmusen & Blackwell, 1994; Wolters & Schuller, 1997). In order to achieve the objectives of the study, there is a need to first conceptualize the role of the government, firm and civil society in game analysis.

Three Players Dynamic Game Model

In this study, it is assumed that all three players those have concern related to environmental issues i.e. the government, firm and civil society are rational participants.

Assumptions and Parameters

1. The firm fulfills its responsibility for the purpose of sustainable environment, the cost of environmental sustainability is C, and in a case the firm doesn't fulfill its environmental responsibility then a firm need to pay a fine to the government with the amount of EC. However, under the normal situation, specifically in the context of developing countries, the fine imposed by the government is lower as compare to the cost of environmental responsibility. So, firms prefer to not satisfy its environmental responsibility and earn (1-E) C where E is coefficient and can be $0 < E < 1$.
2. The government is responsible for the supervision upon the firms operating in the country and if the government accepts bribes it will face the MC penalty imposed by global organizations; firm shall also be penalized by U(1-E)C. The penalty on the government may be imposed by the Kyoto convention that enforces governments to play their role for the purpose of environmental sustainability. When the government doesn't play the role for environmental sustainability it may face the restrictions on exports. Further, if the government doesn't play its role then it will also lose the earning UC that it supposed to earn from the enforcement of laws regarding environmental sustainability, where ($U > 1$). U is the intensity of fine or punishment.
3. The environmental loss cost for the society is K, if the government fulfills its responsibilities, it can be on the benefit side, and the benefit here is C. These benefits government will receive from the firm for environmental sustainability.
4. The role of civil society can be to monitor the behavior of the government and the firm and play their role accordingly. In a case when civil society doesn't play their role, the environmental benefit is 0. However, when the government collects fine, the amount will be utilized for the betterment of environment and society. So, in this case the benefit of society is 2UC.
5. The above are the utilities of the government, firm and civil society. It can be stated here that environmental problems are higher in developing countries due to poor legal frameworks. The strong legal system in developed countries ensures that firms confirms with environmental compliance. However, the problem for the firm is to satisfy environmental responsibility that is a dynamic game process because the government has a choice to introduce or not to introduce the green regulation. Whereas, the firm does not have a choice of fulfilling environmental responsibility and non-compliance with environmental responsibility because when the government introduce the mechanism to sustain the environment, then firm have to follow if firm don't follow the government regulations, the firm will face fine imposed by the government. However, civil

society has a choice to influence or not to influence the green mechanism for environmental sustainability. So, the game process is a significant theoretical tool to achieve the balance situation.

According to the aforesaid assumptions and payoffs, this study develops a model based on a dynamic game theory among the government, the firm and civil society using the dynamic game of complete information. The dynamic game involves the cooperative game between the government, firm and civil society. The specific dynamic game model is as follows:

Behavior of Participants and their Subsequent Probability

1. *The Government*: Introduce green regulations, the probability is (r) and No-green regulation, the probability is (1 - r).
2. *Firm*: Fulfill its environmental responsibility the probability is (p) and do not fulfill environmental responsibility (1 - p).
3. *Civil Society*: if play role for environment sustainability, the probability is (q) and not to play role, the probability is (1 - q).

In order to derive the utility function of each participant, the study used the game tree model to analyze it (Figure 1). Based upon the dynamic game among the government, firm and civil society, a game tree for environmental sustainability is shown in Figure 1.

Using the game tree (Figure 1), the study also used game matrix analysis of the parties to determine the utility matrix as shown in the table 1, from (I) to (VIII) below.

The functions (I) to (VIII) are the utility matrix of all participants. The first part is the utility of the government; the second is firm utility functions and third is the utility functions of civil society.

After the establishment of a game tree and payoff matrix, it is necessary to solve the dynamic game to reach equilibrium solution. Since the order of the game stage is government – firm– civil society, however, it is necessary to solve the payoff solution of the firm first due to its importance in all process. The optimal solution of the government and civil society are also being obtained. Finally, the optimal solution of all three participants is the equilibrium solution of the dynamic game model.

Equilibrium Analysis

The equilibrium solution of the dynamic game can be solved by the game tree and the income matrix. The equilibrium solution is being analyzed as follows:

Figure 1. Game tree based on dynamic game of complete information

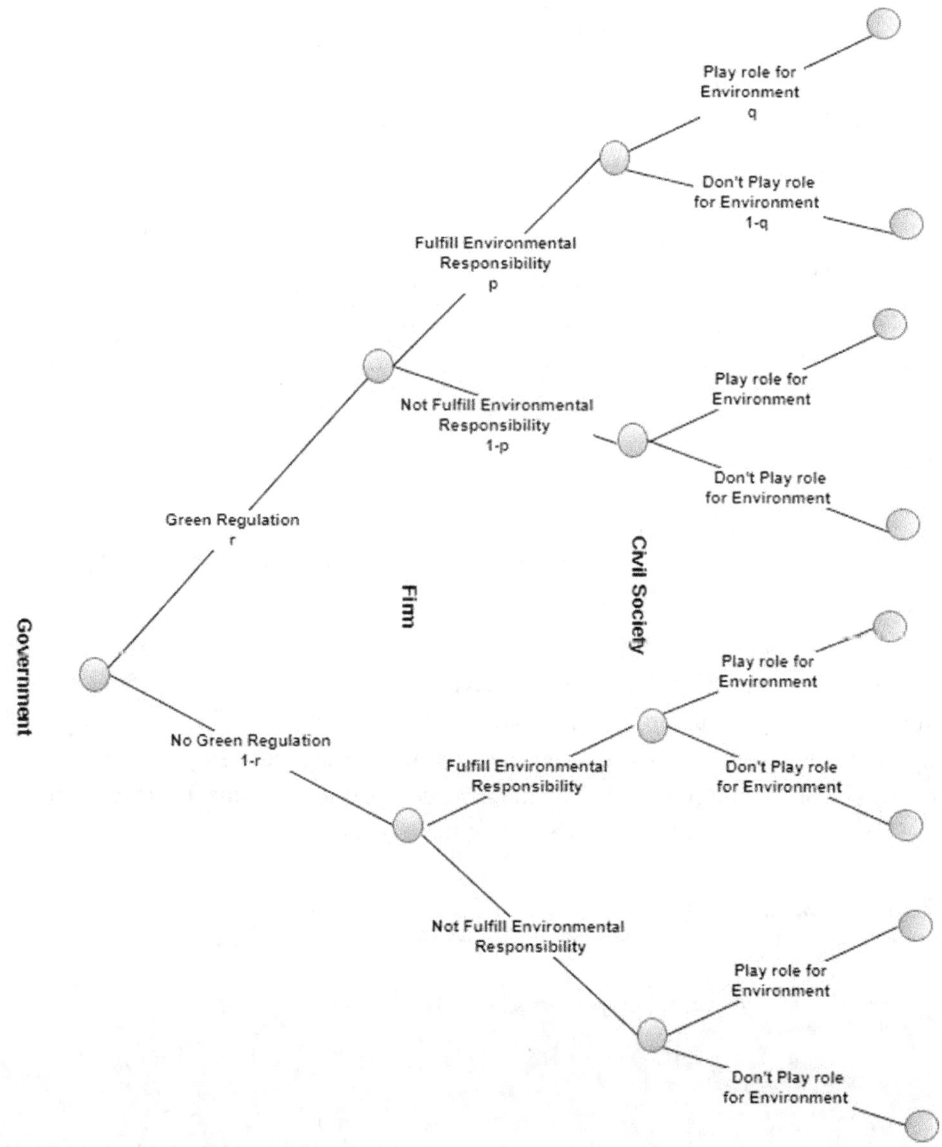

1. The Expected Return/Utility Function of the Firm

Based upon the payoff matrix, behavior of the participants and probability, the expected payoff equation for the firm can be written the as follows:

Table 1. Payoff matrix

	Government	Business	Society
I.	*EC*	(1-*E*)*C*	-*C*
II.	-*MC*	-*U*(1-*E*)*C*	*UC-K*
III.	-*UC*	-*C*	*UC-K*
IV.	*C*	-*C*	-*K*
V.	-*U*(1+*E*)*C*	-*U*(1-*E*)*C*	2*UC-K*
VI.	*C*	-*C*	-*C*
VII.	(1+*E*)*C*	(1-*E*)*C*	-2*C*
VIII.	0	-*C*	0

Equation 1

$$\begin{aligned} Firm = &(1-q)(1-r)(1-p)(1-E)C - q(1-r)(1-p)U(1-E)C - pqrC - q(1-r)pC \\ &-qr(1-p)U(1-E)C - (1-q)rpC + (1-q)r(1-p)(1-E)C - (1-q)(1-r)pC \end{aligned}$$

The above (equation 1) is an expected utility function of the firm which can be simplified based upon the dynamic game theory of complete information to get the optimal value from the function. As the dynamic game theory suggests that partial derivative of the function can be calculated to get the optimal value from the function. Let compute the partial derivative and let them be equal to 0:

Equation 2: Let

$$\begin{aligned} \frac{\partial Firm}{\partial p} = &-(1-q)(1-r)(1-E)C + q(1-r)U(1-E) - qrC - q(1-r)C \\ &- qrU(1-E)C - (1-q)rC + (1-q)r(1-E)C - (1-q)(1-r)C = 0 \end{aligned}$$

Equation 2 is simplified and we get the value as follows in the form of equation 3:

Equation 3

$$q = \frac{1}{(U+1)C}$$

2. The Government's Expectations

Based upon the payoff matrix, behavior of the participants and probability, the expected payoff equation for the government can be written the as follows:

Equation 4

$$\begin{aligned} Government = & -q(1-r)(1-p)MC - pqrUC + q(1-r)pC - qr(1-p)U(1+E)C \\ & +(1-q)rpC + (1-q)r(1-p)(1+E)C + (1-q)(1-r)(1-p)EC \end{aligned}$$

The expected payoff function of the government can be solved by taking partial derivatives to get the optimal value from the function. Let compute the partial derivative and let them be equal to 0:

Equation 5: Let

$$\begin{aligned} \frac{\partial Government}{\partial r} = & -pq + q(1-p)MC - pqUC - q(1-p)U(1+E)C \\ & +(1-q)pC + (1-q)(1-p)(1+E)C - (1-q)(1-p)EC = 0 \end{aligned}.$$

Equation 5 is simplified and we get the value as follows in the form of equation 6:

Equation 6

$$\mathrm{p} = \frac{1}{\mathrm{C}}$$

3. The Expected Payoff for the Civil Society

Based upon the payoff matrix, behavior of the participants and probability, the expected payoff equation for civil society can be written the as follows:

Equation 7

$$\begin{aligned} Civil\,Society = & \, q(1-r)(1-p)(UC-K) + qrp(UC-K) - q(1-r)pK \\ & +qp(1-p)(2UC-K) - (1-q)rpC - 2(1-q)r(1-p)C - (1-q)(1-r)(1-p)C \end{aligned}.$$

The expected payoff function of the civil society can be solved by taking partial derivatives to get the optimal value from the function. Let compute the partial derivative and let them be equal to 0:

Equation 8: Let

$$\frac{\partial\, civil\, society}{\partial q} = rp\left(UC-K\right)-\left(1-q\right)pK+r\left(1-p\right)\left(2UC-K\right)$$
$$+\left(1-q\right)\left(1-p\right)\left(UC-K\right)+rpC-2r\left(1-p\right)C+\left(1-r\right)\left(1-p\right)C=0$$

Equation 8 is simplified and we get the value as follows in the form of equation 9:

Equation 9

$$r=\frac{K}{U+1}$$

Based on the subsequent analysis, authors have accomplished the maximum expected utility/returns of the civil society, the government and firm. Hence, based on the dynamic equilibrium solutions of the firm, the government and civil society, it can be drawn three propositions.

Proposition 1

From equation 3, $q=\frac{1}{\left(U+1\right)C}$, it can be inferred that the intensity of punishment, U, business satisfies the environmental responsibility by paying the cost of C and the probability of civil society to play their role (q) is inversely proportional, When C, U is smaller, the probability q is larger, which reflects that smaller the punishment, greater the role of civil society for environmental sustainability.

Proposition 2

The results from equation 6 indicate that $r=\frac{K}{U+1}$ which means higher the level of punishment, the higher the probability of government intervention, which enhance the role of the government for environmental sustainability. But, it also reflects that the appropriate increase in penalties reduces the role of government intervention.

It means, when environmental cost for society is higher, and then the role of the government is also higher for the implementation of environmental responsibility and punishment.

Proposition 3

The optimal point equation 9 $p = \frac{1}{C}$ shows that if the cost of environmental responsibility is low, the probability of firm to fulfill the environmental responsibility is higher. In real life, the high cost of environmental responsibility is the main reason that the firm does not fulfill the environmental responsibility. Under the high cost, the firm often has the phenomenon of bribery to the government other means to avoid the environmental responsibility.

DISCUSSION

Based upon the subsequent analysis, this paper argues that to promote the environmental responsibility, policy maker worldwide should design and enforce green regulations. However, for the enforcement of green regulations, governments should engage civil society and increase their role for supervision. The coordination with the civil society and giving them right to monitor the environmental responsibility of the government and firm is proposed in the context of developing countries. Because in the context of developed countries; the enforcements of laws are already strong, which ensures that firm follow the environmental regulations. As discussion from the subsequent part has highlighted that the government and firm are the main participants that are creating worse conditions for the environment. However, due to poverty and low level of education in developing countries, the common public is unaware and can't be influential for environmental sustainability. Thus, the role of civil society should be strengthened to cope with environmental issues in these countries. Moreover, the distribution of environmental responsibility cost should be reasonable. The above analysis has indicated that the probability of the firm to fulfill their environmental responsibility depends upon the environmental cost. In real life, the high cost of environmental responsibility is the main reason that the firm does not fulfill the environmental responsibility. Under the high cost, the firm often has the phenomenon of bribery to government officials and use other means to avoid the environmental responsibility in the context of developing countries. The government collects taxes from a common man and also from the factories in developed and developing countries. Therefore, the governments in developing countries should compensate firms similar to developed countries, so that firms can bear environmental responsibility happily and don't try to find unacceptable ways to avoid environmental responsibility.

As the government regulates and control the environmental system in the country, so its role is the most crucial for environmental sustainability. So, it is the responsibility of the government to play the role for environmental sustainability. The governments in developing countries should not only introduce green regulations, but it should also facilitate the firms like firms are facilitated in developed countries. Furthermore, it can be seen that if the firm has the realization of environmental responsibility, the pressure of social supervision will be smaller; the supervision cost is also low. Thus, from the perspective of strengthening environmental responsibility, the important means can be to promote environmental responsibility, such as the governments in developing countries can emphasize on publicity and education related to environmental responsibility. Moreover, the governments in developing countries can encourage firms to play their role for energy saving and reduction in carbon emissions. Further, in order to stimulate the enthusiasm of firms to fulfill their environmental responsibilities, the government can make firms realize that environmental responsibility helps to improve their corporate reputation. It is conducive to long-term sustainable competitiveness of firms to upgrade, so it can also reduce the supervision of the government and civil society efforts. When the firm realizes to fulfill its environmental responsibility, it will be the win-win situation for all players.

CONCLUSION AND POLICY IMPLICATIONS

The decision-making process seeks consensus among scholars and practitioners in providing an outline/ mechanism to cope with global environmental issues. Policy makers worldwide prefer the mechanisms that provide quick and reliable roadmap for policy formulation. However, there is also a consensus that there is no mechanism available that can help in policy formulation to cope with environmental issues in the context of developing countries. Therefore, this study has been designed to put forward a game theory-based mechanism that outlines the role of the government, firm and civil society to ensure the environmental sustainability for future generations. In this study, we used a dynamic game theory of complete information to understand the equilibrium solutions for the government, firm and civil society to provide the basis for policy makers to understand how and when the government, firm and civil society must intervene and play a vital role for environmental sustainability. Based upon the equilibrium analysis, the study has highlighted that when the punishment for non-compliance with environmental responsibility is smaller, the role of civil society would be higher for environmental sustainability. On the other hand, when the environmental responsibility cost is higher, then the role of the government is also higher for the implementation of environmental responsibility and to ensure the punishment. However, we found from model analysis that if the cost is low, the probability of firm to fulfill environmental responsibility is higher. In real life, the high cost of environmental responsibility is the main reason that the firm does not fulfill environmental responsibility. Under the high cost, the firm often has the phenomenon of bribery to government institutions and other means to avoid environmental responsibility.

The study provides valuable suggestions for policy makers worldwide. Based upon the equilibrium analysis and scientific literature, this study has highlighted that there is a poor framework of legislation for environmental issues in the context of developing countries as compare to developed countries. Therefore, the first implication from this study is that policy makers worldwide should emphasize on introducing green regulations to cope with environmental issues and also the government should ensure the implementation of green regulations. Moreover, the government should strictly penalize the firms those are not following the green regulations. For this purpose, the government can also engage the civil society to ensure the implementation of green regulation. But, the government must include the role of civil society in green regulations; otherwise it would not be easy to engage civil society for the purpose of environmental sustainability.

In case of developing countries worldwide, only few industries have their own recycling plants to recycle the wastage of their factories. For example, in the context of Pakistan, a developing country, the government should take strict actions to their emissions and wastage according to the NEQS under the 1997 Act to sustain the environment for future generations. If any industry violates the rules of the country and create harmful conditions for the environment, should be punished with heavy fine. In addition, creating public awareness through education can also improve environmental conditions in worldwide.

REFERENCES

Abid, M., & Ashfaq, A. (2015). CPEC: Challenges and opportunities for Pakistan. *Journal of Pakistan Vision, 16*(2), 142–169.

Abid, M., Schilling, J., Scheffran, J., & Zulfiqar, F. (2016). Climate change vulnerability, adaptation and risk perceptions at farm level in Punjab, Pakistan. *The Science of the Total Environment, 547*, 447–460. doi:10.1016/j.scitotenv.2015.11.125 PMID:26836405

Ahmed, K., Shahbaz, M., Qasim, A., & Long, W. (2015). The linkages between deforestation, energy and growth for environmental degradation in Pakistan. *Ecological Indicators*, *49*, 95–103. doi:10.1016/j.ecolind.2014.09.040

Ali, M., & Athar, M. (2008). Air pollution due to traffic, air quality monitoring along three sections of National Highway N-5, Pakistan. *Environmental Monitoring and Assessment*, *136*(1), 219–226. PMID:17385053

Anwar, J. (2016). Analysis of energy security, environmental emission and fuel import costs under energy import reduction targets: A case of Pakistan. *Renewable & Sustainable Energy Reviews*, *65*, 1065–1078. doi:10.1016/j.rser.2016.07.037

Azam, M., & Khan, A. Q. (2016). Urbanization and environmental degradation: Evidence from four SAARC countries—Bangladesh, India, Pakistan, and Sri Lanka. *Environmental Progress & Sustainable Energy*, *35*(3), 823–832. doi:10.1002/ep.12282

Azizullah, A., Khattak, M. N. K., Richter, P., & Häder, D.-P. (2011). Water pollution in Pakistan and its impact on public health—a review. *Environment International*, *37*(2), 479–497. doi:10.1016/j.envint.2010.10.007 PMID:21087795

Bajari, P., Hong, H., & Ryan, S. P. (2010). Identification and estimation of a discrete game of complete information. *Econometrica*, *78*(5), 1529–1568. doi:10.3982/ECTA5434

Bernauer, T., & Gampfer, R. (2013). Effects of civil society involvement on popular legitimacy of global environmental governance. *Global Environmental Change*, *23*(2), 439–449. doi:10.1016/j.gloenvcha.2013.01.001

Bernauer, T., Gampfer, R., Meng, T., & Su, Y.-S. (2016). Could more civil society involvement increase public support for climate policy-making? Evidence from a survey experiment in China. *Global Environmental Change*, *40*, 1–12. doi:10.1016/j.gloenvcha.2016.06.001

Betsill, M. M., & Corell, E. (2008). *NGO diplomacy: the influence of nongovernmental organizations in international environmental negotiations*. Mit Press.

Cheffins, B. R. (1997). *Company law: theory, structure, and operation*. Oxford, MA: Clarendon Press.

CIA. (n.d.). The World Factbook — Central Intelligence Agency Retrieved August 03, 2017, from https://www.cia.gov/library/publications/the-world-factbook/geos/pk.html

Colman, A. M. (2013). *Game theory and its applications: In the social and biological sciences*. Psychology Press.

Coutinho, M., & Butt, H. K. (2014). *Environmental Impact Assessment Guidance for Coal Fired Power Plants in Pakistan.* National Impact Assessment Programme, *IUCN Pakistan*.

Daud, M., Nafees, M., Ali, S., Rizwan, M., Bajwa, R. A., Shakoor, M. B., ... Murad, W. (2017). Drinking Water Quality Status and Contamination in Pakistan. *BioMed Research International*. PMID:28884130

Drews, S., & Van den Bergh, J. C. (2016). What explains public support for climate policies? A review of empirical and experimental studies. *Climate Policy*, *16*(7), 855–876. doi:10.1080/14693062.2015.1058240

Edenhofer, O., Steckel, J. C., Jakob, M., & Bertram, C. (2018). Reports of coal's terminal decline may be exaggerated. *Environmental Research Letters*, *13*(2), 024019. doi:10.1088/1748-9326/aaa3a2

Epstein, M. J., & Buhovac, A. R. (2014). *Making sustainability work: Best practices in managing and measuring corporate social, environmental, and economic impacts*. Berrett-Koehler Publishers.

Farooqui, S. Z. (2014). Prospects of renewables penetration in the energy mix of Pakistan. *Renewable & Sustainable Energy Reviews*, *29*, 693–700. doi:10.1016/j.rser.2013.08.083

Fiedler, P. E. K., & Zannin, P. H. T. (2015). Evaluation of noise pollution in urban traffic hubs—Noise maps and measurements. *Environmental Impact Assessment Review*, *51*, 1–9. doi:10.1016/j.eiar.2014.09.014

Finch, K. C., Snook, K. R., Duke, C. H., Fu, K.-W., Tse, Z. T. H., Adhikari, A., & Fung, I. C.-H. (2016). Public health implications of social media use during natural disasters, environmental disasters, and other environmental concerns. *Natural Hazards*, *83*(1), 729–760. doi:10.100711069-016-2327-8

Gao, Y., Li, Z., Wang, F., Wang, F., Tan, R. R., Bi, J., & Jia, X. (2018). A game theory approach for corporate environmental risk mitigation. *Resources, Conservation and Recycling*, *130*, 240–247. doi:10.1016/j.resconrec.2017.12.009

Grigorescu, A. (2007). Transparency of intergovernmental organizations: The roles of member states, international bureaucracies and nongovernmental organizations. *International Studies Quarterly*, *51*(3), 625–648. doi:10.1111/j.1468-2478.2007.00467.x

Harijan, K., Uqaili, M. A., Memon, M., & Mirza, U. K. (2011). Forecasting the diffusion of wind power in Pakistan. *Energy*, *36*(10), 6068–6073. doi:10.1016/j.energy.2011.08.009

Khalil, H. B., & Zaidi, S. J. H. (2014). Energy crisis and potential of solar energy in Pakistan. *Renewable & Sustainable Energy Reviews*, *31*, 194–201. doi:10.1016/j.rser.2013.11.023

Khan, M. A., Khan, J. A., Ali, Z., Ahmad, I., & Ahmad, M. N. (2016). The challenge of climate change and policy response in Pakistan. *Environmental Earth Sciences*, *75*(5), 412. doi:10.100712665-015-5127-7

Li, S., & Xu, J. (2018). Game analysis among the central government, local governments, and firms in China's environmental pollution governance. *Paper presented at the IOP Conference Series: Earth and Environmental Science.* 10.1088/1755-1315/153/6/062007

March, R. G. (2016). Detection of Forest Mortality from the 2011 Texas Drought and Examination of Environmental Drivers.

McVea, H., & Charalambu, N. (2014). Game theory and sovereign wealth funds. *Journal of Financial Regulation and Compliance*, *22*(1), 61–76. doi:10.1108/JFRC-12-2012-0049

Mehdi, M. R., Kim, M., Seong, J. C., & Arsalan, M. H. (2011). Spatio-temporal patterns of road traffic noise pollution in Karachi, Pakistan. *Environment International*, *37*(1), 97–104. doi:10.1016/j.envint.2010.08.003 PMID:20851468

Myerson, R. B. (2013). *Game theory.* Harvard university press.

Önder, S., & Akay, A. (2015). Reduction of Traffic Noise Pollution Effects by Using Vegetation, Turkey'Sample. *Journal of Engineering and Economic Development*, *2*(2), 23.

Osborne, M. J., & Rubinstein, A. (1994). *A course in game theory.* MIT press.

Poudel, H. N. (2016). State of Human Development: SAARC and Nepal. *Economic Literature*, *12*, 50–55. doi:10.3126/el.v12i0.14887

Poundstone, W. (1993). *Prisoner's Dilemma/John von Neumann, Game Theory and the Puzzle of the Bomb.* Anchor.

Qudrat-Ullah, H. (2015). Independent power (or pollution) producers? Electricity reforms and IPPs in Pakistan. *Energy*, *83*, 240–251. doi:10.1016/j.energy.2015.02.018

Qudrat-Ullah, H., & Karakul, M. (2007). Modelling for policy assessment in the electricity supply sector of Pakistan. *International Journal of Energy Sector Management*, *1*(3), 240–256. doi:10.1108/17506220710821125

Qureshi, S. (2010). The fast growing megacity Karachi as a frontier of environmental challenges: Urbanization and contemporary urbanism issues. *Journal of Geography and Regional Planning*, *3*(11), 306.

Rasmusen, E., & Blackwell, B. (1994). *Games and information*. Cambridge, MA: Blackwell.

Roseland, M. (2000). Sustainable community development: Integrating environmental, economic, and social objectives. *Progress in Planning*, *54*(2), 73–132. doi:10.1016/S0305-9006(00)00003-9

Rosenbloom, S. (2001). Sustainability and automobility among the elderly: An international assessment. *Transportation*, *28*(4), 375–408. doi:10.1023/A:1011802707259

Sassi, P. (2016). Built Environment Sustainability and Quality of Life (BESQoL) Assessment Methodology. In Engaging stakeholders in education for sustainable development at university level (pp. 21-32). Cham, Switzerland: Springer.

Schelling, T. C. (2010). Game theory: A practitioner's approach. *Economics and Philosophy*, *26*(1), 27–46. doi:10.1017/S0266267110000040

Severo, E. A., de Guimarães, J. C. F., & Dorion, E. C. H. (2017). Cleaner production and environmental management as sustainable product innovation antecedents: A survey in Brazilian industries. *Journal of Cleaner Production*, *142*, 87–97. doi:10.1016/j.jclepro.2016.06.090

Severo, E. A., de Guimarães, J. C. F., Dorion, E. C. H., & Nodari, C. H. (2015). Cleaner production, environmental sustainability and organizational performance: An empirical study in the Brazilian Metal-Mechanic industry. *Journal of Cleaner Production*, *96*, 118–125. doi:10.1016/j.jclepro.2014.06.027

Severo, E. A., & Guimarães, J. C. F. D. (2015). Corporate environmentalism: An empirical study in Brazil. *International Journal of Business and Globalisation*, *15*(1), 81–95. doi:10.1504/IJBG.2015.070225

Sheikh, M. A. (2009). Renewable energy resource potential in Pakistan. *Renewable & Sustainable Energy Reviews*, *13*(9), 2696–2702. doi:10.1016/j.rser.2009.06.029

Shi, G.-M., Wang, J.-N., Fu, F., & Xue, W.-B. (2017). A study on transboundary air pollution based on a game theory model: Cases of SO2 emission reductions in the cities of Changsha, Zhuzhou and Xiangtan in China. *Atmospheric Pollution Research*, *8*(2), 244–252. doi:10.1016/j.apr.2016.09.003

Tariq, M. I., Afzal, S., Hussain, I., & Sultana, N. (2007). Pesticides exposure in Pakistan: A review. *Environment International*, *33*(8), 1107–1122. doi:10.1016/j.envint.2007.07.012 PMID:17765971

Valasai, G. D., Uqaili, M. A., Memon, H. R., Samoo, S. R., Mirjat, N. H., & Harijan, K. (2017). Overcoming electricity crisis in Pakistan: A review of sustainable electricity options. *Renewable & Sustainable Energy Reviews*, *72*, 734–745. doi:10.1016/j.rser.2017.01.097

Van Rooy, A. (2004). *The global legitimacy game: Civil society, globalization and protest*. Springer. doi:10.1057/9780230000957

Wakeel, M., Chen, B., & Jahangir, S. (2016). Overview of energy portfolio in Pakistan. *Energy Procedia*, *88*, 71–75. doi:10.1016/j.egypro.2016.06.024

Waseem, A., Arshad, J., Iqbal, F., Sajjad, A., Mehmood, Z., & Murtaza, G. (2014). Pollution status of Pakistan: A retrospective review on heavy metal contamination of water, soil, and vegetables. *BioMed Research International*. PMID:25276818

Wolters, H., & Schuller, F. (1997). Explaining supplier-buyer partnerships: A dynamic game theory approach. *European Journal of Purchasing & Supply Management*, *3*(3), 155–164. doi:10.1016/S0969-7012(97)00011-7

Yamin, F. (2001). NGOs and international environmental law: A critical evaluation of their roles and responsibilities. *Review of European, Comparative & International Environmental Law (Northwestern School of Law)*, *10*(2), 149–162.

Zaman, K., Abdullah, A., Khan, A., Nasir, M. R. M., Hamzah, T. A. A. T., & Hussain, S. (2016). Dynamic linkages among energy consumption, environment, health and wealth in BRICS countries: Green growth key to sustainable development. *Renewable & Sustainable Energy Reviews*, *56*, 1263–1271. doi:10.1016/j.rser.2015.12.010

Zaman, K., Abdullah, I., & Ali, M. (2017). Decomposing the linkages between energy consumption, air pollution, climate change, and natural resource depletion in Pakistan. *Environmental Progress & Sustainable Energy*, *36*(2), 638–648. doi:10.1002/ep.12519

Zameer, H., & Wang, Y. (2018). Energy production system optimization: Evidence from Pakistan. *Renewable & Sustainable Energy Reviews*, *82*, 886–893. doi:10.1016/j.rser.2017.09.089

Zameer, H., Wang, Y., Yasmeen, H., Mofrad, A. A., & Saeed, R. (2018). A game-theoretic strategic mechanism to control brand counterfeiting. *Marketing Intelligence & Planning*, *36*(5), 585–600. doi:10.1108/MIP-02-2018-0053

Zhang, R., Andam, F., & Shi, G. (2017). Environmental and social risk evaluation of overseas investment under the China-Pakistan Economic Corridor. *Environmental Monitoring and Assessment*, *189*(6), 253. doi:10.100710661-017-5967-6 PMID:28477273

Zürn, M. (2004). Global governance and legitimacy problems. *Government and Opposition*, *39*(2), 260–287. doi:10.1111/j.1477-7053.2004.00123.x

This research was previously published in the International Journal of Agricultural and Environmental Information Systems (IJAEIS), 10(2); edited by Francois Pinet and Petraq Papajorgji, pages 82-97, copyright year 2019 by IGI Publishing (an imprint of IGI Global).

Chapter 4
Reverse Logistics and Solid Waste:
Challenges for the National Waste Policy (PNRS) in Brazil

Hermes de Andrade Júnior
Catholic University, Portugal

ABSTRACT

This chapter promotes a selection of works collected that seek to analyze the need and the evolution of reverse logistics into the context of the National Policy on Solid Waste in Brazil. Nineteen years of intensive discussion have been held until the legal framework for the implementation of Agenda 21 of 1992 on the environmentally sound management of solid waste could be announced. The principle of shared responsibility for the product lifecycle, which reaches manufacturers, importers, distributors and traders, consumers, and holders of public solid waste management services, is the central theme of the law and undoubtedly innovates on the issue, placing Brazil alongside countries such as those of the European Union and Japan. However, a serious problem that distances them is to achieve large population densities with the benefit of municipalization of the process of control of urban waste. The rate of effective management of solid wastes is relatively low at the municipal level compared to the countries mentioned.

DOI: 10.4018/978-1-7998-0441-3.ch004

Copyright © 2020, IGI Global. Copying or distributing in print or electronic forms without written permission of IGI Global is prohibited.

INTRODUCTION

In the twentieth century, due to industrial revolution and technology development, consumption patterns of the people, all over the globe, have changed. The use of natural resources and goods has increased a lot. Due to this, huge quantities of different types of solid wastes are produced every day creating an enormous problem of their disposal. It is now recognized that proactive management is required to deal with this problem, i.e., it is required to reduce the generation of solid waste, effective collection of solid waste and utilization of solid waste rather than concentrating on disposal alone. Thus, solid waste management involves management of activities associated with generation, storage, collection, transfer and transport, reuse and recycling, processing and disposal which should be environmentally compatible, adopting to the principles of economy, aesthetics, and energy conservation.

Over the past 25 years, Europe has moved ahead, ordering the entire waste management process in the Old Continent and issuing dozens of European Community directives. And, notably in Europe, there has been considerable scientific and technological progress in the field of waste management.

Brazil's neighbors Uruguay, Argentina, Peru and Chile were able in the meantime to edit their General Waste Laws. In the wake of these achievements, the quality of life, the urban environment, society and economy were privileged. The United Nations Conference on Environment and Development (Rio-92) adopted a program of action for sustainable development, now known as Agenda 21. With regard to solid waste, the document recognizes that unsustainable patterns of production and consumption are causes of environmental degradation, and a significant change of this model is indispensable in order to minimize the generation of waste at all stages cycle of the product or service.

According to Agenda 21, the environmentally management of solid waste must go beyond its deposit or use by safe methods, which requires the participation of the whole society. To this end, public power in all spheres, together with the business sector, consumers and other segments of society, must act by encouraging the introduction of new environmentally healthy products and services, recycling and reuse of the product consumed and the inputs used in processes of production and also by reducing the waste in the packaging of the products.

The debate about a national solid waste management and management policy goes in the 1980s, started from a proposal presented in the Brazilian Federal Senate, which had specific provisions on health service waste. During the long process of the matter in Parliament (11 years, until the Agenda 21 and then, 8 more until signing The National Policy of Solid Waste, totaling 19 years), the project started to incorporate different issues related to solid waste and, gradually, a legislative proposal based on the principles established in Agenda 21 was consolidated.

In this chapter, solid residues are approached from the perspective of the national policy of solid waste in Brazil and its process, with a focus on understanding conceptual details of the management, control, tools and integrated participation proposal, presenting some difficulties in its implementation and with emphasis on the treatment and management of municipal solid waste.

GENERAL CATEGORIES OF SOLID WASTE (SW)

"Based on the source of contributors, four common categories of solid waste are: a) Municipal waste[1], b) Industrial waste, c) Agriculture and Animal waste and d) Hazardous waste" (Nag & Vizayakumar, 2005, p.11).

MUNICIPAL WASTE (MSW)

According to Nag & Vizayakumar (2005), there are several derivations of meaning and context to consider. Useless, can not be used, not wanted. These are some of the meanings of the waste dictionary. Generally, readers are driven by such concepts, as if only part of the solid waste meant municipal solid waste (MSW). "In addition to MSW, there is a substantial contribution in terms of solid industrial waste of variable composition and solid waste from thermoelectric power plants and activities related to agriculture. A few words, very frequently being used, are refuge, trash, garbage, rubbish, etc. Refuge, has a broader meaning like waste and is used as its synonym. Trash is generally used for the most combustible components of throwaway materials (paper, rag, wood chips, toys, etc.)" (Nag & Vizayakumar, 2005, p.12).

A Municipal solid waste (MSW) that mainly consists in food wastes, commonly called garbage, are prone to decompose and they originate from food products of animal and vegetable origin, arising out of preparation, processing, handling, catering, and eating. "Rubbish is combustible and non-combustible rejected materials other than those mentioned above. The combustible portion (trash) consists of paper, cardboard, textiles, plastics, rubber, etc. The non-combustible portion consists of glass, ceramics, metals, etc. Ashes and cinders originate mainly from coal, firewood, and burnt residues of other combustible materials." As the authors say, "construction and demolition wastes include wide varieties of materials, mostly non-combustible in nature and come from civil works of construction, remodeling, repair works and demolition of building structures and others that include broken pieces of bricks, stones, plasters, dirt, sand, wooden articles, metal pieces, electrical parts, etc." Special wastes are "uncommon materials accumulated from unpredictable and infrequent sources, i.e., abandoned vehicles, dead animals, limbs, blood, etc. from hospitals and that found from street sweepings." (Nag & Vizayakumar, 2005, pp.12-13).

DAMAGES FOR HEALTH: SOME PATHOGENS AND DISEASES THAT COME FROM URBAN SW[2]

Waste management is critical in the municipal administration's agenda, including the preservation of public health.

According to Epstein (2015), there are health concerns about the placement of materials in the soil and subsoil, as well as the inability to manage and treat solid waste. It contains pet stools, contaminated human tissues, contaminated food residues (e.g., chicken with Salmonella) and other contaminated sources. In the United States, most MSW are landfilled. Incineration or heat recovery systems are few, "approximately 200 biosolid composting facilities in the United States and the largest facility is the Inland Empire composting facility located in Chino, California". MSW composting facilities are few, but food waste collection and composting are increasing. He says "Approximately 154 municipalities are collecting food waste. On June 17, 2013, Mayor Bloomberg of New York indicated that the city will initiate food waste composting as being done in several other cities" (Epstein, 2015, p.5).

In developed countries, pathogens and the diseases resulting from them are predominantly waterborne or foodborne. There have been relatively few illnesses reported in the literature from the disposal and management of sewage sludge, biosolids, or municipal solid waste (MSW). The populations most likely to be exposed are workers. Even diseases from direct contact or exposure during wastewater treatment, collection of MSW, or disposal of MSW are rare. Hygiene, sanitary conditions, and the education of workers are high. However, the potential exposure to diseases by workers can result from inhaling bioaerosols and having respiratory complications. Bioaerosols produced during composting of sewage sludge or MSW has been a concern for workers and populations residing near composting facilities. The bioaerosols of predominant concern have been aspergillus fumigatus and endotoxins. Numerous studies have been conducted on this issue in the United States and United Kingdom (Epstein, 2015, preface, *Op. Cit.*).

Epstein quickly addresses the issue of the emission of gaseous pollutants from untreated solid waste. For example, in the simple domestic use of daily urban life, we have a lot of junk. In fact, garbage means destruction and disposal of the kitchen, market, stores, warehouses, and may contain various putrescible residues. It requires careful handling as it attracts and produces flies and other insects. It also generates unpleasant odor due to rapid fermentation. Garbage includes all non-putrescible residues except ashes and consists of combustible and non-combustible substances such as cans, paper, brushes, metals, etc. It becomes a nuisance when sprayed by wind or careless handling. Solid wastes may contain pathogens, such as viruses, bacteria, parasites, fungi and actinomycetes. For example, aspergillus fumigatus can infect humans directly through the respiratory system. Pathogens in the air can land on crops and eventually infect food and humans. In Epstein's review, Ritter et al. (2002) stated that, on a global scale, pathogenic contamination of drinking water represents the most significant health risk to humans. Contaminated water resulting from landfill leachate or other disposal methods may enter potable water sources in groundwater or surface water. Pathogens can move through the soil (eg through cracks, wormholes, etc.) and eventually end up with groundwater and drinking supplies (Ritter et al., 2002). Sobsey (1978) found enteric viruses in landfill leachate. Contaminated water can directly affect the food supply. Pathogens may enter the soil from sewage and sewage applications (unless disinfected by composting, heat treatments or limes), but Teschke et al. (2010) have indicated that studies of endemic diseases in various water or sewage systems are rare. (Epstein, 2015).

MSW IN THE REALITY OF EACH ONE

The context of developing countries is definitely still not very encouraging, perhaps alarming in comparative terms. Epstein (2015) says "that in developed countries, most MSW is either landfilled or incinerated. In the United States, over 50% is landfilled, whereas in Europe the majority is incinerated". Landfills are regulated "to reduce or prohibit groundwater contamination, and daily landfill cover reduces the dispersion and dissemination of pathogenic bioaerosols". Recycling, reuse, source separation, and mechanical separation are increasing but often depend on markets for such items as paper, cardboard, glass, and plastics. "In developing nations and emerging industrial nations with large populations, much of the MSW ends up in uncontrolled dumps". These sites are open, and scavengers roam them to seek reusable items, often in the simple quest for day-to-day survival, without speculating on technical subjectivities. "Consequently, workers or scavengers are exposed to pathogens, resulting in diseases. Not only are the individuals exposed, but also they bring pathogens into the home and public places on their clothing and shoes. Furthermore, MSW is often left in streets and not removed for days, harboring vectors" (Epstein, 2015, p.39).

SW IN LATIN AMERICA AND THE CARIBBEAN

The review below is from Espinoza and other analysts (2010) of the Pan American Health Organization (PAHO). The governments of Latin America and the Caribbean (LAC) face the problem of SW management and its impact in the context of global concern for economic, environmental and social services. Topics such as adequate final disposal of waste, the difficult and complex reality of informal workers, minimization and recycling, climate change and carbon markets, and the role of national, regional and municipal authorities in planning and regulating services are discussed in this paper: Integrated research report.

According to Espinoza *et al.* (2010), "between 2001 and 2008, the population of the LAC increased from 518 to 588.6 million people, while urban residents spent 405.7 to 468.8 million people (78.3% to 79.6% of the total), bringing the claimant population of public cleaning services for 63.1 million people". On the other hand, after the crises suffered in early 2000, "the socio-economic situation of the region experienced a strong improvement between 2002 and 2008, when GDP *per capita* increased by 23.2%". Despite the subsequent global crisis, the socio-economic indicators of poverty, unemployment, inequality and human development index improved during the period. However, "LAC remains the region with the highest inequality in the world". (Espinoza *et al.*, 2010, p.24).

The Regional Evaluation of Solid Waste Management in Latin America and the Caribbean (AVAL), held in 2010 by Espinoza *et.al.*, estimated that the generation *per capita* of solid waste in Latin America and the Caribbean "reached 0,63kg / person / day, while the solid waste reached 0.93 kg/person/day. Indicators *per capita* obtained for the region imply an approximate daily urban generation of 295,000 tons of household solid waste (HSW) and 436,000 tons of MSW". Espinoza *et al.* (2010) complement the services increased its coverage during the last decade: "A full coverage sweeping in LAC increased by 10% to 82.3%, with 5.56 manual street sweepers and 0.17 mechanical sweepers per 10,000 inhabitants. The total cover about 91% of the residues are scanned manually and the remaining 9%, mechanically". In turn, the AVAL estimated that the "residues collection coverage in LAC also increased by more than 10 percentage points, to 93.4%. In six countries of the region (Argentina, Brazil, Chile, Colombia, Uruguay and Venezuela), it can be said that the coverage of the collection of household waste is practically universal". The collection frequency is daily for "just over 45% of Latin Americans, while almost 53% of the population receiving the service from 2 to 5 times a week, and nearly 2% weekly". "The burning and uncontrolled disposal of RS" increases as decreases the frequency of collection. "It is an average of 1.31 vehicles for every 10,000 inhabitants, of which 33% are over 10 years old and 57.8% are compactor vehicles". (Espinoza *et al.*, 2010, p.24).

According Espinoza *et al.* (2010), among the main waste treatment alternatives before its final disposal may be mentioned compost, recycling and heat treatment, including, in some cases, the energy utilization. Waste management is more feasible from their separation, and the formal screening in picking stations is almost nonexistent. The informal scavenging, in turn, is widespread, but it is not known precisely the quantities recycled. Composting, used in numerous opportunities, could get a boost with the use of carbon credits.

The EVAL 2010 estimated that "the waste of 54.4% of the inhabitants of Latin America and the Caribbean are disposed in landfills, one significant increase over the 22.6% recorded in 2002. At the same time, the use of landfills to heaven open decreased from 45.3% to 23.3%". (Espinoza *et al.*, 2010, p.24). Perhaps the main reasons for this phenomenon are due to the strong momentum that some countries have given to rules, forcing the closure of landfills and uncontrolled landfills and specifically defining the technical characteristics that must have a suitable solution for disposal. One should also highlight the environmental sensitivity of making some governments and their political will to enforce the rules, carbon credits and modifications of incurred operating practices and the adoption of regional schemes of disposal which seem to be the main drivers the remarkable improvement of final disposal in the period.

Espinoza *et al.* (2010) conclude that the municipalities have very little experience in dealing with this problem and find a way to solve it. The formalization as part of the municipal management system is a valid solution, however still very incipient in the region.

SW IN BRAZIL

Article 30 of the 1988 Federal Constitution establishes that the municipality is responsible for legislating on matters of local interest; organize and provide, directly or under a concession or permit system, public services of local interest, including collective transportation, which are essential; to promote, where appropriate, adequate land-use planning through planning and control of land use, urbanization and urban land occupation.

Thus, urban cleaning, understood as logistics for the management of solid waste (collection, transport, packaging and final destination), is a service that falls within the constitutional requirements. It is the responsibility of the municipality to organize and provide public services of local interest, such as collection and proper destination of all garbage generated.

But, as Espinoza *et al.*, the rapid population growth in recent decades and its concentration in urban areas, has led to an increase in demand for public services in general, including the SW management. Rapid urban growth occurred mainly in the outskirts of towns or marginal areas where the needs are numerous, which stressed the importance not only to increase the coverage of services, but also to carry out an adequate risk management, socially and environmentally sustainable. This urbanization process is observed, from before, intensely in Brazil. The generation of solid waste is the most important indicator to scale the different services of its management, in addition to predicting the difficulties to be found in the processes and it serves, therefore, as parameter for the decision making with regard to the planning of the systems of collection and final disposal. As Espinoza et al. evaluated, Brazil has a weak image on the municipal solid waste management. The published results showed that the percentage of the municipal budget, at most 5%, was for the management of solid waste or for urban sanitation, with the great majority of municipalities with a population of less than 50,000 inhabitants. In the pursuit of urban sustainability and eco-efficiency in cities, inadequate urban waste management implies significant direct and indirect social and environmental costs, which further affect the suburbs. Thus, adequate waste management is included as complementary information to meet the objectives related to the sustainability of the human environment (Espinoza *et al.*, 2010).

Brazilian municipalities would have until 2014 to close the disposal of their waste in dumps, as determined by the NPSW. However, there were still 2,906 landfills to be erected in Brazil, distributed in 2,810 municipalities in 2012 (IPEA, 2012). According to another source, at 2012 only 766 (13.76%) of the 5,565 Brazilian municipalities, had selective collection systems managed by city halls (Cempre, 2012).

Considerations of analysts and entities of the pressing situation of solid waste in the reality of Brazilian territorial space and population, especially in cities, where the perspective of analysis and monitoring of the implementation of public policy is clearer, shows a sophisticated and integrated set of legal, systemic and administrative apparatuses, but as difficulties in practice in various sectors. I quote here the electronics sector, in the work of studying reverse logistics (RL) to a brazilian modelling for computers and cell phones, articles widely consumed in urban life, Demajorovic, Augusto & Souza (2016), advise an intricate waste management mechanism that has not yet been established (Table 1). The Brazilian government tends to become active in defining and creating fiscal, credit and financial incentives, but not in inspecting the Management Organization (MO); this is going to under the responsibility of the industry managing all RL for Waste Electric and Electronic Equipment (WEEE). The operating organization will be hired in the Brazilian market and the steps in the RL process may or may not be outsourced. Besides the MO, another organization will be created to audit all the companies in the reverse process, as well as the results obtained. Brazilian consumers should be responsible for returning cell phones or computers cost and incentive-free at drop-off centers. An embedded fee to cover

Table 1. International models and potential Brazilian model

	Switzerland	Sweden	China	India	Brazil
Government	Active	Participative	Not Active	Not Active	Active
Industry	Total Management	Transportation/ Recycling	Ships WEEE		Total Management
Informal S.			Excluded	Pre-processing	No training
Model	EPR PRO	EPR Gov/PRO	Consortium StEP	Empa Program	SPC
Financing	TAR	Industry	Self Sustained	Self Sustained with pre-paids to waste pickers	Embedded fee to cover RL Costs
Performance	9Kg/p.c//year High consc.	16 kg/p.c/year High consc.	Lacks Indicators	Lacks Indicators	Lacks Indicators
Consumer	Pays ARF Returns	return			Pays embedded fee and returns

Source: Demajorovic, Augusto & Souza (2016)
Note: SPC = Shared Producer Responsibility, EPR = Extended Producer Responsibility

RL costs will be charged at product purchase, which the industry will transfer to the MG. After disposal of equipment at drop-off retail points, the MG will collect and transport it to a site, where it will be sorted by brand and pre-stored. After that, it will be transferred to trans-shipment centrals for de-characterization and recycling, decontamination and elimination of final rejects.

THE NATIONAL POLICY ON SOLID WASTE (NPSW) OF BRAZIL: SUMMARY

The Brazilian NPSW recognizes the polluter pays principle, introduced by the Organization for Economic Co-operation and Development (OECD), responsible for the fundamentals of the economic aspects of environmental policies. It imposes on natural or legal persons, governed by public or private law, the payment of the costs of measures to eliminate contamination or reduce it to the limit set by standards that ensure the quality of life.

This principle arose in the Brazilian legal system in 1981, through Law 6.938 (National Environmental Policy, article 4, item VII), and later was ratified by the Federal Constitution of 1988, establishing environmental protection and consecrating a new dimension to the State Democratic system of law, which must be subject to ecological principles.

It also establishes as an environmentally balanced environment the right of everyone and imposes on governments the obligation to implement policies aimed at that end. One of the many challenges proposed during the legislative debates in Brazil about a solid waste policy was to agree on a solid waste management and management law that were limited to establishing general guidelines applicable throughout the national territory, with respect to the constitutional competencies of the States, Federal District and Municipalities, and Autonomy of the federated entities. The management and management of solid waste are concepts that are different from each other. The definition of management is more comprehensive and refers to the set of actions that encompasses the planning and coordination of all stages of management, from the perspective of sustainable development. Management already understands the actions related to the operationalization and can be restricted to only one type of waste.

The adoption of state laws, the proactivity of various industrial segments in the field of post-consumer responsibility, the recognition of the economic opportunities arising from the use of solid waste and the increasing pressure of society have favored the construction of a consensus around a law that establishes the minimum guidelines for solving one of the most serious urban environmental problems. Inadequate disposal of solid waste, in addition to threatening the quality of the environment and public health, is a response to serious social problems, such as the inhuman and degrading conditions of those living in the dumps, including children.

The Legislative Bulletin 15 of 2012 of the Federal Senate of Brazil does its analysis, as it shows that, on August 2, 2010, Law 12,305, which established the National Policy on Solid Waste (NPSW), signed after a long time and after a long and tiring debate. This effort was attended by representatives of organs from the three governmental spheres, from the most diverse business segments and retail trade, from environmental and consumer organizations, to recyclers of reusable and recyclable material collectors and specialists in Solid Waste Management. This legal text brings innovative precepts, such as shared responsibility for the product life cycle and the reverse logistics system; in summary, the post-consumer responsibility of the productive sector.

The principle of shared responsibility for the product lifecycle, which reaches manufacturers, importers, distributors and traders, consumers and holders of public solid waste management services, is central to the law and undoubtedly innovates in the issue, placing Brazil alongside countries such as the European Union and Japan.

In the context of shared responsibility, manufacturers, importers, distributors and traders are obliged to include, among other measures, the collection of the remaining products and residues after use - post-consumer responsibility - as well as their subsequent environmental disposal, adequate in the case of products subject to a reverse logistics system, independently of the public urban cleaning and solid waste management service.

In turn, reverse logistics is characterized by a set of actions, procedures and means to enable the collection and restitution of solid waste to the business sector for reuse, in its cycle or in other production cycles, or other final disposal environmentally adequate, being an instrument of economic and social development.

The Brazilian NPSW establishes which products are submitted, immediately, to the regime of reverse logistics: agrochemicals, their residues and packaging; batteries; tires; Lubricating oils, their residues and packaging; fluorescent lamps, sodium and mercury vapor and mixed light and electronics and its components. For other products and non-discriminated packaging, reverse logistics will be extended as required by law regulation or the sectoral agreements and the terms of compromise signed between the public power and the business segment.

The law also establishes the order of priority to be observed in the management and management of solid waste, such as: non-generation, reduction of quantity and volume generated, reuse, recycling, treatment of solid waste and final disposal of waste (Decree 7,404, of December 23, 2010). Under the law, tailings and solid waste are also distinct concepts. The definition of tailings is more restricted and refers to solid wastes which, after all treatment and recovery possibilities have been exhausted, have no alternative other than the final environmentally appropriate disposal, i.e. safe and orderly distribution in landfills.

Solid waste plans are also among the main instruments regulated by law, including national, state and municipal plans. The local public power - Federal District and Municipalities - is responsible for the integrated management of solid waste generated in their respective territories.

The elaboration of state and municipal plans is a prerequisite for States, the Federal District and Municipalities to access the resources of the Union, or controlled by it, destined to the management of solid waste. Municipalities that opt for joint solutions for solid waste management, as well as those that implement selective collection with the participation of recyclable and recyclable material collectors will have priority in accessing federal resources.

The Solid Waste Act also includes specific provisions on selective waste collection in the broadest sense, including municipalities' responsibilities, within the framework of integrated solid waste management plans, once it is the constitutional competence of these federated entities the actions Management of solid waste. The norm also establishes that municipal public power may institute economic incentives, in the form of municipal law, to the consumers that participate in the selective collection system.

It also clarifies the responsibilities of certain types of generators - industry, commerce, health, mining, public sanitation, among others - in the elaboration and implementation of a solid waste management plan specific to the sector, and links these plans to licensing environmental.

In addition to the above rules, the law recognizes the role of collectors in the framework of a federal law and strengthens the performance of their cooperatives; Creates the National System of Information on the Management of Solid Waste; Establishes special rules for hazardous waste; Fixes fences as to the destination or final disposal of wastes or wastes; Expressly prohibits the import of hazardous solid waste and tailings for any purpose and foresees that the final environmental disposal of the tailings should be implemented within four years of the entry into force of the law, i.e. the legal term for the end of the "dumps" is 2014, which by 2017 this does not it occurred.

As can be seen, the idea that permeates the entire PNRS Law is to reduce the amount of material subject to final disposal - in order to add value to the waste generated and to allocate as little as possible to the landfill that can no longer be used - And thus reverse the current logic of solid waste management.

The new law of solid waste - whose provisions apply in conjunction with Law 11,445, of January 5, 2007, establishing national guidelines for basic sanitation - integrates and complements the legal system of environmental protection, putting up alongside important laws, such as the National Policy on the Environment, Water Resources and Environmental Crimes. It also has the merit of bringing to the legal text a set of infralegals acts often challenged for not having support in law in the strict sense.

Law 12,305, 2010, as seen, it incorporates guidance of Agenda 21 and, although not go solve a sudden the wounds resulting from the accumulation of solid waste is the first step inducing a new and modern management model, in favor of environmentally sound, economically viable and socially just conditions - principles of sustainable development.

DESCRIPTION OF THE CONCEPTUAL ELEMENTS OF BRAZILIAN NPSW FOR THE LOGISTICS OF SOLID WASTE

It is common to define a productive system as a process capable of transforming input materials / services into goods / services as outputs. However, this model inherited from the industrial revolution produces, in addition to goods / services, solid waste that in turn produces "greenhouse" gases that accumulate in the atmosphere and which have been provoking global climate change since that time.

In this context, the variable "environment" was considered as strategic in the productive process, changing the traditional "production-consumption" relationship to "production-consumption-sustainability". This new relationship imposes the need to manage the variable "solid waste" in the pre-consumption activities of the good and / or pre-provision of the service, as well as in post-consumption and / or post-delivery activities. This requires planning such activities as embodied in action plans.

The NPSW emphasizes the concept of planning (plan) and the concept of integrated management (shared responsibility) in activities related to solid waste management.

The use of logistics, with its holistic and systemic characteristics, implies integrated management of activities, from the planning stage, through the implementation of the actions to the provision of the service and / or distribution of the good. In this way, logistics management and management occupy a strategic intermediary position among those involved in the application of NPSW.

The term management is used in NPSW. According to Article 3 of Law 12.305 / 10 (NPSW), integrated solid waste management is set of actions aimed at finding solutions to the solid waste in order to consider the political, economic, environmental, cultural and social, with social control and under the premise of sustainable development. The holistic and systemic characteristics of the solid waste management process can be inferred from this concept.

It is important, initially, to have the concept of solid waste and its classification.

Solid waste is all material, substance, object, or well disposed resulting from human activities in society, whose final destination is carried, it is proposed to proceed or is obliged to carry, in solid or semi-solid and gases in containers and liquids whose peculiarities make it unfeasible to be placed in the public sewage system or water bodies, or require technical or economically unviable solutions in view of the best available technology (Article 3 of Law 12,305/10).

According to Article 13 of Law 12,305 / 10, solid wastes are classified according to their origin and their dangerousness.

ORIGIN

- Urban Solid Waste (USW). They are divided into two interesting and specific categories for analysis, unlike the one pointed out as generic by Nag & Vizayakumar (2005). Household Wastes (HSW), which are those resulting from domestic residential household activities and the Urban Cleaning Wastes (UCW), which are those derived from sweeping, cleaning of public places and roads, and other urban cleaning services;
- Waste from commercial establishments and service providers generated in these activities;

- Waste from public basic sanitation services generated in these activities, except Urban Solid Waste;
- Industrial waste generated in production processes and industrial facilities;
- Waste of health services generated in health services;
- Civil construction waste generated in construction, renovation, repair and demolition of civil construction works, including waste resulting from the preparation and excavation of land for civil works;
- Waste generated in agricultural and forestry activities, including inputs;
- Waste from transport services, originating in ports, airports, customs, road and rail terminals and border crossings;
- Mining wastes arising from mining research, extraction or processing.

CONCERNING HAZARD

- **Hazardous Waste:** Those presenting a significant risk to public health or public health or to environmental quality due to their flammability, corrosivity, reactivity, toxicity, pathogenicity, carcinogenicity, teratogenicity and mutagenesis characteristics.
- Non-hazardous waste, which is not previously classified.

The characteristics of the solid waste will be detailed below in the Outbound Logistics activities for the PNRS application.

It is worth mentioning, among the guidelines of the NPSW, the concept of "aggregation of socioeconomic value" to "solid waste" and "aggregation of socio-environmental value" to "reject" through an environmentally adequate final destination.

ENVIRONMENTALLY APPROPRIATE FINAL DESTINATION

Disposal of waste including reuse, recycling, composting, recovery and energy use or other destinations accepted by the competent bodies of systems of environment (SISNAMA), health (SNVS), and water (SUASA), including the final disposal, observing specific operational rules in order to avoid damages or risks to public health and safety and to minimize adverse environmental impacts (Article 3 of Law 12,305 / 10).

Two categories of solid waste are inferred from this concept. A category that includes solid waste with technical and economical feasibility for reuse, recovery, recycling, composting and energy recovery and a category that includes those solid wastes without technical-economic feasibility called "tailings", whose destination is their "environmentally appropriate disposal" in landfills.

DISPOSAL OF SW WITH TECHNICAL-ECONOMIC FEASIBILITY: THE REUSE

Process for the use of solid waste without its biological, physical or physic-chemical transformation, observing the conditions and standards established by the competent bodies of SISNAMA and, if applicable, SNVS and SUASA (Article 3 of Law 12,305 / 10). It is to use a material again before discarding it.

RECYCLING

Process for the transformation of solid waste that involves the change of its physical, physical-chemical or biological properties, with a view to transformation into inputs or new products, observing the conditions and standards established by the competent bodies of SISNAMA and, if applicable, the SNVS And SUASA (Article 3 of Law 12,305 / 10).

That is, it is to transform the materials into raw material to start a new cycle of production-consumption-disposal. It begins with the selective collection in the separation and collection of recyclable solid waste, from its origin.

There is a distinction between pre-consumer recycling and post-consumer recycling. Pre-consumption recycling refers to waste generated in production processes. Post-consumer recycling refers to waste arising from the use of a good.

COMPOSTING AND ENERGY UTILIZATION

Composting is the process of decomposition of organic matter through aerobic digestion (presence of oxygen), that is, in the presence of air and water it is digested by microorganisms and it becomes a compound used to improve soil quality. It is the "recycling of organic waste", transforming them into fertilizer and energetic use of the gas emitted. It also begins with the selective collection in the separation and collection of organic waste.

In addition to the formation of the compound, solid waste, when decomposed, generates gases, in particular methane (CH4), which is toxic and highly flammable, and carbon dioxide (CO2) which, together with other gases present in the atmosphere, produce the greenhouse effect, contributing to global warming. The decomposition of the organic matter together with the action of the rainwater also generates a liquid of dark coloration, of unpleasant smell, toxic, and high power of contamination, because it can infiltrate in the soil contaminating the superficial and subterranean waters. This liquid is called "pergolated liquid", "leached" or "leachate".

Among the actions foreseen in Decree 7,390/ 10, which regulates the National Policy on Climate Change, is the expansion of the supply of energy from renewable sources such as bioelectricity.

Bioelectricity can be generated by recovering and destroying methane gas in suitable facilities in order to increase energy efficiency. Biogas, produced by the degradation of organic solid waste, can be converted into a form of energy recovery such as electricity, steam, fuel for boilers or stoves, vehicular fuel or to supply pipelines.

The energy utilization of organic solid waste on a large scale, by biodigestion that eliminates methane and generates organic compost, is increasingly used in countries with advanced environmental management.

DISPOSAL OF SOLID WASTE WITHOUT TECHNICAL-ECONOMIC FEASIBILITY: REJECTS

Tailings are solid wastes that, after having exhausted all the possibilities of treatment and recovery by available and economically viable technological processes, present no possibility other than the final environmentally appropriate disposition (Article 3 of the Law 12,305 / 10).

THE ENVIRONMENTALLY APPROPRIATE FINAL DISPOSAL

Orderly distribution of tailings in landfills, observing specific operational norms in order to avoid damages or risks to public health and safety and to minimize adverse environmental impacts (Article 3 of Law 12,305 / 10).

The following types of landfill are used: Landfill (open pit), controlled landfill, landfill, composting units, sorting and composting units, incineration units and landfills in areas floods. The most common modalities are: the dump or dump, the controlled landfill and the landfill.

1. **Garbage / Dump:** An inadequate form of final disposal of waste and tailings, which consists of the discharge of the material into the soil without any control technique or measurement.
2. **Controlled Landfill:** An inadequate form of final disposal of wastes and wastes, in which the only precaution is to cover the mass of waste and tailings with soil.
3. **Landfill:** It consists of the technique of disposal of municipal solid waste in the soil, without causing damage or risks to public health and safety, minimizing environmental impacts, which uses the principles of engineering (soil sealing, gas drainage, rainwater and leachate) to confine solid wastes to the lowest permissible volume, covering them with a layer of soil at the conclusion of each working day or at smaller intervals if necessary (See ABNT Metrological System, NBR 8,419 / 92).

From the study of Law 12,305 / 10 (Art. 3 and Art. 9) and Decree 7,404 / 10 (Art. 12 and Art. 13) it can be inferred that for the purpose of carrying out the activities of environmentally adequate disposal of solid wastes. Economic use of waste and final disposal of waste) requires prior segregation of solid waste, according to its constitution or composition, through the implementation of a selective collection system and reverse logistics system.

SELECTIVE COLLECT

It is the collection of solid waste previously segregated according to its constitution or composition (Article 3 of Law 12,305 /10). It is a process of selection of solid waste (garbage), involving two distinct steps: separation of the solid waste (litter) at the source (segregation) and collection.

A relevant aspect for Solid Waste Management is the support for the productive inclusion of recyclable and recyclable waste pickers, prioritizing the participation of cooperatives of pickers or other forms of association of recyclable and recyclable material collectors made up of individuals with low incomes. For the economic and financial sustainability of these services, Law 8,666 / 93 allows, in its Article 24, the waiver of bidding for the hiring and remuneration of associations or cooperatives of collectors of recyclable materials.

It is important to highlight the destination for associations and cooperatives of collectors of recyclable materials discarded by the organs and entities of the federal public administration directly and indirectly, segregated in the generating source.

Decree 5,940/06 establishes this separation and destination of recyclable waste, considering the following concepts (Article 2): a) Selective solidary collection: collection of discarded recyclable waste, separated at the generating source, for disposal to associations and cooperatives of recyclable waste pickers; b) Discarded recyclable waste: materials that can be returned to their productive cycle, rejected by the organs and entities of the direct and indirect federal public administration. This selection involves separating the solid waste according to the characteristics of the material. It will be treated in the outbound logistics part.

REVERSE LOGISTICS (RL)

Reverse logistics was historically associated with product recycling activities and environmental aspects (Kopicki, Berg and Legg, 1993; Kroon and Vrijens, 1995; Stock, 1992), thus becoming important in companies due to the pressure exerted by stakeholders related to environmental issues, as appointed another authors (Hu, Sheu & Haung, 2002). This literature is still studied in Brazil, together with the process and possibility of well implementing the NPSW.

Reasons for using reverse logistics are due to increasingly strict environmental legislation. The manufacturer's responsibility for the product is expanding. Therefore, it is not sufficient to reuse and remove scraps that are directly part of your own production process, the manufacturer is being held responsible for the product until the end of its useful life. During the eighties, reverse logistics has gained importance in the operations of the companies (Bowersox, Closs and Helferich, 1986), whether due to recalls carried out by the company itself, responsibility for the correct disposal of dangerous products after use, defective products and returned for exchange, expiration The period of validity of the products or withdrawal of the purchase by consumers.

It is not difficult to think of logistics as managing the flow of products from procurement points to customers. On the other hand, there is also a reverse logistics channel that must be managed. From the logistic point of view the life of a product does not end with its delivery to the customer. The products become obsolete, damaged or damaged and are taken to their points of origin for repair or disposal.

NPSW defines reverse logistics as an instrument of economic and social development characterized by a set of actions, procedures and means to enable the collection and restitution of solid waste to the business sector for reuse in its cycle or in other production cycles, or other environmentally appropriate final destination (Article 3 of Law 12,305 /10 and Article 13 of Decree 7,404 /10).

The reverse logistics system should be structured and operationalized through sectoral agreements, regulations issued by the Government and by terms of commitment (Article 15 of Decree 7,404 / 10).

Sectoral Agreements: Are acts of contractual agreements signed between the government and manufacturers, importers, distributors or retailers, aiming at the implementation of shared responsibility for the product's life cycle (Art. 19 of Decree 7,404 / 10). They may be initiated by the Government or by manufacturers, importers, distributors or merchants (Article 20 of Decree 7,404 /10). The minimum items of these agreements are set forth in Article 23 of Decree 7,404 / 10.

THE RESPONSIBILITY SHARED BY THE PRODUCT LIFE CYCLE

Set of individualized and chained allocations of manufacturers, importers, distributors and traders, consumers and holders of public services of urban cleaning and solid waste management, to minimize the volume of solid wastes and wastes generated, as well as to reduce the impacts Caused to human health and environmental quality resulting from the product life cycle (Article 3 of Law 12,305 /10).

PRODUCT LIFE CYCLE

A series of stages that involve the development of the product, the obtaining of raw materials and inputs, the production process, the consumption and the final disposal (Article 3 of Law 12,305 /10). Regulation: Reverse logistics may be implemented directly by regulation, issued by a decree issued by the Executive Branch (Article 31 of Decree 7,404 / 10). Terms of appointment: The Public Authorities may enter into terms of commitment with manufacturers, importers, distributors or merchants, in order to establish a reverse logistics system (Article 32 of Decree 7,404 /10).

SOLID WASTE MANAGEMENT PLANS

From the NPSW study, one can infer the importance given to the planning process for the public and agents involved in the management and management of solid waste.

The Ministry of Environment (MMA), through the Department of Water Resources and Urban Environment (SRHU), prepared the manual Guide for Development of Solid Waste Management Plan 2011(see www.mma.gov.br, op.cit), which can be inferred that solid waste management planning should involve the formulation of guidelines, which establish strategies, establish targets, define programs and projects, and specify the actions to be taken in the solid waste streams.

The manual suggests a basic questionnaire for each planning item:

- **Formulation of the guidelines (WHAT?):** What are the specific guidelines that should be met by the plan?
- **Determination of Strategies (HOW?):** What are the implementation strategies (legal, facilities, equipment, monitoring and control mechanisms) necessary to fulfill the plan?
- **Establishment of Goals (WHAT and WHEN?):** What are the results to be achieved and what are the deadlines to be met?
- **Definition of the Programs / Projects and Specification of the Actions (WITH WHO?): What** are the public and private agents involved and what actions are necessary to carry out the management?

The manual reinforces that the characterization of the planning items allows estimating the financial resources required for its execution, indicating, for consultation, the publication "Integrated Management of Solid Waste, Fascicle 3°, Sources of Financing", carried out by Bank of Brazil, in Partnership with MMA and the Ministry of Cities.

In it, the sources of financing are broken down into two categories: reimbursable financing sources and sources of non-reimbursable financing. Reimbursable financing sources are: BNDES/PMI / Multi-Sectoral Projects Integrated Urban Environmental Sanitation and Water Resources. Non-reimbursable financing sources are: BNDES/ Social Fund, National Fund for the Environment (FNMA), Climate Fund, FUNASA/ environmental sanitation program for up to 50 thousand municipalities, the Ministry of Cities / National Secretariat of Environmental Sanitation / Municipal Program of Solid Waste and the Ministry of Justice / Diffuse Direct Fund.

Later, in 2012, the MMA produced a guidance manual for the preparation of solid waste management plans. The publication provides guidance to the plans of solid waste can be prepared in accordance with the guidelines set forth by Law 12,305 /10 and by the National Solid Waste Plan, taking into account the specificities and diversity that characterize the Brazilian urban network, avoiding the creation of ready and repetitive models. It also aims at empowering the different segments of society interested in the management of solid waste.

Importantly, management and solid waste management should be subject to the following order of priority: non-generation, reduction, reuse, recycling, solid waste treatment and disposal environmentally sound tailings (Art. 9 of Law 12,305 /10). Therefore, there is a hierarchy in the actions to perform.

Another consideration worth mentioning is the encouragement of consortiums solid waste management. The federal entities that choose to establish micro-regions (states, Art. 16 of Law 12,305 /10) and consortium solutions (municipalities, Art. 18 of Law 12,305 /10) will have priority access to federal funds.

In Article 45 of Decree 7,404 / 10 are set the following solid waste plans: the National Solid Waste Plan, Solid Waste the State Plans, the micro-regional Plans Solid Waste and Waste Plans Solid metropolitan or agglomerations Urban Regions the Intermunicipal Solid Waste Plans, Municipal Plans Solid Waste Integrated Management and Solid Waste Management Plans.

NATIONAL PLAN FOR SOLID WASTE

The Ministry of Environment (MMA) along with the International Council for Local Environmental Initiatives (ICLEI - Brazil) in its publication "Waste Management Plans Solid: Guidance Manual", Brasilia, 2012, states that the preparation of the Plan national was the first step in the planning of solid waste management in the country and establishes with time horizon of twenty, guidelines, scenarios, goals and programs of action, it is expected to review every four years, harmonized with the Multi-year Plans Investment (PPA) of the Federal Government.

Another publication of the MMA, with support from the Institute of Applied Economic Research (IPEA), is the manual "National Plan for Solid Waste (Preliminary Version, 2011)", the final version was scheduled for the second half of 2012. This documentation expressed concepts and proposals that reflect the interface between the various sectors of the economy making compatible economic growth and environmental preservation with sustainable development, featuring the application of concepts and logistics chain management / supply chain.

The plan, according the Manual contains guidelines, strategies and targets, indicating which actions are necessary to achieve national goals and priorities to adopt, with quantitative and qualitative information, conforming sectoral agreements and reverse logistics. It plays a guiding role in the development of the other plans.

The content of the Plan consists of the following items:

- Diagnosis of the current situation of different types of waste (Chapter 1 of Law 12,305 /10);
- Macroeconomic and institutional settings (Chapter 2 of Law 12,305 /10);
- Guidelines targets for the proper management of waste (Chapters 3 and 4 of Law 12,305 /10).

The Plan highlights that will be prioritized in access to federal funds other entities of the Federation to establish micro-regions (states, Article 16 of Law 12,305 /10) and inter-municipal consortium solutions (municipalities, Article 18 of Law 12,305 / 10). The process for the establishment of micro and consortia is to identify territorial arrangements between municipalities in order to share services / activities of common interest. Thus, economies of scale are generated to maximize human resources, infrastructure and financial resources.

When referring to the inter-relationship between different levels of government and between the private sector, scavengers and cooperatives, the Plan applies logistics concepts and creates supply chain management network, leading to the chain of training/reverse network supplies of solid waste to an economic destination (reuse, repair, recycling, composting and energy recovery), adding economic value to waste for final disposal or environmentally sound (landfills), adding environmental value to reject. It is important to note that one of the fundamental NPSW instruments for the Plan is the National Information on Solid Waste Management System (SINIR)[3].

REVERSE LOGISTICS (RL) AND REGULATION OF NATIONAL POLICY OF SOLID WASTE

Precisely, this is the strength of a leading expert on the subject in Brazil. Leite, as president of Reverse Logistics Council in Brazil (CLBR) states that the Reverse Logistics is no longer the unknown area a few years ago, jumping in the eyes of any observer its growing importance to the world of business in general. He understands a problem becomes an opportunity when reverse logistics is well planned and executed. Innovative and strategically astute companies have created resources and attention focused on the gains from this opportunity (Guarnieri, 2016). Himself, in his classic work, Reverse Logistics - Environment and Competitiveness (Leite, 2003) commented briefly on the history of the logistics business by setting logistics as one of the oldest and inherent human activities in that its primary mission is to provide goods and services generated by society, in places, in time, in quantity and quality require users.

The Brazilian Logistics Association conceptualized logistics as a process of planning, implementing and controlling the flow and efficient storage and low cost of raw materials, work in process, finished products and related information, ranging from the point of origin to point of consumption in order to meet customer requirements (Daher, 2006).

Liva *et al.* (2002) address three types of reverse logistics: reverse logistics after-sales; RL post-consumer packaging and reverse logistics. According to these authors, the reverse logistics of aftermarket deals with the logistics flow and related logistics information after sales of goods, unused or little use and that are returned. They are products that can return to the business cycle adding them commercial value and are sent to recycling or to a final destination in the reuse of impossibility. Already reverse logistics post-consumer operationalize the physical flow and the corresponding information from consumer goods discarded by society at end of life or used with reusability, as well as industrial waste which return to the business cycle or the production cycle by reverse - specific distribution channels. Regarding the reverse logistics packaging, although frame in the reverse logistics of post-sales or post-consumer it is considered a separate category. Now the distribution to increasingly distant markets means that there is an increase in packaging costs, which affects the final cost of the product. Depending on the type of product and distribution have been the primary packaging, secondary, tertiary, quaternary and the fifth level, which is the containerized unit or special packaging for shipping over long distances (Liva *et al.*, 2002).

According to Leite (2003), reverse logistics after-sales would be the specific area of activity that deals with the addressing and operation of the physical flow and related logistics information after sales of goods, unused or little use, which for various reasons, return to the different links in the direct distribution chain. It is defined also as the area of operation of reverse logistics which equates and also operationalize the physical flow and the corresponding information from consumer goods discarded by society in general, returning to the business cycle or the production cycle through channels specific reverse distribution.

In another point of view, would point to the reverse logistics could be understood from the operational and strategic perspective, becoming more holistic in their concerns in the disposal or use of inhibitors of reverse chains. The strategic perspective refers to the reverse logistics business decisions in the macro environment, consisting of the society and local communities, governments and competitive environment. Would comment further that the operational perspective refers to decisions involving the use of the main logistics tools applied to reverse logistics (Leite, 2003).

In an article published (Guarnieri, 2016), he analyzes the impact that the new legislation is bringing the market logistics reverse, expanding the opportunities for the provision of all categories and specialties and general business services involving equating the return products, under the responsibility of the productive chain of the same.

Also emphasizes that its relevance is observed both in addressing not yet consumed goods and products that require technical assistance, as in products already consumed or used. In a business vision, a solution to achieve economic benefits, the provision of services, adapt to legislation on corporate image defense, etc.

The inclusion of Reverse Logistics in specific chapters in this law, even with its own definitions, demonstrates the importance given to the operational and logistical equation of this return, revealing its complexity and making it an integral part of many waste plans to be edited by the Federation, states, municipalities and companies involved in generating them.

Over the last few years would act companies in Brazil seeking to integrate new activities in the field of reverse logistics, carrying out joint venture, seeking new reuse technology products, specializing in typical activities of reverse logistics, new companies looking for investments in the areas of promising business among other businesses move. It can be said that there is a certain "race" to better prepare for this avalanche of business that is coming true.

Although NPSW specifically involves the products used (post-consumer in the proper terminology), the fact that some of the product segments and its identified packaging this legislation their high degree of obsolescence or marketing term, high demand for technical assistance, among other reasons causes Reverse Logistics aftermarket (which deals with the return of not consumed product or service conditions) is also directly involved with this legislation.

On December 23, 2010, the Presidency of the Republic issued Decree 7,404 regulating this legislation defining the ministries and government departments involved in the implementation of NPSW through the creation of the Interministerial Committee of the National Policy on Solid Waste and the Steering Committee, as well as the functions thereof. Broadly speaking, one can understand that the Interministerial Committee (Article 3) will respond by implementing reverse logistics and the Steering Committee (Article 33) "will have" a consultant role in defining methodologies, criteria and assess the logistics systems reverse presented. Both committees will be able to invite representatives of the business society of similar entities and create specific technical bodies. In general, I also agree that this decree is a start of the planning phase of the reverse logistics systems implementation to be effected in the country in the coming years. The decree explains the justifiable concern to define very precisely the various systems implementation and operation of reverse logistics that may be adopted by the productive chain of the product or packaging, business sectors or companies. It distinguishes, therefore, the following systems (Article 15): "sectoral agreements", "regulations issued by the Government" and "terms of engagement".

Leite analyzes that Chapter III of Title III of the decree details the various types of implementation of the Reverse Logistics. The system through "sectoral agreement" is a contractual act between a supply chain (manufacturers, importers, distributors or dealers) of products and packages aimed at the implementation of reverse logistics, with the initiative of the production chain or the Public Power. It consists basically of the main aspects of a reverse logistics deployment program, namely: product definition and object packages, participation of the various links in the production chain including the consumer, information required throughout the process, collection, storage, transport, and final destinations, penalties, among other details. The system "Regulation issued by decree of the Public Power" is a government initiative and the "Terms of Commitment" are specific agreements that may be considered cases of sectoral agreements exceptions. Analysis of the decree made by Leite presents a series of reflections and observations. Incomplete, it emphasizes further that progress in NPSW decisions will bring to Brazil great progress in all activities related to reverse logistics, providing increase in the quantities to be treated in the various supply chains and as a result large business and professional opportunities in these various areas.

FINAL

The disposal of solid waste is one of the extremes of an extensive chain, which co-mingles with the exploitation of natural resources, goes through the manufacturing industry and ends up in the manufacture of consumer goods. This process is stimulated by a sophisticated marketing and advertising system to ensure the permanence and growth of the production system. Disposal, the last point in this sequence, is not a mere accident, but is predicted from the outset, as product obsolescence is strategically programmed by the industry.

It is known that the NPSW in some reverse channels, structured course by market law and, in this case, the implementation of reverse logistics depend exclusively on companies in the sector. Here it is stated that the costs added from the collection of post-consumer products, to the reintegration into the production cycle outweigh the economic advantages of reuse them to replace the original raw materials or the new use of assets still in use conditions.

The social concern and the sustainability of NPSW is expressed in the charge of public urban cleaning services and solid waste management should prioritize the hiring of productive collectors of recyclable materials formed by individuals from low-income organizations. The complexity of the issue and the difficulty of reaching consensus among the various sectors involved are the determining factors.

Faced with this problem, it is worth mentioning the fact that the dumps have a term of end, provided for in law for the year 2014. Unfortunately, this has not yet occurred. There is also an abundance of open-air garbage dumps, unprepared to be treated. It is said that in Brazil there are still open sewers. How and when to expect that, the municipality which is the thin capillarity of the state, can have its proper waste treatment? It seems impossible, for the observed, that without a network of cooperators, companies and volunteers and the people responsible for their waste, to put into practice solid waste management, helping in the reverse of the supply chain and in the beauty described in the proposal gives NPSW.

As for shared responsibility for the life cycle of the products, it was noted that this mechanism also requires public urban cleaning services to establish a separate collection system, the articulation with the economic and social agents in the development of measures to enable the returning to the production cycle of the reusable and recyclable solid waste, the implementation of systems for composting solid organic waste and the development of ways to use the produced compound.

The social impact of the new law is ensured when, during the same said device ensures that the holder of public urban cleaning services and solid waste management should prioritize the organization and functioning of cooperatives or other forms of association of recyclable material collectors and recyclable formed by individuals of low income people as well as their hiring for the selective collection activities, reverse logistics, recycling waste and operation of composting systems, waiving public bidding.

Governance mechanisms, essential to the success of the initiative, are guaranteed by the implementation of the National Information on Solid Waste Management System (SINIR) provided for in the Act of shared competence between the Union, the states and municipalities.

The prohibitions and sanctions, the new regulations some advances in relation to the rules already in force, except for the reference to a ban on imports of hazardous waste and waste and solid waste whose characteristics cause damage to the environment and animal and public health nurse and plant health, although for treatment, renovation, reuse, recycling or recovery.

Chapter hazardous waste does not present major changes in today's current regulations (CONAMA, in environment and ANVISA resolutions, in health), except for the permission given to the licensing agencies to demand by the hazardous waste operators hiring liability insurance for damages caused to the environment or public health, subject to the rules on coverage and maximum employment to be established by regulation.

Although it is not yet possible to draw a model of results, the expected success of the reverse logistics of solid waste, from the entry into force of the National Waste Policy, is promising and is closely linked to the concept of sustainability in its economic, social and environmental. The NPSW, together with the National System of Basic Sanitation Information (SNIS) establishes a consistent regulatory framework, allowing, by the general guidelines of the law, a global orientation, respecting the limits of discretionary legislative powers of each federal entity. There is the conviction that the set of instruments of this NPSW, especially the shared responsibility for the life cycle of products and reverse logistics, will be a big factor modifier in Brazilian corporate cultures, particularly in the industrial segments, noticeable a large field business diversification, especially for companies operating in urban cleaning and waste treatment sector. In this context, also is expected to a strong increase in recycling industries, considering the performance spaces will be created downstream of the production chains, where to direct the flow of recoverable waste collected.

By the way, according to authors Demajorovic, Augusto & Souza (2016), Brazil's huge territory makes collecting waste even more difficult outside big urban areas, due to the logistics costs involved. We should keep in mind that the volume and value of product raw materials are essential to offer gains of scale and to ensure the financial feasibility of RL activities. In Brazil, the present technology only allows the recycling of components with lower aggregated value. Computer boards can only be sorted and packaged for reconditioning overseas, thus transferring to other countries the biggest share of the value generated with the activity.

It is important to emphasize too that practices listed in the proposal of the substitute are indicative of a strong commitment to social issues, by creating mechanisms that generate sources of income, benefiting a large number of people who do not have access to the formal labor market, with its promising inflow from operating activities of the reverse chain of recycling. For collectors, the substitute innovates as the first federal law to recognize and assign privileges to these workers.

This, in addition to providing important information, open additionally spaces for a future deepening of matter, being timely and enriching for the management science the development of studies in the field, with a greater degree of coverage in the logistics applied to the reverse flow waste in Brazil and other countries involving public officials and other experts with experience in the issue of waste management in the area of business strategy and logistics.

REFERENCES

Bowersox, D. J., Closs, D. J., & Helferich, O. K. (1986). *Logistical Management: A systems integration of physical distribution, manufacturing support, and materials procurement*. New York: MacMillan Pub Co.

Brazilian Association of Technical Standards (ABNT). (1992). *NBR 8419/92* (Presentation of urban solid waste landfill projects - procedure). Retrieved Jan 04, 2013, from http://www.cetesb.sp.gov.br/wp-content/uploads/sites/36/2015/03/base3.pdf

Brazilian Federal Senate. (2012). *Bulletin of legislative 15 (Opinion on the national solid waste policy)*. Brasília, Brazil: Federal Senate.

Brazilian Institute of Geography and Statistics (IBGE). (2000). Quality and efficiency of sanitation services. In *Atlas of sanitation*. Brasilia, Brazil: Ministry of Cities. Retrieved Jun 28, 2017, from http://www.ibge.gov.br/home/estatistica/populacao/atlas_saneamento/

Brazilian Institute of Geography and Statistics (IBGE/MUNIC). (2011). *Profile of Brazilian Municipalities*. Brasília, Brazil: Ministry of Cities. Retrieved Mar 22, 2017, from http://www.ibge.gov.br/home/estatistica/economia/perfilmunic/2011/

Business Commitment to Recycling (CEMPRE). (2012). *Business Commitment for Recycling: radiographing the selective collection*. Retrieved Mar 17, 2017, from http://www.cempre.org.br/ ciclosoft_2012.php

Daher, C. E., Silva, E., & Fonseca, A. (2006). Reverse logistics: opportunity to reduce costs by managing the integrated value chain. Brazilian Business Review, 3(1), 57-72. doi:10.15728/bbr.2006.3.1.5

Demajorovic, J., Augusto, E. E. F., & Souza, M. T. S. (2016). Reverse logistics of e-waste in developing countries: Challenges and prospects for the brazilian model. *Ambiente & Sociedade*, *19*(2), 117–136. doi:10.1590/1809-4422ASOC141545V1922016

Epistein, E. (2013). *Disposal and Management of Solid Waste: Pathogens and Diseases*. Boca Raton, FL: Taylor & Francis CRC Press.

Espinoza, P. T., Arce, I. N., Daza, D., Faure, M. S., & Terraza, H. (2010). *Regional evaluation on urban solid waste management in Latin America and the Caribbean* (Research Report 2010). IDB, PAHO and AIDIS. Retrieved mar 14, 2017, from https://publications.iadb.org/handle/11319/3286

Guarnieri, P. (2016). Reverse Logistics in Brazil: Advancements and perspectives (Interview with Paulo Roberto Leite). *Gestão, Inovação e Sustentabilidade RE:view*, *2*(1), 265–270.

Hu, T. L., Sheu, J. B., & Haung, K. H. (2002). A reverse logistics cost minimization model for the treatment of hazardous wastes. Transportation Research Part E, 38, 457-473. doi:10.1016/S1366-5545(02)00020-0

Institute of Applied Economic Research (IPEA). (2012). *National Solid Waste Plan: diagnosis of urban waste* (Research Report). Brasília, Brazil: Ministry of Economics. Retrieved Jul 23, 2015, from http://www.ipea.gov.br/agencia/images/stories/PDFs/relatoriopesquisa/121009_relatorio_residuos_solidos_urbanos.pdf

Kopicki, R., Berg, M., & Legg, L. L. (1993). *Reuse and recycling: Reverse logistics opportunities*. Oak Brook, IL: Council of Logistics Management.

Kroon, L., & Vrijens, G. (1995). Returnable containers: An example of reverse logistics. *International Journal of Physical Distribution and Logistic Management, Bradford*, *25*(2), 56–68. doi:10.1108/09600039510083934

Leff, E. (2008). Citizenship, globalization and post-modernity. In *Knowing environmental: sustainability, rationality, complexity* (6th ed.; pp. 118–132). Petrópolis, Brazil: Vozes.

Leite, P. R. (2003). *Reverse logistics, environment and competitiveness*. São Paulo, Brazil: Prentice Hall.

Liva, P. B. G., Oliveira, W. S., & Pontelo, V. S. L. (2002). *Reverse Logistics - I*. Retrieved from http://limpezapublica.com.br/textos/logistica_reversa_01. pdf

Ministry of the Environment. (1981). *National Policy on the Environment* (Extracts from Law 6938/81). Brasília, Brazil: MMA. Retrieved Mar 13, 2015, from http://www.mma.gov.br/estruturas/sqa_pnla/_arquivos/46_10112008050406.pdf

Ministry of the Environment. (2010). *National Solid Waste Policy* (Law 12,305 /10). Brasília, Brazil: MMA. Retrieved Jun 10, 2017, from http://www.mma.gov.br/port/conama/legiabre.cfm?codlegi=636

Nag, A., & Vizayakumar, K. (2005). *Environmental Education and Solid Waste Management*. New Delhi: New Age International.

Presidency of the Republic. (1988). *Constitution of the Federative Republic of Brazil of 1988*. Brasília, Brazil: Civil House. Retrieved mar 20, 2017, from http://www.planalto.gov.br/ccivil_03/constituicao/constituicao.htm

Presidency of the Republic. (1993). *Decree 5,940 of October 25, 2006* (Institutes the separation of discarded recyclables). Brasília, Brazil: Casa Civil. Retrieved Mar 05, 2015, from http://www.planalto.gov.br/ccivil_03/_ato2004-2006/2006/decreto/d5940.htm

Presidency of the Republic. (1993). *Law 8,666 of June 21, 1993* (Normalizes bids and contracts). Brasília, Brazil: Casa Civil. Retrieved Mar 07, 2015, from http://www.planalto.gov.br/ccivil_03/leis/L8666cons.htm

Presidency of the Republic. (2007). *Lei 11445, of January 5, 2007* (Establishes National guidelines for basic sanitation). Brasília, Brazil: Casa Civil. Retrieved Mar 21, 2015, from http://www.planalto.gov.br/ccivil_03/_ato2007-2010/2007/lei/l11445.htm

Presidency of the Republic. (2010). *Decree 7404, of December 23, 2010* (Regulation of Law 12, 305/10). Brasília, Brazil: Casa Civil. Retrieved Mar 21, 2015, from http://www.planalto.gov.br/ccivil_03/_ato2007-2010/2010/decreto/d7404.htm

Presidency of the Republic. (2010). *Decree 7,390, of December 9, 2010* (Regulates the National Policy on Climate Change). Brasília, Brazil: Casa Civil. Retrieved Mar 12, 2015, from http://www.planalto.gov.br/ccivil_03/_ato2007-2010/2010/decreto/d7390.htm

Ritter, L., Solomon, K., Sibley, P., Hall, K., Keen, P., Mattu, G., & Linton, B. (2002). Sources,path-ways, and relative risks of contaminants in surface water and ground water: A perspective prepared for the Walkerton inquiry. *Journal of Toxicology and Environmental Health. Part A.*, *65*, 1–142. doi:10.1080/152873902753338572 PMID:11809004

Sobsey, M. D. (1978). Field survey of enteric viruses in solid waste landfill leachates. *American Journal of Public Health*, *68*(9), 858–864. doi:10.2105/AJPH.68.9.858 PMID:28677

Stock, J. R. (1992). *Reverse logistics.* Oak Brook, IL: Council of Logistics Management.

Teschke, K., Bellack, N., Shen, H., Atwater, J., Chu, R., Koehoom, M., ... Isaac-Renton, J. L. (2010). Water and sewage systems, socio-demographics, and duration of residence associated with endemic intestinal infectious diseases. *BMC Public Health*, *10*(1), 1–24. doi:10.1186/1471-2458-10-767 PMID:21162734

ENDNOTES

1. For purposes of work adequacy, only the concepts and characteristics of the first will be used, by directing the study to the reality of this vector of environmental problems.
2. Although it is not the main theme, it is worth inserting the contribution of Epstein, 2015, Op.cit, due to the terrible consequences that the lack of treatment of urban and domestic waste brings to health, a problem perhaps even greater than the problem of incipient process of elaboration and conduct of a public policy for solid waste, as is the case of Brazil.
3. More information can be found at www.sinir.gov.br. Its design involves the National System of Environment Information (SINIMA) and the National System of Basic Sanitation Information (SNIS), being fed with information from the States, Federal District and Municipalities.

This research was previously published in the Handbook of Research on Supply Chain Management for Sustainable Development edited by Ulas Akkucuk, pages 282-304, copyright year 2018 by Business Science Reference (an imprint of IGI Global).

Chapter 5
Environmental Reporting and Accounting:
Sustainability Hybridisation

Radiah Othman
Massey University, New Zealand

Nirmala Nath
Massey University, New Zealand

Fawzi Laswad
Massey University, New Zealand

ABSTRACT

This chapter examines the context in which 'accounting-sustainability hybrids' were constructed in anticipation of the Environmental Reporting Act (ERA) passed in 2015 in New Zealand. Using governmentality perspectives, the researchers consider how the key individuals in local government discursively articulate the 'sustainability programmatic' – sustainability policies that deploy the Act as a regulative intervention, the mediating instruments (reporting medium), and the hybridisation of accounting and sustainability (accounting-sustainability hybrids). The chapter draws on archival materials and qualitative survey of 90 key individuals from all 78 local authorities. The analysis showed that local mediating instruments were aligned with ERA and sustainability programmatics, resulting in modification of the existing hybrids and the creation of new ones. The chapter demonstrates how legislation facilitates action from a distance.

DOI: 10.4018/978-1-7998-0441-3.ch005

Copyright © 2020, IGI Global. Copying or distributing in print or electronic forms without written permission of IGI Global is prohibited.

INTRODUCTION

Accounting-sustainability[1] processes and practices are regarded as central to embedding sustainability within public service organisations. They act as intermediaries between government programmes of reform and local service imperatives (Thomson, Grubnic, & Georgakopoulos, 2014, p. 453). Accounting-sustainability practices are considered to be hybrids of accounting with different aspects of the sustainability programmatic (Thomson et al., 2014, p. 455). For instance, full cost accounting (Bebbington et al., 2001; Herbohn, 2005) can be seen as a hybrid between conventional costing, sustainability science and environmental economics.

In New Zealand (NZ), the central government has introduced policies aimed at fostering greater sustainability in public service delivery. In 2014, the Environmental Reporting Bill was introduced by the Environment Minister, Amy Adams, and was later passed as a new law (the ERA) in September 2015. The ERA sets a broad framework to scope the reporting of one of the five environmental domains (air, freshwater, land, marine, atmosphere and climate), with the cycle beginning with freshwater in mid-2016 (MFE, 2016b). The framework requires a synthesis report, with analysis of air, freshwater, land, marine, atmosphere and climate trends and interactions, to be published every three years. The Resource Management Act (RMA) 1991 has also been amended to make local governments' (LG[2]) responsibility in collecting data and reporting[3] the state of their local environment explicit. Collectively, these policies and acts are referred to as programmatics. Programmatics are high-level discourses with purposive attempts to organise and reorganise institutional spaces, their routines, rituals and procedures, and the conduct of actors in specific ways (Dean, 1999, p. 32; Raco & Imrie, 2000; Thomson et al., 2014). As such they can be seen as part of many central government's strategies and interventions for LGs to transform their processes and practices.

In regard to environmental sustainability programmatics, in NZ the local authorities are required to report on the state of the environment under the RMA, and now the ERA has added on a new framework for local authorities to report on. In absence of specific standards and criteria of reporting the LG are left to interpret both Acts (considered as regulative sustainable programmatics) and other programmatics relevant to the public sector to construct their accounting-sustainability hybrids and to select mediating instruments (Kurunmäki & Miller, 2011, p. 222) to embed environmental sustainability. Mediating instruments are vital for constructing points of common reference between environmental sustainability reform and the local authorities' processes and practices. Mediating instruments can be local (internally generated) or non-local (generic solution by external body, such as GRI reporting). Thomson et al. (2014) advocate that understanding how the local hybrids programmatic are constructed is critical for the understanding of the hybridisation of accounting and sustainability.

The main aim of this chapter is to analyse the context in which accounting-sustainability hybridisation emerged. Specifically, the researchers examine how the local programmatic has been interpreted, prioritised and constructed at LG level, and then the researchers analyse how the local authorities selected a suitable mediating instrument, and the accounting-sustainability hybrids. This research contributes to the accounting-sustainability literature by providing insights on how regulation facilitates actions from distance and the importance of mediating instruments and hybrids being grounded in local contexts in order to support and facilitate change (Thomson, Grubnic, & Georgakopoulos, 2014, p. 473; Gray, 2010; Lehman, 2001).

CONTEXT OF STUDY

There are 78 local authorities in NZ. Eleven are regional councils and 67 are territorial authorities. The Auckland Council (a unitary authority), 54 district councils and 12 city councils are collectively known as territorial authorities. Regional and territorial authorities are defined by the scope of their responsibilities. According to www.localcouncils.govt.nz (2016), the LG sector employs over 28,000 staff. LG's contribution to NZ's gross domestic product (GDP) for the year ending March 2014 was 3.5% ($8.13 billion) of the total GDP ($229.7 billion). This indicates the importance of LG in the NZ economy.

LG is an interesting research site because in terms of proximity it is much closer to the community and thus has a greater role in influencing local communities towards sustainability (Potts, 2004; Ball, Broadbent, & Jarvis, 2006, p. 260). Carnegie and West (2010, p. 111) claim that LG has a capacity to influence and change the behaviour of their constituents.

PREVIOUS RESEARCH

There is a distinct lack of comprehensive research on sustainability reporting and accountability within the LG segment (Ball & Grubnic, 2007; Gray, Dillard, & Spence, 2009; Williams, Wilmshurst, & Clift, 2011). Research in this area would enhance an understanding of the nature of sustainability accounting and accountability (Ball & Grubnic, 2007, p. 257) and the processes undertaken by the local authorities in this regard.

The majority of previous studies on sustainability reporting by local authorities have examined the nature and extent of sustainability, reporting performance and motivation for reporting (e.g. Goswami & Lodhia, 2014; Joseph, Pilcher, & Taplin, 2014; García-Sánchez, Frías-Aceituno, & Rodríguez-Domínguez, 2013; Greco, Sciulli, & D'onza, 2012). These studies have been criticised as having little to do with sustainable development (Bebbington & Larrinaga, 2014; Unerman & Chapman, 2014, p. 385) advocate more research that constructively but critically engages business and organisations.

The roles of key individuals in the LG sector in driving the sustainability initiatives and the consequences of regulatory regime have been under-researched. Bebbington and Larrinaga (2014, p. 404) propose research that stimulates ongoing participation with those who are affected by (and who affect) a sustainable development problem situation. In Australia, Farneti and Guthrie (2009) noted that key individuals were the drivers of sustainability initiatives, including the reporting and the reports. In NZ context, Bellringer, Ball, and Craig (2011) interviewed key individuals in five NZ local authorities responsible for preparing the sustainability report. Their study concluded that even though such reporting was mainly for internal stakeholders, it also served as a tool to demonstrate leadership, public accountability and financial responsibility to the external parties. By comparison, in Australasia, research that critically engages LG and the individuals affected by sustainability regulations is limited (Tregidga & Milne, 2006; Bellringer et al., 2011).

Another strand of research in this area combines the accounting and sustainability disciplines. This combination of disciplines has triggered research into accounting-sustainability hybrids such as corporate social reporting (Gray, Owen, & Maunders, 1988), the sustainable balanced scorecard (Figge, Hahn, Schaltegger, & Wagner, 2002) and carbon accounting (Ball et al., 2009; Hopwood, 2009). Miller, Kurunmäki, and O'Leary (2008, p. 943) define hybrids as new phenomena produced out of two or more elements normally found separately. Accounting is itself a hybrid that was formed and reformed at the margins of other practices and disciplines (Miller, 1998). According to Thomson et al., (2014), accounting-sustainability hybrids aim to steer organisational change in local service delivery practices to be aligned with elements of environmental sustainability programmatics external to LG. These can be acts and other relevant legislations imposed by the central government on the LG. Miller et al., (2008, p. 944) argued that the existing literature has largely neglected the hybrid practices, processes and expertise that make possible lateral information flows and the diversity of hybrids in order to understand better how organisations manage the different types of uncertainties they face.

In view of the above, this study fills the gap by examining how key individuals in LG sector interpret, prioritise and construct their own locally relevant programmatic discourses and how they select and/or construct mediating instruments, thus the embedding of accounting-sustainability hybrids into their local processes and practices in anticipation of the ERA. According to Bebbington and Larrinaga (2014) and Unerman and Chapman (2014), this type of research is critical to sustainable development by providing more robust evidence and understanding of the current policies and practices so that they can be evaluated and critiqued and new ones developed.

THEORETICAL FRAMEWORK

In this study, Foucault's governmentality framework is used to theorise the management and relationships of power dynamics in the LG setting in New Zealand. Foucault (1979, p. 62) describes governmentality as "an ensemble formed by the institutions, procedures, analyses and reflections, the calculations and tactics, that allow the exercise of this very specific albeit complex form of power". Governmentality forms a basis of political thought and action (Raco & Imrie, 2000), characterised by particular ways of thinking about the types of problems that can and should be addressed by various authorities ('problematics') (Rose & Miller, 1992, p. 32) such as threats to environmental sustainability. These ways of thinking are political rationalities or discourses connected to central government programmes (Raco & Imrie, 2000, p. 2190). The programmes therefore normalise the central government's intervention via rules, norms and processes (Rose & Miller, 1992, p. 183) in order to organise and reorganise institutional spaces, routines, rituals and procedures, and the conduct of actors in specific ways (Dean, 1999, p. 32; Foucault, 1979).

Political rationalities and programmes are deployed by the central government in a variety of ways, including legal, architectural, professional, administrative, financial, and judgemental (Dean, 1999, p. 212), and are known as the "technologies of government" (Miller & Rose, 1990, p. 183). They constitute logistical and infrastructure powers, and subsume the moral and political shaping of conduct (Dean, 1999, p. 212). This is to ensure that aspects of the decisions and actions of the individuals, groups, organisations and populations are aligned with the regulated authoritative criteria (Dean, 1999, p. 212). Raco and Imrie (2000, p. 2201) assert that a Foucauldian approach enables them to link the mentalities, norms, aspirations and actions of members of the population with the objectives and techniques of government. They use Foucault's concept of governmentality to argue that the recent shift towards a 'right and responsibilities' agenda in urban policy is part of broader transformations in the rationalities and techniques of the UK government.

Foucauldian governmentality theory, though limited, has been used to frame studies in accounting. Specifically, it has been used in examining discounted cash flow analysis (Miller & Rose, 1990), auditing practices (Power, 1994), healthcare costing

(Preston, Chua, & Neu, 1997), and the role of accounting techniques in colonisation and genocide (Neu, 2000). However, its use in the sustainability literature has been limited to the development of accounting-sustainability practices (Thomson et al., 2014). It was not until recently that this notion of governmentality became part of the vocabulary of sustainability research (Everett, 2004; Neu, 2006; Spence & Rinaldi, 2014) in addressing one of the criticisms in the literature that sustainability research has been "decoupled from the factual organisational actions" (Laine, 2009, p. 1047). Unerman and Chapman (2014, p. 386) suggest that new research is needed to engage with and critique practice and policy so as to examine whether or not power structures are in need of replacement, reformation or redirection.

Within the context of LG, the governmentality framework is useful to understand aspects of, and constraints to, local authorities for substantive sustainable development and transformation (Thomson et al., 2014). In this line of research, governmentality has been used as a means of analysing and seeking to make sense of the government's sustainability framework and policies. Thomson et al. (2014) provide evidence of how financial resource constraints in UK public services and the onus of meeting tough service delivery targets have limited the effectiveness of accounting-sustainability hybrids. Spence and Rinaldi (2014) use governmentality theory to further develop Dean's (2009) four analytics of government to examine how specific accounting practices, in the context of a new regime of sustainability practices instituted by a supermarket, were effective in shaping the governance of a supply chain. Argento, Grossi, Jããskelãinen, Servalli, and Suomala (2016) apply the governmentality framework and its biopolitical implications to explore how municipalities in the City of Helsinki drive smart city programmes.

Miller and Rose (1995) advocate that the governmentality framework is particularly useful for directing attention to the heterogeneous sites through which governmentality is exercised and the ways in which technologies of government facilitate action at a distance. Specifically, Miller and Rose (1995, p. 8) explain that the term 'technologies' suggests a particular approach to the analysis of the activity of ruling, one which pays great attention to the actual mechanisms through which authorities of various sorts have sought to shape, normalise and instrumentalise the conduct, thought, decisions and the aspirations of others in order to achieve the objectives they consider desirable. Regulative environmental reporting can be seen as a technology of the government that has certain target populations (cf. Foucault, 1991, p. 92). By employing this governmentality framework, the researchers are able to examine and understand the ways in which NZ central government's sustainability policies are operationalised and translated into local authority practices, particularly in how the LGs known to have limited capacities cope with the central government's rationalities and techniques of governmentality and how their practices have been shaped and reshaped in this regard. The researchers' study is similar to Kurunmaki and Miller (2011), who studied the concepts of flexibility introduced in the Health Act 1999 in the UK by exploring the early experiences of those experimenting with inter-organisational co-operation; in the researchers' case, it is sustainability policies via the introduction of ERA in 2015.

RESEARCH METHODOLOGY

This qualitative study analysed data collected from primary and secondary sources. The primary data were analysed from an online qualitative survey in August 2015. Qualitative survey has been used in other disciplines, such as biology (Stenius & Cunnington, 1972), anthropology (Carter, 2002), education (Debski & Gruba, 1999) and psychology (Kane, 2008). Wolcott (1990, p. 1) emphasised that "the real mystique of qualitative inquiry lies in the process of using data rather than in the processes of gathering data". Guba and Lincoln (1998) suggest qualitative survey may be useful in a positivist or post-positivist project (including ontological realism and epistemological objectivism), but it could also be performed in the context of critical theory or constructivist projects. Jansen (2010) suggests that it is the analysis that would determine whether a study is a qualitative or quantitative. Unlike the quantitative survey, the qualitative type of survey aims at determining the diversity of some topic of interest within a given population instead of frequencies, means or other parameters (Jansen, 2010). The emphasis is establishing meaningful variation within that population.

The aim of a qualitative survey analysis is to explore meanings and experiences (Fink, 2003, p. 61); in the researchers' case, it is exploring the key individuals' articulation of their daily work practices in order to reveal the everyday doings of the local authorities in terms of environmental sustainability initiatives. In order to enlarge the coverage, Jansen (2010) suggests that one might want to reconsider the selection of dimensions and categories. In our study the data are framed closely to the themes generated from governmentality studies i.e. the programmatic/discursive; the practices, processes and instruments to which they are intrinsically linked, and the professional or 'expert' knowledge through which they are made operable (Rose & Miller, 1992; Kurunmaki & Miller, 2011, p.221). This multi-level of analysis by using governmentality themes has been successfully used in accounting research and other domains in the context of modernising government reforms, but it has been little used to examine the inter-relations between large-scale policy reforms and the more localised redesign of management control practices and organisational arrangements for service delivery (Kurunmaki & Miller, 2011, p. 222). Thus, this study aims to fill this gap.

The qualitative survey targeted 318 key individuals working in all 78 local authorities in NZ. The respondents were asked about their roles, knowledge and experience with environmental sustainability initiatives, including sustainability reporting and ERA. Due to the qualitative nature of the survey, respondents were given one month to complete the survey. The progress of the survey was monitored online. Anonymity was assured and the response rate was 28.3% (90 respondents) (Table 1). Fifty-one percent of the respondents were directly involved in deciding what sustainability information should be reported. The rest were involved in various roles such as making recommendations to management and the council, planning, managing assets, and drafting the environmental sustainability plan and strategy.

The secondary data were obtained from archival analysis of published documents, information from each local authority, the Ministry for the Environment, and other related websites. The documents reviewed included public and stakeholders' submissions, ministers' speeches, press releases and announcements such as "Environmental Reporting Bill First Reading", "New Environmental Reporting Act Passed", "About the Environmental Reporting Act 2015" and "Departmental Report on Environmental Reporting Bill 2014", which are all available from the NZ central government website, www.beehive.govt.nz. Governmentality studies emphasise the importance of performing distinct levels of analysis and their inter-relations (Rose & Miller, 1992; Kurunmaki & Miller, 2011, p. 221). The documents provide vital information of the central government's political rationalities and discursive nature of its intervention in the LG boundaries, as discussed below.

Table 1. Respondents' Position in Council

Respondent no	Position	Type of council	Code
1	Unit Manager in Strategy and Planning	City	CC11
2	Manager	City	CD12
3	Chief Executive Officer	City	CD21
4	Chief Finance Officer	City	CH11
5	Divisional Manager, Strategy and Planning	City	CH21
6	Chief Executive Officer	City	CH22
7	Principal Advisor, Strategy and Planning	City	CH23
8	General Manager, Strategic Services	City	CH24
9	Director of Environmental and Planning Services	City	CI11
10	Chief Executive Officer	City	CI12
11	Planning Manager	City	CN11
12	Corporate Service	City	CN12

continues on following page

Table 1. Continued

Respondent no	Position	Type of council	Code
13	Group Manager – Strategy & Environment Group	City	CN21
14	Senior Manager, Strategy	City	CP11
15	Project Manager	City	CP21
16	Sustainability and Community Development	City	CT11
17	Chief Operating Officer	City	CW11
18	Department Manager	District	DA11
19	Group Manager Customer and Community Services	District	DB11
20	Planning and Regulatory Manager	District	DC11
21	GM Strategic Planning and Policy	District	DF11
22	District Planning Manager, Strategic Planning and Policy	District	DF12
23	Shared Services Science Manager	District	DG11
24	Senior Planner	District	DG21
25	Infrastructure and Community Facilities Manager	District	DG31
26	Group Manager Strategy and Development	District	DH11
27	Strategy Manager	District	DH12
28	Policy Analyst (Strategic Planning Department)	District	DH21
29	Chief Executive Officer	District	DK11
30	District Planner	District	DK12
31	Chief Executive Officer	District	DM11
32	Environment and Sustainability Advisor	District	DM21
33	Environment and Sustainability Advisor	District	DM22
34	Planning Leader	District	DN11
35	Finance and Corporate Services Manager	District	DO11
36	Chief Financial Officer	District	DO21
37	Policy Advisor	District	DR11
38	Chief Executive Officer	District	DR21
39	Corporate Service Manager	District	DS11
40	Group Manager	District	DS21
41	Chief Financial Officer	District	DS22
42	Group Manager, Environmental and Community	District	DS23
43	Group Manager, Policy and Operations	District	DS31
44	Manager, Environmental Services	District	DS32
45	Group Manager Environmental Services	District	DT11
46	Strategy Manager	District	DT21

continues on following page

Table 1. Continued

Respondent no	Position	Type of council	Code
47	Community Development Manager	District	DT31
48	Corporate Services Manager	District	DT32
49	Corporate Planning Manager	District	DT41
50	Team Leader, Strategic Policy	District	DW101
51	Group Manager, Regulatory	District	DW11
52	Group Manager, Strategic Development	District	DW12
53	Policy Manager	District	DW21
54	Chief Executive Officer	District	DW22
55	Manager, Finance Business Support	District	DW23
56	Chief Executive Officer	District	DW31
57	Finance Manager	District	DW41
58	Corporate and Community Planner	District	DW42
59	Environment and Sustainability Advisor	District	DW51
60	Chief Executive Officer	District	DW61
61	Water Services and Waste Manager	District	DW71
62	Principal Planner, Deputy Strategy and Development Manager	District	DW81
63	Deputy Chief Executive Officer	District	DW82
64	Senior Policy Analyst	District	DW83
65	Group Manager, Planning, Community and Environment	District	DW91
66	Chief Sustainability Officer	Regional	RA11
67	General Manager	Regional	RB11
68	Policy Analyst	Regional	RM11
69	Strategy Manager	Regional	RM12
70	Group Manager, Environmental Management	Regional	RM13
71	Manager, Environmental Monitoring and Reporting	Regional	RN11
72	Group Manager Regulatory Services	Regional	RN12
73	Chief Executive Officer	Regional	RO11
74	Strategy and Corporate Planning Manager	Regional	RS11
75	Manager, Policy and Strategy	Regional	RT11
76	Senior Manager – Strategic Planning	Regional	RU1
77	Chief Financial Officer	Regional	RU10
78	Manager	Regional	RU11
79	General Manager, Strategy	Regional	RU2

continues on following page

Table 1. Continued

Respondent no	Position	Type of council	Code
80	Senior Planner	Regional	RU3
81	Manager, Environment and Community	Regional	RU4
82	Chief Financial Officer	Regional	RU5
83	Manager, Strategy and Community Planning	Regional	RU6
84	Chief Executive Officer	Regional	RU7
85	Senior Manager, Strategy	Regional	RU8
86	Chief Executive Officer	Regional	RU9
87	Director, Community and Services	Regional	RW11
88	Manager, Sustainability Projects	Regional	RW12
89	Corporate Sustainability Coordinator	Regional	RW21
90	Chief Executive Officer	Regional	RW21

FINDINGS AND DISCUSSION

Political Rationalities and Local Authorities' Responses

Political Rationalities of the Central Government

The secondary data were used to examine the rationalities of the central government and how they were constructed and reproduced through the deployment of ERA to promote environmental sustainability in NZ (Foucault, 1979, p. 92).

On 5 March 2014, the (then) Environment Minister, Amy Adams, announced that NZ was the only OECD country that did not have a statutory requirement to regulate the state of the environment reporting. ERA was intended to address this anomaly and deliver the National Party's 'Bluegreen' policy commitment. She rationalised ERA by linking it to the wider agenda of the nation's economy,

In New Zealand our economy is dependent on our environment, and our environmental protection efforts rely on us having a strong economy. (MFE, 2016a)

When ERA was passed as the new Act on 24 September 2015, the new Environment Minister, Nick Smith, stressed that the new law was to strengthen NZ's clean and green brand:

The new Environmental Reporting Act will back up our clean, green brand with authoritative and independent information on the state of our environment. It will tell us where we match up, where we don't and give regular updates so that we can track long-term changes ... this new law requiring fair and accurate reports on the state of the environment reflects the increased importance New Zealanders put on our natural wealth. (MFE, 2016b)

From the economy, the environmental reporting agenda was then tied to the image branding and natural wealth importance. The Environment Minister also announced that:

It (ERA) proposed the following key changes: improve the consistency of environmental monitoring statistics at local levels, create a statutory obligation in New Zealand to require regular and independent state of the environment reporting. (MFE, 2016c)

The wordings of "local levels" and "statutory obligations" marked the responsibilities of the local authorities to the environmental sustainability agenda, which had been previously associated with the wider national agenda of economy, image branding and natural wealth, those of which fell primarily under the jurisdiction of the central government. It can be inferred that the new agenda of environmental sustainability via environmental reporting was suffused with language (and practices) to actually operationalise (extend) the central government's governing of the LG, in pursuit of their overall objectives (Dean, 1999). Dean (1999) predicted that this is how the government reinvent themselves: focusing on rebuilding and operationalising the intended groups by including them in the process of achieving their own objectives. As such, the regulated technology, in this case (ERA), was used to direct conduct of the LG to instantiate the authorities of the central government (Rose & Miller, 1992; Argento et al., 2016) and to structure the conduct of policy recipients in ways that could be regulated, controlled, calculated, and shaped to meet specific ends (Dean, 1999, p. 11).

ERA is the most recent programmatic discourse impacting on the governance of and by NZ LG. Under the Resource Management Act 1991, the amendments to section 360 are aimed at explicitly regulating local authorities' responsibilities for "on-the-ground" monitoring and reporting the environment of its region or district (MFE, 2016d, p. 47). Even though section 35(2)[4] stipulates that this responsibility is to the extent that is appropriate for enabling the local authority to effectively carry out its functions, ERA would further constrain small local authorities' resources in better serving their communities, especially in generating the required statistics to be measured against national and international standards (MFE, 2016d, p. 48). Even though the Minister asserted that "The intent of the Bill is that data already collected by local government and other agencies is used to inform reporting, and that the regime should not create any requirement to generate information that is not currently collected" (MFE, 2016d, p. 8), only six local authorities submitted (2%) out of the 323 submissions received by the LG and environment committee during that time. This raised concern as to whether the local authorities were aware of the implications that ERA might have had on their operational processes and practices that prompted this study to approach the key individuals responsible for sustainability agenda in each local authority. This is discussed in the next section.

Seven submitters (mainly from the LG) were concerned that the Bill did not acknowledge the existing local authorities' monitoring and reporting responsibilities (under section 35 of the RMA) and the interplay between national level reporting and the compliance functions of local authorities, which use the same standards to measure compliance (MFE, 2016). In its submission, the Otago Regional Council stressed LG needed sufficient time to plan for any additional resourcing required for data collection and to allow for changes in scope and process (MFE, 2016d, p. 39).

In order to alleviate these concerns, it was suggested that LG collaborate with the Ministry for the Environment to make sure that the data they collect meet their monitoring and reporting obligations and are aligned with national reporting needs. This collaboration was intended to benefit the LG environmental and reporting systems and, ultimately, the national reporting regime (MFE, 2016d, pp. 48–9). In this aspect, collaboration was discussed in a sideline with obligations to balance between 'rights' and 'responsibilities'. According to Raco and Imrie (2000), the ways in which these are institutionalised are of critical importance to the future of the politics. The landscape of the LG's moral and political conduct is changed, shaped and reshaped by modifying the accountability structure and requiring more information. The literature has shown that regulation has been found to be effective in sealing public sector organisation's commitment to report and make their performance visible (Adams, Muir, & Hoque, 2014). In NZ's case, an open reporting was introduced by the Environment Minister, as stated below.

Open reporting is a powerful tool for improving performance. The Fiscal Responsibility Act 1994 requiring open reporting of public accounts played a pivotal role in improving the management of New Zealand's finances. The objective of this new Environmental Reporting Act is to reduce the debate over where New Zealand has environmental problems and to help focus the Government, councils and communities on finding long-term, sustainable solutions (MFE, 2016b).

Here, the discourse explicitly links not only the reporting tool to performance improvement but also extends the responsibilities to the communities in finding solutions to environmental threats. In this case, the environmental threat is the 'problematic' in Foucault's term. This indicates that the experts in the political domain rather than the scientific field played a part in swaying the agenda more readily towards influencing behaviour than concentrating on precise measurement and calculation (Thomson et al., 2014; Ascui & Lovell, 2011). In the UK, Thomson et al. (2014, p. 461) noted that this was done to secure individual staff members' engagement throughout the processes.

Local Government Responses to Central Government's Political Rationalities

The obvious proximity of the LG, which is closer to the communities (than the central government), was perceived as having the capacity to influence and change the behaviour of their constituents towards sustainability (Potts, 2004; Ball et al., 2006; Carnegie & West, 2010). One respondent rightfully stated that:

Council should be looking at its environmental footprint to improve its environmental performance. Lead the way for the community, reduce its negative impact on the environment, set priorities for doing this, i.e., having a sustainability strategy... [Project Manager, City Council, CP21]

ERA could be viewed as a mobilisation of the central government's power through the development of "voluntary and coercive exercise of power upon the subjectivity of those to be empowered" (Cruikshank, 1994, p. 35). Foucault (1983, p. 213) perceives the state as a major form of power, which he describes as both individualising and totalising, in this case totalising the responsible parties for the achievement of the ruling government's bluegreen policy (individualising). What was clear was that ERA was the deployment technique of the central government's powers, mirroring discourses of control and monitoring (Hastings & McArthur, 1996; Raco & Imrie, 2000) and the conduct of others, as explained by Dean (2009, p. 18).

... and therefore any more or less calculated and rational activity, undertaken by a multiplicity of authorities and agencies, employing a variety of techniques and forms of knowledge, that seeks to shape conduct by working through the desires, aspirations, interests and beliefs of various actors, for definite but shifting ends and with a diverse set of relatively unpredictable consequences, effects and outcomes.

ERA was introduced as a Bill in Parliament in March 2014 and passed as a law in September 2015. The researchers' primary data were collected in August 2015 immediately prior to the passing of the new Act, with the notion that the local authorities were already cognisant with the Bill. This was also to rule out concerns that low submissions from the LG during the consultation process might indicate that they were not aware of the implications of ERA, if enacted. Other possible reasons for the low submission could also be that the majority of the local authorities were not surprised about ERA either. After all, as they are part of the government, they are categorised as regulatory hybrid organisations, which have limited capacities to respond to (or contest) voluntary programmatic such as sustainability, especially if they are backed up with specific regulatory changes (Thomson et al., 2014, p. 471, Kurunmäki, 2004, Lapsley & Wright, 2004, Laughlin, 2007, Llewellyn & Northcott, 2005). The low submission did not necessarily indicate acceptance.

By probing the respondents in the survey, the researchers noticed that the respondents' reactions towards ERA and the political rationalities of the central government were quite mixed. Only 12 respondents were clearly in support of the new regulation. Their justifications were mainly about the ability of fronting the sustainability issues to the public prominence to increase public awareness of sustainability issues. They also thought ERA would bring more clarity on the concepts, specific metrics of measurements and reporting focus. As indicated by a Group Manager, Planning, Community and Environment, District Council [DW91]:

I consider this to be a valuable piece of legislation, which would ensure council's [sic] play their part in environmental reporting. It would encourage consistency in what is measured and improve national and regional data availability and enable communities to track changes.

None of these 12 responses were consistent with the political rationalities of the government on re-branding NZ image as a solution to the sustainability issues or for a better economy. In fact, financial implication and burden of reporting were the major concerns highlighted by the majority of the respondents who were not in favour of ERA, as indicated in the following quotes.

While it would get NZ in better step with other OECD members, there appear to be three issues that would affect our council. First, independent reporting is an important component of the bill being a success, as it would ensure more robust data from which to make better decisions and prevent any potential political influence to scientific information. Second, access to reported information appears uncertain. Having access to all data collected on our council's area would bolster our own data and be useful in informing what and how we report more locally. Finally, the issue of topic selection may restrict what kinds of information are collected, potentially influencing the type and regularity of reporting information we might be able to use (District Planner, District Council, DK12].

Reporting is only one aspect. The main priority is to do the monitoring to get the information to report on. [...] yet another regime of reporting assumes that Councils will have systems in place, people available, time available, and funding available to do reporting that may well duplicate some of the existing reporting formats. We are heading into "death by measurement"... The costs of undertaking this monitoring and reporting are being shifted from the taxpayer to the ratepayer, with no scale of cost being provided. Regional councils are sorting the government's priority of fresh water at significant costs to the ratepayers... The government's demands placed on local government over air quality improvements should provide a lesson to the government – it is heading to a failure. Depending on the domain being reported on in the six month cycle, local authorities will be stretched as more often than not, the timing of the reporting demands clash with the core business reporting cycles. [Manager, Environmental Monitoring and Reporting, Regional Councils, RN11]

The tensions in both quotes represented the obvious operating challenges at local practices neglected by the central government at national level. This resistance could be the pinnacle as many LG might not only risking legal implications for non-compliance, but perhaps their own sustainability in the future – the "real" risk of failure (Miller & Rose, 1992, p. 14). The legislative intervention extended the local authorities' responsibilities in monitoring and reporting of the environment so as to be aligned with the regulated authoritative criteria (Dean, 1999; Foucault, 1979; Rose & Miller, 1992).

Further analysis of the responses seemed to suggest that the respondents wanted a balance between economic interest and long-term environmental concerns (Thomson et al., 2014; Boston & Lempp, 2011) as indicated by the following respondent.

Sustainability is a corner stone of the council's overall strategic approach. We want a future that provides for sustainable economic and population growth, prudent planning to address pressure on our natural resources, allowing for growth while protecting the productive capacity of rural areas. These priorities are constantly challenged by pressure for economic development. Sustainability reporting needs to be regularly reinforced and supported by information that keeps it at the forefront of council's strategic concerns. [Group Manager, Strategy and Development, District Council, DH11]

For many small district councils in New Zealand, ERA inevitably signifies uncertainties, risking their own sustainability in the pursuit of the central government for a more clean, green and sustainable environment. Without proper support, the imposition of the central government's aspirations via ERA intervention could lead to an adverse impact to the country as a whole.

The Emergence of Accounting-Sustainability Hybrids

The conflicts identified above were not unprecedented. Prior to ERA, the local authorities already had their own localised sustainability strategies embedded into their local practices due to RMA and LGA reporting. Thus, it would be interesting to examine to what extent ERA would bring changes to the local practices. The respondents were asked to provide their insights on what had been initiated in the local authorities in relation to sustainability. They were asked about the relevant sustainability issues to their council, what processes were currently in place, their extent of involvement, and current challenges in reporting sustainability issues. These inquiries were carried out in order to better understand how the respondents interpret, prioritise and construct sustainability in their current processes and practices. These processes were then traced as to the intended outcome – the emergent accounting and sustainability hybrids – as depicted in the following diagram [Figure 1].

The Interpretation, Prioritisation and Construction of Local Hybrid Programmatic

Local hybrid programmatics are used to frame and legitimise any organisational transformation (Thomson et al., 2014). The responses showed that the top three sustainability issues identified as relevant to their councils were environmental management, water management and climate change. The issues selected were context-specific, as indicated below:

… climate change affects sea levels. Our town is very low lying and prone to surface flooding when ground water is high, which is affected by tide. [Finance and Corporate Service Manager, District Council, DO11]

Figure 1.

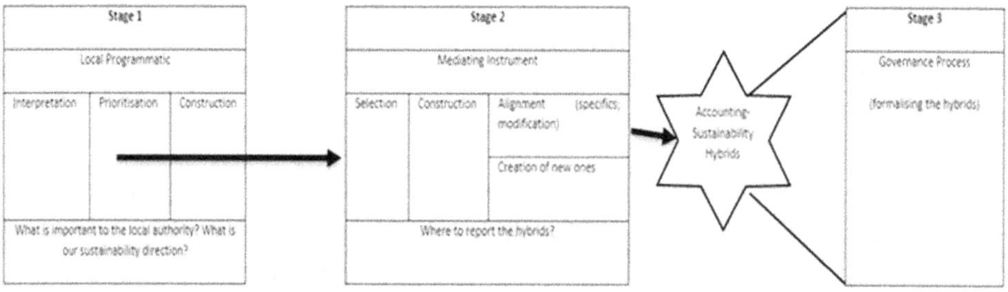

The least relevant issue identified by the respondents was human rights (13%). In NZ, human rights issues were not significant to be addressed at both local and central government levels. NZ played an important role in the development of the Universal Declaration of Human Rights (UDHR) and in the 1945 United Nations Conference on International Organisation in San Francisco, which marked the start of the nation's self-perception as a human rights leader and thus explained the choice.

As the following quotes indicate, the responses also revealed the link between the elements of the sustainability programmatic with local strategic discourses, in terms of performance and accountability.

All Council [sic] should evaluate their performance and their community's wellbeing through a sustainability lens. [Group Manager, Planning, Community and Environment, District Council, DW91]

We ought to be able to demonstrate to our ratepayers that we are achieving our corporate community values of being socially responsible, economically progressive and environmental champions. [Corporate and Community Planner, District Council, DW42].

These links revealed instances where a nexus of other programmatic discourses on local discourses was intertwined and filtered the activities and initiatives to be undertaken by the councils[5] in relation to sustainability. In the UK public sector, Thomson et al. (2014) found that sustainability agenda being linked to programmatic discourses, such as new public management and austerity measures.

There was a general consensus that the councils' responsibilities are also sustainability related:

A lot of what local government does has a direct impact on the overall sustainability of the environment we are charged with managing and the communities we are charged with representing. [Shared Services Science Manager, District Council, DG11]

Thus, few respondents expressed in their concerns the risk of non-action in regard to addressing sustainability issues. As noted by Thomson et al. (2014), the risks could be potentially catastrophic. Nonetheless, a large number of the respondents considered sustainability issues as a combination of both risk and opportunity. They did acknowledge that given the magnitude of the pressures that they were facing now (in terms of environmental sustainability), they would be exposed to an unprecedented level of risk if they failed to act soon or not do enough. The opportunities to transform the shift towards a higher level of sustainability were welcomed by the respondents, as highlighted below:

The Local Government Act requires local governments to progress sustainable development and look after the future needs of their communities. Reporting is a mechanism for ensuring this occurs. [CEO, City Council, CD21]

This has been highlighted as part of our Long-Term Plan that the Councillors are interested in this going forward. For example, all new projects should be reviewed for sustainability e.g., solar panels. [Senior Planner, Regional Council, RU3]

Notwithstanding the above, the other quarter of the respondents view ERA as a high risk in various aspects. These respondents emphasised that the inability to meet the required reporting criteria would put them in the spotlight, which in turn would distract and divert their attentions from service delivery performance and, ultimately, become a political football. Further, they were aware of the strong lobbyist NZ groups who knowingly refused to compromise other than a pure sustainable approach. One respondent summarised that there is always some risk in overstating the position, even for minor items, if not followed through, which would annoy those critical to the LG without appreciating the risk they are facing [Strategy Manager, District Council, DT21]. All these were in addition to the major concerns of the financial implications the additional reporting would cost.

The localised success of the central government's sustainability programmatic discourse depends on the ability of local authorities to prioritise their actions in order to initiate this change locally (Thomson et al., 2014) to prioritise. Prioritisation involves consideration and balancing the needs of various stakeholders with competing and contradictory programmatic discourses (Van Helden, Aardema, ter Bogt, & Groot, 2010). However, being in the public sector domain, the local authorities have limited capacities, freedom and power (e.g. Hyndman & Connolly; Kurunmäki, 2004; Kurunmäki & Miller, 2006; Power, 2007). In his own words, a Senior Manager, Strategy in a City Council [CP11] elaborated:

Significant increase in costs and requirements on the Council – can't really see a lot of benefits actually. We are one of the 7 poorest Councils in the country and I suspect our community would like us to do better on managing the environment rather than meeting reporting compliance requirements. But in the end the top priority is always the legal requirements.

At operational level, the local authorities indeed had to prioritise actions in order to have their own hybrids so that any change could be legitimised without losing their identity, responsibilities, priorities, governing practices and permitted actions

(Thomson et al., 2014, p. 456). This is made possible by local authorities choosing the appropriate mediating instruments in which the hybrids would be reported on. This is discussed in the next section.

The Selection and Construction of Accounting-Sustainability Mediating Instruments

Mediating instruments operate as both means of representation and means of intervention, connecting with, yet remaining distinct from, the object of intervention (Kurunmäki & Miller, 2011, p. 222). In our research context, the mediating instruments are the medium of reporting in which the local councils make their accounting-sustainability hybrids visible. In this context, accounting means making things calculable as they become visible, knowable, thinkable and governable (Thomson, 2014). Thomson et al. (2014) suggest that without the possibility of calculating the other, there would be very little possibility of accounting hybridisation. Thus, due consideration of the mediating instruments is vital as it determines where and in which format they would make their commitment to sustainability become reflected and therefore accountable.

Table 2 summarises a total of 26 different types of mediating instruments currently in use (E) by the local authorities in reporting their commitment to sustainability. Some are purely regulative (R) in nature, as prescribed by LGA and RMA, and some are the combination of both regulative and voluntary (V) formats. The instruments indicated as new (N) are those that the respondents commented that they would use in the near future.

The researchers' further analysis of the responses and the mediating instruments used revealed that there were three strategies commonly undertaken by the local authorities that were "localised" into the local systems and practices. They were either (i) aligned or (ii) modified to be consistent with the regulative hybrids. Only a few mediating instruments were found to be newly created for reporting and monitoring purposes. In this category, the hybrids were (iii) "formalised" and replaced the existing systems and practices. The first two categories are similar to what Thomson et al. (2014) termed as a "local to local" and "non-local to local" trajectory respectively. Unlike prior research into accounting hybridisation, there was no evidence of a "local to non-local" accounting hybridisation process in our study.

For the first strategy, the majority of the respondents narrated that the aim was to report their commitment to sustainability in the annual reports, (current) state of environmental reporting to the Ministry for the Environment and long-term planning, including consultation report, quarterly and management reports. The main reason was indicated by the CEO of District Council [DW31] below.

Table 2. Selected Mediating Instrument

	Instrument				#
	R	V	E	N	
Long-term and Annual Plan	X		X		3
Long-term Plan including consultation report	X	X	X		1
Annual Report	X		X		2
Annual Report + Long-term Plan (LTP)	X		X		1
Special report on sustainability		X		X	1
Newsletter / Good News stories in local newspaper		X		X	2
"General Community", led by LTP and Annual Cycle	X	X	X		1
State of Environment (SOE) Reporting	X		X		4
Cultural Monitoring Report		X		X	1
Resource management policy and monitoring and evaluation plan	X		X		2
Council activity reporting	X		X		1
Quarterly reporting	X		X		1
Management Report		X	X		1
Report to MFE National Monitoring System		X		X	2
Short report card (every 6 months) and 5-yearly Report	X	X	X		1
Global Reporting Initiative (GRI)		X	X		2
Total	26				

R-Regulative V-Voluntary E-Existing N-New

...First, I think it's important to appreciate that reporting on sustainability is multi-faceted ... generally, our Council's sustainability reporting is probably incidental to reporting on particular activities, for example reporting around waste management and provision of drinking water. The information presented is often around customer satisfaction in the delivery of the service, and in regard to drinking water – mandatory drinking water standards.

The second strategy is a modification to the existing local systems and practices. This strategy includes the art of making use of the previously reported information to suit the purpose, as in this commentary by a Policy Analyst, Strategic Planning Department, District Council [DH21].

In the past, we have reported on sustainability through the annual report and the four well-beings. As a policy analyst/strategic planner I would assist in the preparation of this report and in teasing out the information from Asset/Activity Managers on what was sustainable in this regard.

This modification strategy also targeted specific issues of sustainability, depending upon calculative risks of that specific region, the ones that they 'can directly influence/control' as indicated below.

[I am involved in] ... differentiating between regional sustainability (not specifically our function/role) and areas of sustainability we can directly influence/control. [Manager, Sustainability Projects, Regional Council, RW12]

The above commentaries showed that the mediating instruments had been 'localised' by aligning and modifying them to be consistent, not only to suit the local context but also to be consistent with the regulative programmatic such as RMA and LGA. The literature suggests that alignment and/or modification is necessary, otherwise it can distort the aspirations of the intended reform or change (Broadbent & Guthrie, 2008; Broadbent, Jacobs, & Laughlin, 2001; Hyndman & Connolly, 2001; Kurunmaki et al., 2003).

The third strategy is more radical than the first two. As explained by Miller et al. (2008), "hybridising is a continually inventive process". With this strategy, the new mediating instrument was 'formalised' onto the local systems. Previous studies claimed that the adoption and implementation of a new accounting hybrid could substantively alter the organisation (Thomson et al., 2014) and trigger a wider series of changes (see Armstrong, 2006; Miller & O'Leary, 1993), though a total radical transformation might not be possible in a regulative environment. Any changes had to be somewhat linked to their regulative hybrids or the local authorities could risk violating their compliance obligations, as narrated in the following quotes.

We are currently developing a sustainability framework which will include reporting criteria. [Planning Manager, City Council, CN11]

The Annual Report is an obvious opportunity however readership of this is likely to be low. A special report on sustainability may be resource intensive and we would need to assess potential reach. More regular snappy communication tools such as newsletter but be more effective at engagement and education. [Sustainability and Community Development Advisor, City Council, CT11].

Only one respondent reported that the council used an international benchmark, Global Reporting Initiatives (GRI) in addition to their mandatory reporting [Corporate Sustainability Coordinator, Regional Council, RW21]. The other council intended to use GRI in its reporting in the future [Chief Operating Officer, City Council, CW11]. Two respondents revealed that they had "advanced" their reporting framework to include an on-line monitoring tool.

Council has developed an on-line monitoring tool. Officers report to Council quarterly and annually on progress with strategies. [Senior Manager, Strategy, City Council, CP11]

Staff produce a report which covers social, economic and environmental wellbeing. Staff are also responsible for reporting on climate change and strategies covering the 3 wellbeing. Staff developed and maintain an on-line monitoring tool. [Senior Manager, Strategy, District Council, DH12]

Thomson et al. (2014) suggest that the selection of the mediating instruments was used to add impetus to the sustainability agenda and help engage with the staff. The respondents indicated that the various internal processes were facilitated by selection process, including frequent intra- [e.g. Team Leader, Strategic Policy, District Council, DW101; Corporate Service Manager, District Council, DS11] and inter-organisational communications, such between regional councils [Senior Planner, District Council, DG21]. By contrast, Argento et al. (2016) found that inter-organisational collaboration might not evident due to absence of incentives and no control structures for it.

Examples of [the hybrids] that respondents reported are listed in Table 3. Similar to the Thomson et al. (2014) findings, the accounting-sustainability hybrids were formed (and reformed) in the intersections and repeated interactions among programmatic discourses, resource constraints, political priorities, local accounting practices, the organisation's legal responsibilities, accountability relationships, existing assemblage of governing practices, existing knowledge of the sustainability programmatic, and everyday service delivery (p. 471). In NZ, the central government specified a set of required performance measures to be reported in the Long-term Plan and Annual Plan, as these are the main indicators that the councils had to agree and work upon. As this respondent explains,

Yes. Currently we report on aspects of sustainability but not an overall sweet of sustainability indicators. However, this is largely driven from central government and its focus on core services. [General Manager, Strategic Planning and Policy, District Council [DF11]

The hybridisation of accounting with sustainability was seen as imperative to showcase local authorities' commitment to the community in addition to compliance responsibility. The accounting-sustainability hybrids disclosed by the respondents were context-specific in nature, which raised the question of the suitability of the standardised new reporting framework to be introduced by ERA.

Table 3. Examples of accounting-sustainability hybrids in NZ local authorities

Water	Surface water quality and quantity, KPIs, benchmarking
	Coastal water quality, KPIs, benchmarking
	Groundwater levels and quality, KPIs, benchmarking
	Energy
	Energy use/efficiency, budgeting, KPIs
Air	Air quality, benchmarking
Land	Waste minimisation, KPIs
	Sustainable land use monitoring, KPIs
	Soil health, KPIs
	Coastal erosion impact analysis
	Carbon accounting
Social	Housing affordability monitoring [Cost (rental) /income; cost to own a house/income]
	Economic sustainability of small communities, budgeting
Performance	Financial sustainability benchmarking

A Chief Financial Officer, Regional Council [RU10] narrated the emergent accounting-sustainability process,

Once meaningful, reliable measures were developed, they could provide further information and insight into an organisation's risk profile thus, it was important for local authorities to have a long-term community perspective rather than profit oriented drivers (although local authorities should be run in a business-like manner with regard to fiscal probity and discipline).

As suggested by Miller et al. (2008), "hybridising is a continually inventive process, in which proliferation and multiplication is the norm" (p. 96). This was evident in the case of NZ local authorities. A Chief Sustainability Officer, Regional Council [RA11] reflected on the continuous improvement process in his council.

We take the balanced scorecard approach with KPIs established before my position was established. We're working to modify/improve the current sustainability metric. Also, we've determined what should be part of our annual report and have been producing a sustainability-specific spread in the annual report. Finally, our sustainability office will be choosing metrics to benchmark against other cities and reporting that in our quarterly external/internal report.

Similar to Thomson et al. (2014), the researchers' findings seemed to suggest that the accounting-sustainability hybrids were driven by the locally defined problems, as indicated in the following quote.

...I advised councillors of the parameters to measure and reported on, that best fitted our functions, within the context of our region's circumstance. [CEO, Regional Council, RW21]

The hybrids (environmental data and sustainability indicators) were also chosen based on work streams and outcomes sought for the community [District Planner, District Council, DK12). The researchers expected the hybrids to trigger intra-organisational dynamics at local levels (Miller et al., 2008).

Previous studies, such as Gray et al. (1995), found a tension between environmentalists' view of environmental accounting and the use of this practice by an organisation. This signals a potential conflict between accounting and sustainability quarters. Though this was not evident in our study, the Chief Financial Officer and Finance Officer did not appear as the champions or as the drivers of the sustainability agendas described below.

The [CFO] role includes being able to balance the cost of measuring with the benefit – and the relationship to other KPIs already included in the statement of service performance. It is also about selecting key measures that can influence strategy and behaviour versus lots of measures that obscure the key measures. [Chief Financial Officer, City Council, CH11]

Being involved in this process would aid in providing a financial perspective on the key elements of a sustainability report. [Finance Manager, District Council, DW41]

This finding is consistent with Gray et al. (1995) and Larrinaga-Gonzalez et al. (2001), who noted little or no role by the accountants in forging sustainability change. In fact, the analysis revealed that heavy involvement was traced back to the sustainability-related quarter with its scientist and engineering staff pushing for the agenda.

The researchers asked the respondents about their roles and extent of involvement in deciding what sustainability information should be reported. Their responses revealed vigorous processes and discourses at various levels of the organisations.

Part of a team which developed the council's strategic framework based on sustainability principles. This has in turn influenced the reporting back on progress. [Strategy Manager, District Council, DH12]

Gouldson and Bebbington (2007, p. 7) claim that the management is enlisted to "do the governing" as multilevels and multi-actors with authority and responsibility dispersed within broader networks (Black, 2002). The respondents seemed to view the processes as critical in gaining feedback and commitment of various groups within the organisation before an environment and sustainability plan and/strategy was presented to the councils for approval.

...This [My] role requires facilitation of the various areas...being part of the Council Leadership team I can ensure that the various aspects of Council are aware of what is required in this space [sustainability]. [CEO, Regional Council, RU7]

From governmentality perspective, this open and participative decision-making process gives the embedding of sustainability some internal credibility (Spence & Rinaldi, 2014, p. 441) to be used as a means through which top management can control and negotiate the "conception of environment" (Gray et al., 1995, p. 233). It also legitimises sustainability governance of the sustainability programmatic at the local level. In this way, it serves more at legitimising purpose rather than transforming decisions (Owen, Swift, Humphrey, & Bowerman, 2000; Owen, Swift, & Hunt, 2001; Owens, 2000).

Decisions of this nature have been made by our senior management team and the policy team. This type of reporting needs to be looked at from several angles by people that have a good overview of the issues. When it comes to the stats required this is when affected areas/teams should be consulted. [Manager, City Council, CD12]

Ultimately, the sustainability governance agenda is underpinned by the visibility of the sustainability plan and/or strategy through the endorsement of the councils. This sustainability plan/or strategy document represents the visible artefact of governmentality, which consists of documented policies giving de facto authority to the local authority to pursue the sustainability agenda. It was a consequence of how the sustainability programmatic at the central government level had managed to be operationalised by the local authorities. Our analysis did not reveal any evidence of a deep organisational change or fundamental modus operandi of the local authority towards greater sustainability. However, the researchers concur with Gray et al. (1995) and Larrinaga-Gonzalez et al. (2001) that the sustainability programmatic triggered increased transparency and some changes in organisational discourse, which facilitated actions from a distance.

CONCLUSION

The aim of the chapter was to gain an understanding on how the central government's sustainability programmatics in the form of strategies and policies interact with LG processes and practices. The researchers analysed responses from 90 key individuals from all NZ local authorities immediately prior to the passing of the new law, the Environmental Reporting Act 2015. The NZ sustainability programmatics were the 'technologies' of the central government and were intended to change local organisations and society to act sustainably, consistent with NZ's clean and green image, and to address the issue of being the only country in the OECD not having a statutory requirement for its environmental reporting and monitoring.

The researchers found that the evolution of accounting-sustainability hybrids in NZ local authorities to be consistent with the governmentality framework, which requires the local authorities to interpret, prioritise and construct their local programmatic discourses. This is followed by the processes of constructing and embedding hybrids of accounting and sustainability into their governing process. These evolutionary processes require the alignment with the external regulative (e.g. LGA and RMA) and sustainability programmatics, and to a certain extent the modification of existing and/or creation of new hybrids. The NZ local authorities either adopt a pragmatic approach by localising sustainability programmatics or a radical approach by replacing the old mediating instrument and formalising it into the local processes and practices. The accounting-sustainability hybrids disclosed by the respondents were context-specific in nature, which raised questions of the suitability of the standardised new reporting framework to be introduced by ERA. Regardless, the researchers anticipate that NZ local authorities might not require a radical transformation approach in the future in regard to ERA, as the evidence infers that they are structurally ready, with local processes and practices that have been shaped and reshaped, even prior to the enforcement of the new Act.

Overall, the researchers' findings seem to suggest that the introduction of an Act is a necessary driving force to initiate change in relation to environmental sustainability in public sector organisations such as the LG. The researchers show how the regulation could facilitate actions from a distance. Nonetheless, the power of the local actors as to how they interpret and select mediating instruments was a precursor to further transformation. It is vital for the local contexts to be considered by the central government before enforcing the new reporting framework since the local actors' discourses are framed by their everyday practices and knowledge. The findings seem to suggest that the framework should be flexible enough to accommodate local contexts as local to non-local mediating instruments would have a better chance of success.

The limitation of the study lies with its inability to offer insights of the actual evolutionary processes of accounting-sustainability hybrids in a specific local authority. This requires a longitudinal case and offers richer data and greater insight into the complex dynamic of external programmatics and local processes and practices. Future research should consider the impact of the new reporting framework to the local processes and practice (post ERA). A study to examine how effectively the organisational accounting-sustainability hybrids translate the sustainability programmatic into local processes and practices would complete the cycle, the holistic framework introduced by Thomson et al., (2014) to observe sequences of inter-connected transformations.

ACKNOWLEDGMENT

The researchers are grateful to Massey University for the funding of this research (MURF Grant No: 1000017839).

REFERENCES

Adams, C. A., Muir, S., & Hoque, Z. (2014). Measurement of sustainability performance in the public sector. *Sustainability Accounting. Management and Policy Journal*, *5*(1), 46–67.

Argento, D., Grossi, G., Jããskelãinen, A., Servalli, S., & Suomala, P. (2016). Governmentality and performance for the smart city. *Paper presented at 8th Asia-Pacific Interdisciplinary Research in Accounting Conference*, RMIT University, Melbourne, Australia, July 13–15.

Armstrong, P. (1994). The influence of Michel Foucault on accounting research. *Critical Perspectives on Accounting*, *5*(1), 25–55. doi:10.1006/cpac.1994.1003

Ascui, F., & Lovell, H. (2011). As frames collide: Making sense of carbon accounting. *Accounting, Auditing & Accountability Journal*, *24*(8), 978–999. doi:10.1108/09513571111184724

Ball, A. (2004). A sustainability accounting project for the UK local government sector?: Testing the social theory mapping process and locating a frame of reference. *Critical Perspectives on Accounting*, *15*(8), 1009–1035. doi:10.1016/S1045-2354(02)00209-5

Ball, A., Broadbent, J., & Jarvis, T. (2006). Waste management, the challenges of the PFI and 'sustainability reporting'. *Business Strategy and the Environment*, *15*(4), 258–274. doi:10.1002/bse.532

Ball, A., & Grubnic, S. (2007). Sustainability accounting and accountability in the public sector. In J. Unerman, J. Bebbington, & B. O'Dwyer (Eds.), *Sustainability accounting and accountability* (pp. 243–265). London: Routledge.

Ball, A., Mason, I., Grubnic, S., & Hughes, P. (2009). The carbon neutral public sector: Early developments and an urgent agenda for research. *Public Management Review*, *11*(5), 575–600. doi:10.1080/14719030902798263

Bebbington, J., Gray, R., Hibbitt, C., & Kirk, E. (2001). *Full Cost Accounting: An Agenda for Action*. London, UK: Certified Accountants Educational Trust.

Bebbington, J., & Larrinaga, C. (2014). Accounting and sustainable development: An exploration. *Accounting, Organizations and Society*, *39*(6), 395–413. doi:10.1016/j.aos.2014.01.003

Bebbington, J., & Thomson, I. (2013). Sustainable development, management and accounting: Boundary crossing. *Management Accounting Research*, *4*(24), 277–283. doi:10.1016/j.mar.2013.09.002

Bellringer, A., Ball, A., & Craig, R. (2011). Reasons for sustainability reporting by New Zealand local governments. *Sustainability Accounting. Management and Policy Journal*, *2*(1), 126–138.

Black, J. (2002). *Critical reflections on regulation. Discussion paper*. London: Centre for the Analysis of Risk and Regulation, London School of Economics.

Boston, J., & Lempp, F. (2011). Climate change: Explaining and solving the mismatch between scientific urgency and political inertia. *Accounting, Auditing & Accountability Journal*, *24*(8), 1000–1021. doi:10.1108/09513571111184733

Broadbent, J., & Guthrie, J. (2008). Public sector to public services: 20 years of 'contextual' accounting research. *Accounting, Auditing & Accountability Journal*, *21*(2), 129–169. doi:10.1108/09513570810854383

Broadbent, J., Jacobs, K., & Laughlin, R. (2001). Organisational resistance strategies to unwanted accounting and finance changes: The case of general medical practice in the UK. *Accounting, Auditing & Accountability Journal*, *14*(5), 565–586. doi:10.1108/EUM0000000006263

Carnegie, G. D., & West, B. (2010). A conceptual analysis of price setting in Australian local government. *Australian Accounting Review, 20*(2), 110–120. doi:10.1111/j.1835-2561.2010.00084.x

Carter, M. (2002). Husbands and maternal health matters in rural Guatemala: Wives' reports on their spouses' involvement in pregnancy and birth. *Social Science & Medicine, 55*(3), 437–450. doi:10.1016/S0277-9536(01)00175-7 PMID:12144151

Cruickshank, B. (1994). The will to empower: Technologies of citizenship and the war on poverty. *Socialist Review, 23*(4), 29–55.

Dean, M. (1999). *Governmentality: Power and rule in modern society*. London: Sage Publications.

Dean, M. (2010). *Governmentality: Power and rule in modern society*. London: Sage Publications.

Debski, R., & Gruba, P. (1999). A qualitative survey of tertiary instructor attitudes towards project-based CALL. *Computer Assisted Language Learning, 12*(3), 219–239. doi:10.1076/call.12.3.219.5715

Farneti, F., & Guthrie, J. (2009, June). Sustainability reporting by Australian public sector organisations: Why they report. *Accounting Forum, 33*(2), 89–98. doi:10.1016/j.accfor.2009.04.002

Figge, F., Hahn, T., Schaltegger, S., & Wagner, M. (2002). The sustainability balanced scorecard – linking sustainability management to business strategy. *Business Strategy and the Environment, 11*(5), 269–284. doi:10.1002/bse.339

Fink, A. (2003). *The survey handbook*. Thousand Oaks, CA: Sage. doi:10.4135/9781412986328

Foucault, M. (1983). The subject and power. In H. Dreyfus & P. Rabinow (Eds.), *Beyond structuralism and hermeneutics* (pp. 206–226). Chicago: University of Chicago Press.

Foucault, M. (1991). *The Foucault Effect: Studies in governmentality*. London: Harvester Wheatsheaf.

Foucault, M. (2008). *The birth of biopolitics: Lectures at the Collège de France, 1978–1979*. Springer.

Frost, G. R., & Seamer, M. (2002). Adoption of environmental reporting and management practices: An analysis of New South Wales public sector entities. *Financial Accountability & Management*, *18*(2), 103–127. doi:10.1111/1468-0408.00147

García-Sánchez, I., Frías-Aceituno, H., & Rodríguez-Domínguez, L. (2013). Determinants of corporate social disclosure in Spanish local governments. *Journal of Cleaner Production*, *39*, 60–72. doi:10.1016/j.jclepro.2012.08.037

Goswami, K., & Lodhia, S. (2014). Sustainability disclosure patterns of South Australian local councils: A case study. *Public Money & Management*, *34*(4), 273–280. doi:10.1080/09540962.2014.920200

Gouldson, A., & Bebbington, J. (2007). Corporations and the governance of environmental risk. *Environment and Planning. C, Government & Policy*, *25*(1), 4–20. doi:10.1068/c0614j

Gray, R. (2010). Is accounting for sustainability actually accounting for sustainability… and how would we know? An exploration of narratives of organisations and the planet. *Accounting, Organizations and Society*, *35*(1), 47–62. doi:10.1016/j.aos.2009.04.006

Gray, R., Dillard, J., & Spence, C. (2009, September). Social accounting research as if the world matters, social accounting research as if the world matters: Postalgia and a new absurdism. *Public Management Review*, *11*(5), 545–573. doi:10.1080/14719030902798222

Gray, R., Owen, D., & Maunders, K. (1988). Corporate social reporting: Emerging trends in accountability and the social contract. *Accounting, Auditing & Accountability Journal*, *1*(1), 6–20. doi:10.1108/EUM0000000004617

Gray, R., Walters, D., Bebbington, J., & Thompson, I. (1995). The greening of enterprise: An exploration of the (non) role of environmental accounting and environmental accountants in organisational change. *Critical Perspectives on Accounting*, *6*(3), 211–239. doi:10.1006/cpac.1995.1021

Greco, G., Sciulli, N., & D'onza, G. (2012). From Tuscany to Victoria: Some determinants of sustainability reporting by local councils. *Local Government Studies*, *38*(5), 681–705. doi:10.1080/03003930.2012.679932

Guba, E. G., & Lincoln, Y. S. (1998). Competing paradigms in qualitative research. In N. K. Denzin & Y. S. Lincoln (Eds.), *The landscape of qualitative research* (pp. 195–220). Thousand Oaks, CA: Sage.

Guthrie, J., & Farneti, F. (2008). GRI sustainability reporting by Australian public sector organizations. *Public Money & Management*, *28*(6), 361–366. doi:10.1111/j.1467-9302.2008.00670.x

Hastings, A., & McArthur, A. (1996). A comparative assessment of government approaches to partnership with the local community. In R. Hambleton & H. Thomas (Eds.), *Urban policy evaluation: Challenge and change* (pp. 175–195). London: Paul Chapman.

Herbohn, K. (2005). Job Satisfaction Dimensions in Public Accounting Practice. *Accounting Research Journal*, *18*(2), 63–82. doi:10.1108/10309610580000676

Herbohn, K., & Griffiths, A. (2008, December). *Sustainability reporting in local government: Systems change or greenwash?* Melbourne: CPA Australia.

Hopwood, A. G. (2009). Accounting and the environment. *Accounting, Organizations and Society*, *34*(3), 433–439. Retrieved June 3, 2016 from http://www.mfe.govt.nz/node/21288/ doi:10.1016/j.aos.2009.03.002

Hyndman, N., & Connolly, C. (2011). Accruals accounting in the public sector: A road not always taken. *Management Accounting Research*, *22*(1), 36–45. doi:10.1016/j.mar.2010.10.008

Jansen, H. (2010). The logic of qualitative survey research and its position in the field of social research methods. *Forum Qualitative Social Research*, *11*(2). Retrieved September 15, 2016 from http://www.qualitative-research.net/index.php/fqs/article/view/1450/2946

Joseph, C., Pilcher, R., & Taplin, R. (2014). Malaysian local government Internet sustainability reporting. *Pacific Accounting Review*, *26*(1/2), 75–93. doi:10.1108/PAR-07-2013-0071

Kane, M. N. (2008). A qualitative survey of the attitudes of Catholic priests toward bishops and ministry following the sexual abuse revelations of 2002. *Pastoral Psychology*, *57*(3-4), 183–198. doi:10.100711089-008-0171-9

Kates, R. W., Clark, W. C., Corell, R., Hall, J. M., Jaeger, C. C., Lowe, I., & .... (2001). Sustainable science. *Science*, *292*(5517), 641–642. doi:10.1126cience.1059386 PMID:11330321

Kurunmäki, L., & Miller, P. (2011). Regulatory hybrids: Partnerships, budgeting and modernising government. *Management Accounting Research*, *22*(4), 220–241. doi:10.1016/j.mar.2010.08.004

Laine, M. (2009). Ensuring legitimacy through rhetorical changes? A longitudinal interpretation of the environmental disclosures of a leading Finnish chemical company. *Accounting, Auditing & Accountability Journal*, *22*(7), 1029–1054. doi:10.1108/09513570910987367

Lapsley, I., & Wright, E. (2004). The diffusion of management accounting innovations in the public sector: A research agenda. *Management Accounting Research*, *15*(3), 355–374. doi:10.1016/j.mar.2003.12.007

Larrinaga-González, C., Carrasco-Fenech, F., Correa-Ruiz, C., Llena-Macarulla, F., & Moneva-Abadía, J. (2002). Accountability and accounting regulation: The case of the Spanish environmental disclosure standard. *European Accounting Review*, *11*(4), 723–740. doi:10.1080/0963818022000001000

Laswad, F., Fisher, R., & Oyelere, P. (2005). Determinants of voluntary Internet financial reporting by local government authorities. *Journal of Accounting and Public Policy*, *24*(2), 101–121. doi:10.1016/j.jaccpubpol.2004.12.006

Laughlin, R. (2007). Critical reflections on research approaches, accounting regulation and the regulation of accounting. *The British Accounting Review*, *39*(4), 271–289. doi:10.1016/j.bar.2007.08.004

Llewellyn, S., & Northcott, D. (2005). The average hospital. *Accounting, Organizations and Society*, *30*(6), 555–583. doi:10.1016/j.aos.2004.05.005

Localcouncils. (2016). *Local government in New Zealand, Internal Affairs Department.* Retrieved May 26, 2016 from http://www.localcouncils.govt.nz/lgip.nsf

MFE. (2016a). *Environmental Reporting Bill first reading.* Retrieved May 30, 2016 from https://www.beehive.govt.nz/speech/environmental-reporting-bill-first-reading

MFE. (2016b). *New Environmental Reporting Act passed.* Retrieved 30 May 2016, from https://www.beehive.govt.nz/release/new-environmental-reporting-act-passed

MFE. (2016c). *About the Environmental Reporting Act 2015.* Retrieved 31 May 2016, from http://www.mfe.govt.nz/more/environmental-reporting/about-act

MFE. (2016d). *Departmental report on the Environmental Reporting Bill 2014.* Retrieved 31 May 2016, from http://www.parliament.nz/resource/en-nz/51SCLGE_ADV_00DBHOH_BILL12994_1_A422639/0fdc955f04b88f24e07faf93769a71f31e230fcf

MFE. (2016e). *Environmental Reporting – submissions received.* Retrieved 31 May 2016, from http://www.mfe.govt.nz/node/17360/

MFE. (2016f). *A framework for environmental reporting in New Zealand.* Retrieved 13 September 2016, from http://www.mfe.govt.nz/more/environmental-reporting/about-act

Milewa, T., Valentine, J., & Calnan, M. (1999). Community participation and citizenship in British health care planning: Narratives of power and involvement in the changing welfare state. *Sociology of Health & Illness, 21*(4), 445–465. doi:10.1111/1467-9566.00166

Miller, P. (1998). The margins of accounting. *European Accounting Review, 7*(4), 605–621. doi:10.1080/096381898336213

Miller, P., Kurunmäki, L., & O'Leary, T. (2008). Accounting, hybrids and the management of risk. *Accounting, Organizations and Society, 33*(7), 942–967. doi:10.1016/j.aos.2007.02.005

Miller, P., & O'Leary, T. (1993). Accounting expertise and the politics of the product: Economic citizenship and modes of corporate governance. *Accounting, Organizations and Society, 18*(2–3), 187–206. doi:10.1016/0361-3682(93)90033-3

Miller, P., & Rose, N. (1990). Governing economic life. *Economy and Society, 19*(1), 1–31. doi:10.1080/03085149000000001

Miller, P., & Rose, N. (1995). Political thought and the limits of orthodoxy: A response to Curtis. *The British Journal of Sociology, 46*(4), 590–597. doi:10.2307/591573

Mulgan, R. (1997). The processes of public accountability. *Australian Journal of Public Administration, 56*(1), 25–36. doi:10.1111/j.1467-8500.1997.tb01238.x

Neu, D. (2000). Accounting and accountability relations: Colonization, genocide and Canada's first nations. *Accounting, Auditing & Accountability Journal, 13*(3), 268–288. doi:10.1108/09513570010334126

Neu, D. (2006). Accounting for public space. *Accounting, Organizations and Society, 31*(4-5), 391–414. doi:10.1016/j.aos.2005.03.001

Owen, D., Swift, T., Humphrey, C., & Bowerman, M. (2000). The new social audits: Accountability, managerial capture or the agenda of social champions. *European Accounting Review, 9*(1), 81–98. doi:10.1080/096381800407950

Owen, D., Swift, T., & Hunt, K. (2001). Questioning the role of stakeholder engagement in social and ethical accounting. *Accounting Forum, 25*(3), 264–282. doi:10.1111/1467-6303.00066

Owens, S. (2000). Engaging the public: Information and deliberation in environmental policy. *Environment & Planning A*, *32*(7), 1141–1148. doi:10.1068/a3330

Potts, T. (2004). Triple bottom line reporting: A tool for measuring, communicating and facilitating change in local communities. In *Sustainability and Social Science: Round Table: Proceedings of Conference*, Sydney.

Power, M. (1994). *The audit explosion (No. 7)*. London, UK: Demos.

Preston, A. M., Chua, W. F., & Neu, D. (1997). The diagnosis-related group-prospective payment system and the problem of the government of rationing health care to the elderly. *Accounting, Organizations and Society*, *22*(2), 147–164. doi:10.1016/S0361-3682(96)00011-6

Raco, M., & Imrie, R. (2000). Governmentality and rights and responsibilities in urban policy. *Environment & Planning A*, *32*(12), 2187–2204. doi:10.1068/a3365

Rose, N., & Miller, P. (1992). Political power beyond the state: Problematics of government. *The British Journal of Sociology*, *43*(2), 173–205. doi:10.2307/591464 PMID:20092498

Spence, L. J., & Rinaldi, L. (2014). Governmentality in accounting and accountability: A case study of embedding sustainability in a supply chain. *Accounting, Organizations and Society*, *39*(6), 433–452. doi:10.1016/j.aos.2012.03.003

Stenius, B., & Cunnington, A. M. (1972). House dust mites and respiratory allergy: A qualitative survey of species occurring in Finnish house dust. *Scandinavian Journal of Respiratory Diseases*, *53*, 338–348. PMID:4569853

Thomson, I. (2015). *How can governmentality help in the quest for pro-sustainability accounting?* [PowerPoint slides]. Retrieved from https://tinyurl.com/l4zj7oz

Thomson, I., Grubnic, S., & Georgakopoulos, G. (2014). Exploring accounting-sustainability hybridisation in the UK public sector. *Accounting, Organizations and Society*, *39*(6), 453–476. doi:10.1016/j.aos.2014.02.003

Tregidga, H., & Milne, M. (2006). From sustainable management to sustainable development: A longitudinal analysis of a leading New Zealand environmental reporter. *Business Strategy and the Environment*, *14*(4), 219–241. doi:10.1002/bse.534

Unerman, J., & Chapman, C. (2014). Academic contributions to enhancing accounting for sustainable development [Editorial]. *Accounting, Organizations and Society*, *39*(6), 385–394. doi:10.1016/j.aos.2014.07.003

Van Helden, G. J., Aardema, H., ter Bogt, H. J., & Groot, T. L. (2010). Knowledge creation for practice in public sector management accounting by consultants and academics: Preliminary findings and directions for future research. *Management Accounting Research*, *21*(2), 83–94. doi:10.1016/j.mar.2010.02.008

Williams, B., Wilmshurst, T., & Clift, R. (2011, September). Sustainability reporting by local government in Australia: Current and future prospects. *Accounting Forum*, *35*(3), 176–186. doi:10.1016/j.accfor.2011.06.004

Wise, N. (1988). Mediating machines. *Science in Context*, *2*(1), 77–113. doi:10.1017/S0269889700000508

KEY TERMS AND DEFINITIONS

Accounting-Sustainability: A range of accounting techniques in relation to social, environmental, ethical and responsibility accounting.

Environmental Reporting Act 2015: An environmental report act which is administered by the Ministry for the Environment that requires regular reports on New Zealand's environment.

Hybrid: A localise reporting format as an outcome of repeated interactions of key individuals in the organisation with the internal and/or external environmental sustainability reporting practices that consistent with the local context, practices and priorities.

Key Individual: An individual that involves in the preparation of the sustainability reporting of the organisation.

Local Government: Local government is the system that consists of members elected by the local communities to represent their interest.

Mediating Instrument: A medium of reporting in which the local councils make their accounting-sustainability hybrids visible.

Programmatics: A collective of government's policies, acts, strategies and interventions.

ENDNOTES

1 We use the phrase 'accounting-sustainability' introduced by Thomson, Grubnic, and Georgakopoulos (2014) to represent a range of accounting techniques previously referred to as social, environmental, ethical and responsibility accounting. Thomson et al. (2014, p. 453) cautioned that the term should not

be interpreted as endorsement of the accounting techniques or consideration of these techniques in facilitating sustainable development.

2 In this chapter, we use local government, local authorities and local councils interchangeably.

3 The Act requires the publication of a report every six months.

4 On 4 September 2013 section 35(2)(a) was replaced by section 7(1) of the Resource Management Amendment Act 2013 (2013 No 63).

5 Prior to ERA.

This research was previously published in the Handbook of Research on Modernization and Accountability in Public Sector Management edited by Graca Maria do Carmo Azevedo, Jonas da Silva Oliveira, Rui Pedro Figueiredo Marques, and Augusta da Conceicao Santos Ferreira, pages 130-158, copyright year 2018 by Information Science Reference (an imprint of IGI Global).

Section 2
Green Technologies and Innovations

Chapter 6
Leveraging Internet of Things to Revolutionize Waste Management

Mirjana Maksimovic
University of East Sarajevo, Bosnia and Herzegovina

ABSTRACT

A continuously growing population and their migration to urban centers consequently leads to waste expansion. The rapidly increasing quantities of waste generated in the cities affect way of human life, environment and planet. Hence, the necessity for smarter, safer, and greener places have never been more urgent. The novel technologies, Internet of Things (IoT) particularly, holds the potential to better manage waste and recycling. The IoT-driven waste management systems positively influence achieving the vision of smart green cities. This article analyzes the role of smart and safe IoT-powered waste management system, highlights its benefits, and possibilities of implementation and evaluation. It is expected that the IoT-based waste management system will deal successfully with an increasing amount of diverse types of waste and through the realization of a smart green city vision will resolve numerous problems related to human health and environmental contamination.

INTRODUCTION

The worldwide human population continuously grows and according to numerous estimates, it will reach 9 billion people by 2050, with 70% of the population living in cities (OECD, 2012). As more people means more waste, the urban residential areas will generate large amounts of diverse types of waste than other areas of residency. In

DOI: 10.4018/978-1-7998-0441-3.ch006

Copyright © 2020, IGI Global. Copying or distributing in print or electronic forms without written permission of IGI Global is prohibited.

other words, with a rising number of people the expansion of waste and its harmful effects on human health and environment have never been more noticeable. It is anticipated that the entire mass of waste generated globally will rise by about 50% over the following few years and that it will hit 2.3 billion tons by the end of 2025 (Future Market Insight, 2017; Martin, 2014). Hoornweg et al. (2013) estimate that by 2100 the growing global urban population will be producing three times as much waste as it does today (exceed 11 million tonnes per day). OECD (The Organization for Economic Co-operation and Development) countries are responsible for almost half of the world's waste, while Africa and South Asia regions produce the least waste (Hoornweg & Bhada-Tata, 2012). Irrespective of the waste type (organic, papers, plastic, glass, metal, e-waste, mixed) and their source (e.g., residential and commercial buildings, industry, hospitals, transport, etc.), waste represent a serious problem of modern society. The smart management of waste is something that must be successfully realized in order to create smart, green and sustainable city. Hence, the realization of smart, safe, green and sustainable urban center demands accomplishing one of the most unglamorous but essential tasks – the collection, transportation, disposal or recycling, and monitoring of waste (Maksimovic, 2017a). The proper waste management should reduce the waste production, ensure waste disposal properly, recycle and reuse disposed products and therefore protect the environment, improve people's safety and increase efficiency. European Commission in its Waste Framework Directive 2008/98/EC defines a goal to be achieved by the end of this decade - 50% of waste produced by households and similar origins must be prepared for reuse and recycling. The ongoing efforts toward waste prevention, waste treatment and other waste management tasks are proofs that the society is becoming aware of the problem and that the efficient managing of materials produced through diverse peoples' activities is essential to avoid their harmful effects over human health and the environment (Waste Management Resources, 2009). In the traditional approaches, there is no information about the waste bins status and collection routes are pre-scheduled. This means that waste collection is blind and not optimized as it uses static routes – the collection route includes collecting waste from bins that are overflowing with garbage, but at the same time includes the bins that are less than half-full. On this way there are many unnecessary operational and environmental costs, what makes these traditional approaches not efficient. On the other side, the recent intensive technology advancements, the establishment of Internet of Things (IoT) concept particularly, contribute to smarter, faster and more efficient management of waste. The IoT-powered waste management system poses information regarding waste bins status and according to them create optimized collection routes what significantly reduce costs and improve citizens' satisfaction. Some positive examples how smart city technology utilization can help in dealing with increasing volume of waste are (Novoseltseva, 2017):

- **San Francisco:** Has an ambitious goal of "zero waste" by 2020 with aggressive recycling while the current waste diversion rate is 80%;
- **Barcelona:** Poses one of the most advanced waste management system (Madakam & Ramachandran, 2015);
- **Copenhagen:** Less than 2% of waste is sent to landfills (half of the produced waste is recycled and most of the waste is used to generate heat for the city's district heating network); and
- **Seoul:** There is no garbage trucks and bins (all household waste goes straight off from individual kitchens through a vast underground network of tunnels to a station where waste is being treated).

Regardless of the solution implemented, each smart system for waste management (established on smart technology) should provide an effective monitoring of waste collection, processing, recycling and disposal avoiding human intervention. This will prevent less pollution in the air and hazardous materials from contaminating the environment, creating healthy and waste ridden environment. In addition, smart systems for waste management reduce the overall operating costs of waste management by almost 50% (Future Market Insight, 2017) and at the same time create more opportunities for all stakeholders (Greentechlead, 2014). According to the Navigant Research reports, the global smart waste collection technology market will grow from 57.6 million USD in 2016 to over 223.6 million USD in 2025, representing a compound annual growth rate of 16.3% (Hardcastle, 2016; Martin, 2014). Since the waste generated represents a major growing concern in today's world, the realization of smart system for waste management has become a hot topic. This paper represents the analysis of its role in realizing smart green urban areas. Hence, the rest of the paper is organized as follows. The IoT influence to waste reduction and principles of IoT-powered waste management system are discussed in second and third section, respectively. The fourth section proposes a customizable IoT-powered waste management solution and highlights its potential benefits and implementation challenges. The last section contains the concluding remarks.

THE IoT INFLUENCE TO WASTE REDUCTION

The waste monitoring and its possible reduction or elimination are of immense importance. The waste crisis is seen as both a challenge and as an opportunity to create novel solutions for efficient waste management (Heinze, 2016). Smart waste management systems are more cost-effective than traditional solutions, their implementation reduces environmental contamination and decrease influence of waste materials on public health and climate changes. With the inclusion of

Figure 1. The goals of green technology (Source: Maksimovic, 2017b)

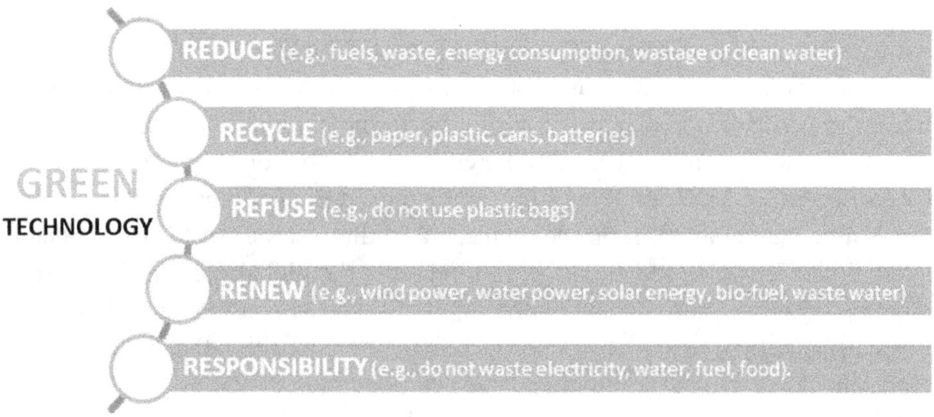

innovative technologies, waste management becomes smarter, more effective and more sustainable. The IoT, as a concept that can connect a variety of numerous physical objects/devices/things to the Internet with the help of sensors and other wireless technologies, enables easier and faster monitoring of waste and pollution generated in numerous sectors (homes, buildings, medicine, industry, construction, transportation, agriculture, etc.) as well as their reduction. The development and implementation of green methods and techniques that satisfy the goals shown in Figure 1., significantly contribute to waste reduction.

The green, environmental or clean technology involves environmental sustainable designing, manufacturing, using and disposing, without harming effects on the environment and human health (Maksimovic, 2017b). The IoT that helps reduction of the greenhouse effect within other industries, and also reduces the effect that IoT itself could have on the environment is known under the term Green IoT (G-IoT). The entire life cycle of G-IoT consists of green design, green manufacturing, green appliance and green disposal/recycling with no or minor impact on the human and the environment (Tracy, 2016). In this way, significant amounts of solid and toxic waste and greenhouse emissions as well as the use of natural and non-renewable raw materials can be significantly decreased or even eliminated. The most present benefits of G-IoT usage are noticeable in the sectors that consume the large amounts of energy and natural resources and that are at the same time the largest sources of greenhouse gas emissions – buildings, industry, and energy sectors. The G-IoT potential to save and more efficiently use natural resources is present in agriculture, forestry, fisheries and aquaculture sectors. The benefits of the G-IoT appliance are present also in transport and tourism sector, and particularly in waste management. Hence, the advantages of G-IoT implementation in numerous sectors of modern society are significant socio-economic and environmental benefits (Maksimovic, 2017b). Obviously, G-IoT development and its widespread use significantly contribute to the realization of smarter, safer and more sustainable places to live.

Alongside numerous benefits technology progress brings in all aspects of our lives, the consequence of electrical and electronic equipment utilization lies in increased quantities of e-waste. Cisco predicts that 50 billion of smart devices will be connected to the Internet by 2020, and if just 10% of IoT devices are trashed per year beginning in 2021, this could mean 2+ billion new e-waste items annually (Vertatique, 2014). Evidently, e-waste represents the global concern as it grows rapidly, 3-5% per year (Cucchiella et al., 2015), and is considered as hazardous waste because they have toxic compounds in their composition that are harmful to human health and to the environment (Veit & Bernardes, 2015). Green technology utilization is expected to significantly reduce harming effects of e-waste.

On the other side, the good news is that waste from electrical and electronic equipment (e-waste) is one of the most controllable parts of the waste streams. E-waste recycling poses following benefits:

- Conservation of rare, critical and valuable natural materials (e.g., gold, silver, aluminum, platinum, copper, iron), as well as glass and plastics (Maksimovic, 2017b);
- E-waste recycling contributes to significant energy efficiency (CFES, 2012). The energy savings consequently lead to the less dependence on finite oil reserves, reduced air pollution and greenhouse gas emissions (Nayab, 2011);
- Repaired and reused electronic products appear as cost-effective and more convenient compared to completely new products (Maksimovic, 2017b);
- E-waste recycling companies strength economies and produce more business opportunities. According to Morgan (2015), the potential revenue from the recycling of e-waste is around 2 billion USD and it is expected to go up to almost 3.5 billion USD by 2020.

New technologies and IoT enhance manners for waste reduction, reuse and recycling alongside reduced costs and better service delivery to citizens. The benefits are also seen in a cleaner and healthier environment through pollutants or hazardous waste removal from the environment. Possible negative impacts of IoT are minimized through the utilization of green technology approaches that eliminate or at least decrease the measure of pollutants or toxic compounds as products of operations (Maksimovic, 2017b).

Greening the waste sector bring numerous environmental and socioeconomic benefits, reduced greenhouse gas emissions, saved energy, natural and valuable resources, new jobs, improved quality of life, etc. At the same time, policy framework, economic incentives and financing are factors that influence the greening waste sector (Elagroudy et al., 2016). However, there is no universal approach and design for smart waste management. Development of effective system through the implementation of advanced technology that will be easy to use and not be expensive is a challenge.

IoT-POWERED WASTE MANAGEMENT SYSTEM

The IoT progress has enabled smarter and more efficient waste management than ever before. Reliability, mobility, user convenience, service continuity and energy efficiency are the advantages of an IoT-based waste management system (Hong et al., 2014). Nevertheless, significant cost savings and increased profit as a consequence of successful waste management are especially worth mentioning. There are numerous smart waste systems developed by companies around the world (e.g. Enevo, Compology, Bigbelly, Sutera, etc) (Musulin, 2015). These systems are Cloud-based and use the smart bins, equipped with the wireless sensors that are able to detect the waste. Weight sensors provide the information about waste weight, while the sensors that are able to measure the bin filling level (e.g. infrared sensor, ultrasonic sensor) are more efficient in the development of smart waste management systems. These sensors can be used together with temperature and motion sensors in order to measure is bin on fire or it has been moved (Norman & Revill, 2016). Zhao et al. (2017) in their work propose to measure fill-level indirectly using a vibrating mini motor placed on the outside surface of waste bins. Irrespective of sensors used, bins capacity status from wireless fill-level sensors are collected in real-time and if the bin waste level is higher than threshold level this information together with the bin location are uploaded to the Cloud and other stakeholders (Maksimovic, 2017a). Separating bins for each category of waste (organic, papers, plastic, glass, metal, e-waste, mixed), the information of waste type is also available alongside information regarding bin waste levels (Aazam et al., 2016). After the receiving information from wireless sensors that measure bin filling level and other multiple sensors and information sources (e.g., GPS - Global Positioning System, smart devices, RFID - Radio Frequency Identification tags) that provides information on bins and garbage track locations, and traffic congestion, the decision about bin collection time and optimal route for waste collection are being generated. Emptying bins when they need to be (some bins are emptied twice a day and others only every few days) leads to improving collection route and hence reduces costs and CO_2 emissions alongside the increased citizens satisfaction. If garbage truck is full, the next step is to determine the optimal route for waste dumping. The accumulated

Figure 2. Waste disposal options

waste can be dumped off in controlled manners, to the landfill or to the incineration (Figure 2). These are the least preferred options, but still the most popularly used techniques for waste disposal.

Landfills are usually found in developing countries, while burning waste at high temperature, known as incineration or combustion, significantly reduce the quantity of disposed waste (Saha et al., 2017). The thermal treatment of waste materials can be utilized for the generation of energy as a fuel (Waste Management Resources, 2009). However, pre-separation of waste simplifies the choice of an adequate method for the dumping of waste. The biological reprocessing or recycling of organic waste materials (food, plant, and paper products) results in forms (e.g. mulch or compost) that can be used for agricultural purposes. Recycling enable changing waste product into new one. Discarded things can be reused through extracting and recovering resources and materials. Another way is to convert them to energy within the variety of useable heat, electricity or fuel. Nevertheless, the most preferred option is to reduce amount of waste, energy consumption, greenhouse gas emission and pollutions and preserve natural resources for future uses (Saha et al., 2017).

According to Norman & Revill (2016), the potential benefits of smart waste management, are more effective city-wide waste collection, reduced costs, round optimization, reduced emissions and pollutions, etc.

The Proposition of a Customizable IoT-Powered Waste Management Solution

The development of smart bins and smart waste management systems can be found in numerous works (Aazam et al., 2016; Boh, 2016; Catania & Ventura, 2014; Dhaya et al., 2016; Giacobbo et al., 2016; Hong et al., 2014; Kokila et al., 2017; Lamichhane, 2017; Likotiko et al., 2017; Lundin et al., 2017; Medvedev et al., 2015; Meghana & Nataraj 2016; Musulin, 2015; Omar et al., 2016; Oralhan et al., 2017; Popa et al., 2017; Ravichandran, 2016; Sai, 2017; Shukla & Shukla 2017; Sinha

& Couderc 2013; Tausif 2015; Tracy 2016; Vrochidou, et al., 2017; Wen et al., 2017; Zhao et al., 2017). Anagnostopoulos et al. (2017) made a detailed survey of existing IoT-based waste management systems, and performed their classification and comparative assessment. The increased interest in this topic is the result of the numerous benefits that smart collecting, processing, and disposition of waste bring. The recent research has resulted in the development of solar-powered smart bins that act as Wi-Fi hotspots. Also, the bins can be equipped with a compactor that after crunching trash enable the increasing of smart bin capacity up to eight times (Boh, 2016). It is significant to highlight that implementation of the pre-separated waste method is of important significance in waste management systems. Ravichandran (2016) introduces the smart dustbin that is capable of distinguishing important materials from waste with the assist of an embedded image scanner. In addition, the proposed solution performs automatic garbage collection from the ground where it circulates. As a part of automated waste collection system, video recognition system for identifying the waste type has been also proposed by Popa et al. (2017). The stakeholders, based on information on the waste type and their collected quantities can perform more effective actions regarding specific waste type. Therefore, the optimization of the number and the placement of bins, as well as the better organization of bins for different types of waste, can be performed.

Irrespective of the approach and ideas, the design and implementation of IoT-powered system for smart management of waste is a complex process, and requires numerous and detailed analysis of all segments. Before the implementation of any system it is essential to consider the cost of sensors implemented, network communications possibilities, database issues, data processing, application choices, data access, and data management (Norman & Revill, 2016). The motive for ongoing efforts in enhancing smart waste management system through the utilization of advanced technologies is the fact that the efficient and on-time waste collection contributes to better health, hygiene, and disposal (Aazam et al., 2016).

In order to build and design a customizable IoT-powered waste management system some consideration should be taken into account. The developed solution should be affordable, easy to install, low power consumption and to pose ability to learn fast since it is based on IoT principles and should make decisions without human intervention. The proposed solution is presented in the Figure 3, and as node components Raspberry Pi 3 and ultrasonic sensor are used.

The Raspberry Pi is chosen as a key building element because it is inexpensive, small, powerful and customizable thanks to its inbuilt support for a large number of input and output peripherals and network communication (Maksimovic et al., 2014). Thanks to its performances, the Raspberry Pi can be easily used as a gateway as well in the proposed approach. The ultrasonic sensor is applied for detecting the level of waste filled in the waste bin. The level is determined according to the distance between sensor and waste (the measurement range is from 2 cm to 4 m). If the RFID tag is added to the node placed on the bin, the garbage truck, which poses the RFID reader, getting closer to the bin communicate with the RFID tag and obtain necessary data. Connecting GPS module to Raspberry Pi 3, the location of waste bin alongside waste level information is also being sent further. The battery problems can be resolved implementing solar battery solutions. The measured data (waste bin level and location) are sent wirelessly through integrated 802.11n wireless LAN, Bluetooth 4.1. or via ZigBee, LoRa, and GSM/GPRS (with the help of modules connected to the Raspberry Pi 3 unit). A gateway receives and forwards data to the NoSQL (Not Only Structured Query Language) database where data are going to be saved and analyzed in order to perform decision making in real-time or near real-time (Figure 3). The database can be filled with other data that can be relevant for the waste collection (i.e. traffic congestion, weather forecast, team shifts). These data collected from another IoT devices and connected objects hold the potential to make solution more effective. Cloud-based applications can receive data from the nodes using different IoT protocols, like MQTT (Message Queue Telemetry Transport) or CoAP (Constrained Application Protocol). Another approach is to implement the Fog computing paradigm at the gateway level what will reduce the amount of data sent to the Cloud and enable faster response of the system. The different parameters and collected information regarding the waste management and their visualizations must be enabled to waste management company and users via GUI (Graphical User Interface).

Figure 3. The proposed solution of IoT-powered waste management system

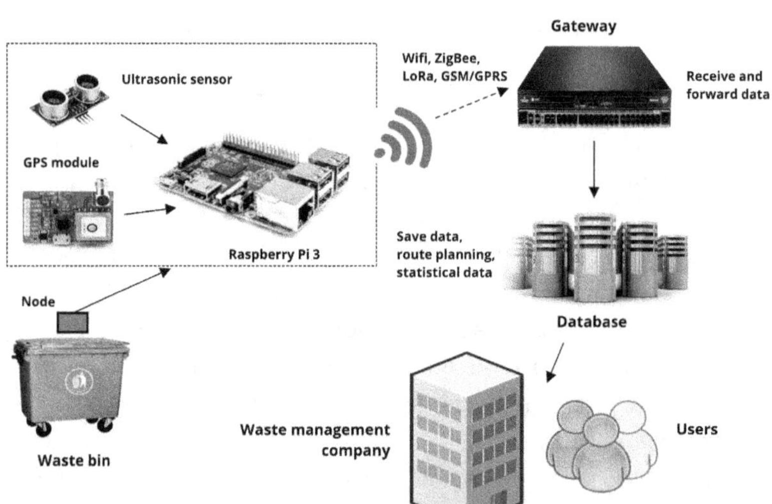

The flowchart of proposed IoT-powered waste management system is shown in Figure 4. Threshold level can be up to 90% bin's capacity in order to don't let bins become overfilled. However, the data regarding bin's waste level can be forwarded at any moment and based on historically collected data predictions when bins will reach a level predefined for performing waste collection can be made. The visualization of data collected in real-time or near real-time is essential to improve decision making in smart waste management systems (Figure 5).

Determining the optimal routes for waste collection (Figure 5) leads to a reduction in the number of garbage trucks on the road what contributes to significant cost savings and maximizes efficiency. Another significant benefit is pollution reduction. For instance, in the case of proposed solution optimal routes can be found using graph theory optimization algorithms. The suitable path is chosen based on applied algorithm on the collected data, graph modeling and graph bi-partitioning for possible routes.

Figure 4. The flowchart of a smart waste management system (Source: Maksimovic, 2017a)

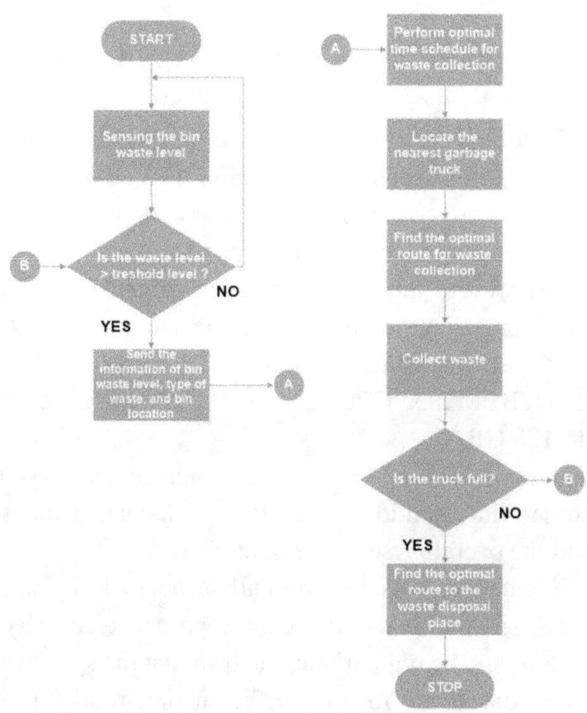

Figure 5. The optimal garbage truck route for waste collection (Source: Maksimovic, 2017a)

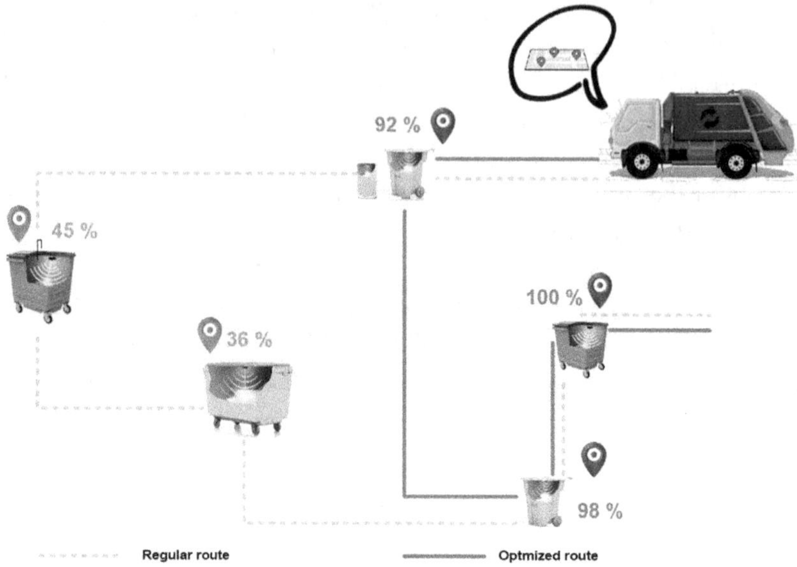

Some of the challenges regarding the proposed solution are:

- **Energy:** Wireless sensor node presented in this paper is consisted of Raspberry Pi unit, ultrasonic sensor and GPS module. In order to operate, collect, measure, analyze and send data, the node requires adequate level of energy. The problems related to energy can be overcome using solar battery approaches. However, the implementation of energy-aware protocols is necessary for increasing the lifetime of node and overall system;
- **Memory and Storage Space:** Raspberry Pi 3 uses a 1.2GHz 64-bit quad-core Broadcom BCM2837 CPU, and has 1GB RAM. External storage can be expanded to 128 GB;
- **Self-Management:** The proposed waste management system should work without human intervention. Therefore, self-management of the node, network, and the overall system is essential;
- **Security:** The data sent to the gateway and further to the Cloud, waste company and end users, should be secured with appropriate security algorithms. In addition, data should be regularly updated so that the system and the end user have the newest data at any moment. Efficient data routing protocols also play a key role in the efficiency of the overall waste management system;

- **Quality of Service (QoS):** As a level of service provided by the waste management system to its users, QoS is an important issue. The proposed waste management system holds a potential to satisfy the required QoS level.

Nevertheless, proposed solution provides an enhanced manner for managing waste in cities. The advantages of presented design of a smart waste management system are the real-time access to information of waste volume, waste type, bin locations, and significantly reduced unnecessary costs (through optimization of routes and waste collection times). However, the more satisfied citizens and their positive experience with advanced systems for waste management are important as well and represents a goal which has to be achieved. Increased eco-consciousness of people will for sure lead to the prevention of waste following the 3R's, reduce, reuse and recycle, therefore reducing negative impact of waste on the environment and human health.

In summary, effective utilization of bins, real-time information of the waste volume, intelligent management of the city's services, cost-effectiveness and optimization of resource utilization, as well as enhanced environmental quality by minimizing potential harmful effects on the environment and human health, are the main benefits IoT-based waste management systems bring. Based on data collected from IoT-driven waste management systems, bin-filling and waste collection/scheduling patterns can be identified. In other words, large quantities of data generated by IoT-based waste management systems enable obtaining valuable insights and making adequate decisions and activities towards the realization of smart, safe, and sustainable cities.

CONCLUSION

The last few decades have witnessed massive growth of urban population, accompanied by waste expansion. Traditional approaches for management of waste are not adequate to deal with the increasing quantities of waste. The novel technology progress has enabled the omnipresence of IoT-powered systems that pose the power to completely revolutionize waste management. Numerous information ranging from waste filling levels to toxic substances is available at any moment from anywhere thanks to the IoT-based waste management system. The availability of various types of data is of immense importance to all parties included in waste management and guides to more successful waste management what is a precondition for the realization of smart, safe, and sustainable urban centers. This paper highlights the benefits of IoT leverage in waste management systems and proposes the customizable IoT-powered waste management solution. Systems like this enable easier waste collection and optimize resource utilization. Innovative waste management systems are based on the 3R's, reduce, reuse and recycle. The IoT inclusion in waste management systems makes them efficient than ever before contributing to significant economic benefits. Utilization of G-IoT based on the green design, production, utilization, and disposal, significantly reduces generated amounts of waste and greenhouse gas emissions leading to the cleaner and safer places to live. In summary, innovative technologies represent a powerful tool for dealing with the challenges modern society face with, such as successful waste management.

REFERENCES

Aazam, M., St-Hilaire, M., Lung, C.-H., & Lambadaris, I. (2016). Cloud-based Smart Waste Management for Smart Cities. In *IEEE 21st International Workshop on Computer Aided Modelling and Design of Communication Links and Networks (CAMAD)*, Toronto, Canada. 10.1109/CAMAD.2016.7790356

Anagnostopoulos, T., Zaslavsky, A., Kolomvatsos, K., Medvedev, A., Amirian, P., Morley, J., & Hadjiefthymiades, S. (2017). Challenges and Opportunities of Waste Management in IoT-enabled Smart Cities: A Survey. *IEEE Transactions on Sustainable Computing*.

Boh, S. (2016). Solar-powered smart bins that act as Wi-Fi hot spots launched at Orchard Road. *Straits Times*. Retrieved August 8, 2017, from http://www.straitstimes.com/singapore/solar-powered-smart-bins-that-act-as-wi-fi-hotspots-launched-at-orchard-road

Catania, V., & Ventura, D. (2014). An Approach for Monitoring and Smart Planning of Urban Solid Waste Management Using Smart-M3 Platform. In *15th Conference of FRUCT association* (pp. 24-31).

CFES. (2012). The environmental and economic benefits of electronic waste recycling. Retrieved July 23, 2017, from http://cashforelectronicscrapusa.com/the-environmental-and-economic-benefits-of-electronic-waste-recycling/

Cucchiella, F., D'Adamo, I., Koh, S. C. L., & Rosa, P. (2015). Recycling of WEEEs: An economic assessment of present and future e-waste streams. *Renewable & Sustainable Energy Reviews*, *51*, 263–272. doi:10.1016/j.rser.2015.06.010

Dhaya, R., Pattabiraman, P., Shahrukh, S., & Aravindh, J. (2016). Smart Waste Management Using Internet of Things (Iota). *Middle East Journal of Scientific Research*, *24*(10), 3358–3361.

Elagroudy, S., Warith, M. A., & El Zayat, M. (2016). *Municipal Solid Waste Management and Green Economy*. Global Young Academy.

Future Market Insight. (2017). *Smart Waste Management Market: Global Industry Analysis and Opportunity Assessment 2015-2025*. Retrieved June 3, 2017, from http://www.futuremarketinsights.com/reports/smart-waste-management-market

Giacobbo, M., Puliafito, C., & Scarpa, M. (2016). The Big Bucket: An IoT Cloud Solution For Smart Waste Management in Smart Cities. In *The 4th International Workshop on CLoud for IoT (CLIoT 2016)*, Vienna, Austria.

Greentechlead. (2014). Annual revenue from smart waste technology to be $6.5 billion by 2023. Retrieved July 16, 2017, from http://www.greentechlead.com/waste-management/annual-revenue-from-smart-waste-technology-to-be-6-5-billion-by-2023-14758

Hardcastle, J. L. (2016). How Smart Waste Collection Can Reduce Costs, Improve EHS Performance. *Environmental Leader*. Retrieved June 28, 2017, from https://www.environmentalleader.com/2016/11/how-smart-waste-collection-can-reduce-costs-improve-ehs-performance/

Heinze, K. (2016). IoT and waste management: Revolutionizing an old industry. Industrial, Smart Cities. *Readwrite*. Retrieved August 1, 2017 from http://readwrite.com/2016/07/22/iot-and-waste-management-revolutionizing-il2/

Hong, I., Park, S., Lee, B., Lee, J., Jeong, D., & Park, S. (2014). IoT-Based Smart Garbage System for Efficient Food Waste Management. The Scientific World Journal.

Hoornweg, D., & Bhada-Tata, P. (2012). *What a waste - A Global Review of Solid Waste Management*. The World Bank.

Hoornweg, D., Bhada-Tata, P., & Kennedy, C. (2013). Waste production must peak this century. *Nature*, *52*, 615-617.

Kokila, J., Devi, K. G., Dhivya, M., & Jose, C. H. (2017). Design and Implementation of IoT Based Waste Management System. *Middle East Journal of Scientific Research*, *25*(5), 995–1000.

Lamichhane, M. (2017). *A Smart waste management system using IoT and blokchain technology* [Master thesis]. ITMO University, Russia.

Likotiko, E. D., Nyambo, D., & Mwangoka, J. (2017). Multi-agent based IoT smart waste monitoring and collection architecture. *International Journal of Computer Science, Engineering and Information Technology*, *7*(5).

Lundin, A. C., Ozkil, A. G., & Schuldt-Jensen, J. (2017). Smart cities: A case study in waste monitoring and management. In *Proceedings of the 50th Hawaii International Conference on System Sciences* (pp. 1392-1401).

Madakam, S., & Ramachandran, R. (2015). Barcelona Smart City: The Heaven on Earth (Internet of Things: Technological God). ZTE Communications, 13(4).

Maksimovic, M. (2017a). Internet of Things driven waste management system for greener city. In *18th Hellenic Forestry Congress "Hellenic Forestry facing major challenges: sustainable forest management, forest cadaster, environmental technologies-networking and nature protection" & International Workshop "Information Technology, Sustainable Development, Scientific Network & Nature Protection"*, Edessa, Greece (pp. 1678-1685).

Maksimovic, M. (2017b). Greening the future: Green Internet of Things (G-IoT) as a key technological enabler of sustainable development. In N. Dey, A. E. Hassanien, C. Bhatt, A. S. Ashour, & S. C. Satapathy (Eds.), *Internet of Things and Big Data Analytics toward Next Generation Intelligence*. Springer.

Maksimovic, M., Vujović, V., Davidović, N., Milošević, V., & Perišić, B. (2014). Raspberry Pi as Internet of Things hardware: Performances and Constraints. In IcETRAN, Vrnjacka Banja, Serbia.

Martin, R. (2014). *Revenue from Smart Waste Technology is Expected to Exceed $42 Billion From 2014-2023*, Navigant Research. Retrieved July 4, 2017, from https://www.navigantresearch.com/newsroom/revenue-from-smart-waste-technology-is-expected-to-exceed-42-billion-from-2014-2023

Medvedev, A., Fedchenkov, P., Zaslavsky, A., Anagnostopoulos, T., & Khoruzhnikov, S. (2015). Waste management as an IoT-enabled service in smart cities. In *Conference on Smart Spaces* (pp. 104-115). Springer. 10.1007/978-3-319-23126-6_10

Meghana, K. C., & Nataraj, K. R. (2016). IOT Based Intelligent Bin for Smart Cities. International Journal on Recent and Innovation Trends in Computing and Communication, 4(5).

Morgan, K. (2015). Is there a future for e-waste recycling? Yes, and it's worth billions - A methodology to help organizations in e-waste management. Retrieved July 18, 2017, from https://www.elsevier.com/atlas/story/planet/is-there-a-future-for-e-waste-recycling-yes,-and-its-worth-billions

Musulin, K. (2015). Refuse revolution: 4 companies transforming the trash bin. *Wastedive*. Retrieved July 2, 2017, from http://www.wastedive.com/news/refuse-revolution-4-companies-transforming-the-trash-bin/405405/

Nayab, N. (2011). Benefits of e-Waste Recycling. *Brighthub.* Retrieved July 4, 2017 from http://www.brighthub.com/environment/green-computing/articles/71375.aspx

Norman, T. & Revill, S. (2016). IoT Investment case Toolkit – Smart waste. IoTUK.

Novoseltseva, E. (2017). Top smart city projects & leaders to watch in 2017, In Technology industry centers. *Apiumhub.* Retrieved November 30, 2017, from https://apiumhub.com/tech-blog-barcelona/smart-city-projects-leaders-barcelona/

OECD. (2012). *OECD Environmental Outlook to 2050: The Consequences of Inaction.* Paris: OECD Publishing.

Omar, M. F., Termizi, A. A. A., Zainal, D., Wahap, N. A., Ismail, N. M., & Ahmad, N. (2016). Implementation of spatial smart waste management system in Malaysia. In *IOP Conference Series: Earth and Environmental Science* (Vol. 37, No. 1, p. 012059). IOP Publishing. 10.1088/1755-1315/37/1/012059

Oralhan, Z., Oralhan, B., & Yiğit, Y. (2017). Smart City Application: Internet of Things (IoT) Technologies Based Smart Waste Collection Using Data Mining Approach and Ant Colony Optimization. *The International Arab Journal of Information Technology, 14*(4), 423–427.

Popa, C. L., Carutasu, G., Cotet, C. E., Carutasu, N. L., & Dobrescu, T. (2017). Smart City Platform Development for an Automated Waste Collection System. *Sustainability, 9*(11), 2064. doi:10.3390u9112064

Ravichandran, S. (2016). Smart Dustbins Capable of Distinguishing Important Materials from Waste Materials. *International Journal of Computer Technology and Applications, 9*(6), 2889–2902.

Saha, H. N., Auddy, S., Pal, S., Kumar, S., Pandey, S., Singh, R., ... Saha, S. (2017). Waste management using Internet of Things (IoT). In *2017 8th Annual Industrial Automation and Electromechanical Engineering Conference (IEMECON)* (pp. 359-363). IEEE. 10.1109/IEMECON.2017.8079623

Sai, P. Y. (2017). IOT Smart Garbage Monitoring System in Cities-An Effective Way to Promote Smart City. *International Journal of Advanced Research in Computer Science and Software Engineering, 7*(2), 99–102. doi:10.23956/ijarcsse/V7I2/0105

Shukla, S., & Shukla, N. (2017). Smart Waste Collection System based on IoT (Internet of Things): A Survey. *International Journal of Computer Applications, 162*(3).

Sinha, A., & Couderc, P. (2013). Smart Bin for Incompatible Waste Items. *In The Ninth International Conference on Autonomic and Autonomous Systems (ICAS)* (pp. 40-45).

Tausif, M. (2015). *Waste Management & Recycling - Kingdom of Saudi Arabia.* Retrieved June 28, 2017, from https://www.linkedin.com/pulse/moving-saudi-arabia-towards-sustainable-future-zero-wast-shaikh

Tracy, P. (2016). Keeping the IoT green by reducing e-waste. *RCR Wireless*. Retrieved July 1, 2017, from http://www.rcrwireless.com/20161004/telco-cloud/green-iot-e-waste-tag31-tag99

Veit, H. M., & Bernardes, A. M. (2015). Electronic Waste: Generation and Management. In H. M. Veit & A. M. Bernardes (Eds.), *Electronic Waste*. Switzerland: Springer International Publishing. doi:10.1007/978-3-319-15714-6_2

Vertatique. (2014). Internet of Things Presents Green ICT Challenges. Retrieved June 15, 2017, from http://www.vertatique.com/internet-things-will-rerquire-more-storage-and-bandwidth

Vrochidou, E., Kagkasidis, N., Koutaliaga, G., & Sgouros, C. (2017). iBIN: Intelligent Monitoring System for Recyclable Materials Using Arduino and the IoT. *International Journal of Innovative Research in Electrical, Electronics, Instrumentation and Control Engineering*, *5*(12), 1–5.

Waste Management Resources. (2009). Retrieved August 1, 2017, from http://www.wrfound.org.uk/

Wen, Z., Hu, S., De Clercq, Beck, M. B., Zhang, H., Zhang, H., Fei, F. & Liu, J. (2017). Design, implementation, and evaluation of an Internet of Things (IoT) network system for restaurant food waste management. *Waste Management*. doi:10.1016/j.wasman.2017.11.054

Zhao, Y., Yao, S., Li, S., Hu, S., Shao, H. & Abdelzaher. T. F. (2017). VibeBin: A Vibration-Based Waste Bin Level Detection System. *PACM Interact. Mob. Wearable Ubiquitous Technol.,* 1(3).

This research was previously published in the International Journal of Agricultural and Environmental Information Systems (IJAEIS), 9(4); edited by Francois Pinet and Petraq Papajorgji, pages 1-13, copyright year 2018 by IGI Publishing (an imprint of IGI Global).

Chapter 7
Analysis of the Nexus Between Smart Grid, Sustainable Energy Consumption, and the Smart City

Luke Amadi
University of Port Harcourt, Nigeria

Prince I. Igwe
University of Port Harcourt, Nigeria

ABSTRACT

Since the 1990s, the field of smart grid has attempted to remedy some of the core development deficiencies associated with power supply in the smart city. While it seemingly succeeds in provision of electricity, it fails to fully resolve the difficulties associated with sustainable energy consumption. This suggests that the future of smart grid analytics in the smart city largely depends on efficiency in energy consumption which integrates sustainability in the overall energy use. This chapter analyzes the nexus between smart grid, sustainable energy consumption, and the smart city.

INTRODUCTION

Cities are the hub of economic development where energy consumption plays a key role. According to the United Nations Conference on Sustainable Development in Rio de Janeiro in 2012, half of humanity lives in cities. The urban population has increased from 750 million people in 1950 to 3,600 million in 2011. It is estimated

DOI: 10.4018/978-1-7998-0441-3.ch007

Copyright © 2020, IGI Global. Copying or distributing in print or electronic forms without written permission of IGI Global is prohibited.

that by 2030 almost 60% of the population of the world will reside in urban areas. The question of green energy and technology has not been adequately resolved in both developed and developing societies.

The strategies to have electricity consumption under efficient control are the basis of smart grid. This involves the control of energy usage based on the process data. The argument is that the process of achieving consensus on the future vision for technology development should lead to the mobilization of different environments around the idea of propagating a smart grid for a smart city. According to the European 2020 Strategy Document (European Commission, 2010b), Europe is suffering a period of structural transformation in the socio-economic framework. The priorities to favor the new model that Europe needs include institutionalization of a smart city dynamics. This includes a variety of activities and lifestyles.

In the United States, the first official effort at defining Smart Grid was provided in the Energy Independence and Security Act of 2007 (EISA-2007). The US Congress gave its approval in January 2007 and it was signed to law by President George W. Bush in December 2007. The EISA (2007) argued that a smart grid is an electrical grid which includes a variety of operational and energy measures including smart meters, smart appliances, renewable energy resources, and energy efficient resources.

Beyond secure and equitable energy consumption among the high-income societies, are the challenges of greenhouse gas emissions which are produced in cities. These appear to be at variance with the smart city prognosis. Equally, a number of studies explore “inclusive grid” within the sustainability contexts. This largely explores the persistent dichotomy between the rich and poor societies. Across Africa, parts of Asia and Latin America power outage points to opposite direction with the smart grid analytics.

This suggests the saliency of engagement with the question of smart grid analytics within sustainability contexts. In particular, this encompasses building solutions with the evolving needs and nature of citizens to accommodate adaptable foundations to shape the future of cities. This is premised on the notion to accommodate a number of evolving interests in the smart city dynamics including clean and green energy consumption, inclusive grid, grid control models etc. Thus, smart grid analytics emerged as basic building block to implement smart city.

Smart cities which could act as catalyzers of the foreseen energy, environmental and social equity are expected to play key roles in order to achieve a sustainable energy consumption. Energy issues relating to sustainable development were discussed at inter-governmental level for the first time at the 9th Session of the Commission for Sustainable Development (CSD-9), held in April 2001. Countries agreed that stronger emphasis should be placed on the development, implementation and transfer of cleaner, more efficient technologies and that urgent action is required to further develop and expand the role of alternative energy sources.

Sustainable consumption debate contends that consumption is a key lever to achieving more sustainable development, on the contrary, that unsustainable consumption patterns are major causes of global environmental deterioration, including the overexploitation of renewable resources and the use of non-renewable resources with their associated environmental impacts (European Environment Agency,2005).

Similarly, technological innovations have reduced the energy and material intensity of most products. However, the increasing volumes of consumed goods have outweighed these gains: Household energy consumption contributes to almost 30% to the total final energy consumption and is, after transport, the second most rapidly growing area of energy use (EEA,2005). Equally, industrial energy consumption also accounts for over 78% of total energy consumption(Barr et al., 2005).

The dynamic and rapid growth of cities will lead to scholarly and research trajectories on the future of smart grid. In particular, cities and urban planners must device new modalities in a sustainable manner that transcends 'meeting the energy needs of the present and future generation' to 'protecting the natural ecosystem where human and non-human species inhabit' .This stand point provides a new set of scholarship that proposes institutionalization of Marxian political ecology, a sustainable development initiative which matches political power and technological resource extraction with ecological equity and justice.

Against this background, this chapter is a foresight research which attempts to resolve the question regarding how smart grid analytics could help smart cities in sustainable consumption. The chapter proposes equitable mobilization of resources with emphasis on sustainability. In a distinct manner, it demonstrates that although increase in green energy producers and consumers may lead to more agencies with ability to provide smart grid analytics. However, that the concomitant challenge of 'efficient energy consumption' points out the saliency of sustainable consumption.

The assumption is that smart grid would be connected with process analysis that could mitigate deleterious energy consumption. To advance this knowledge, it is important to understand how process analysis in energy usage is manifested and how it can be enhanced at various levels in smart cities (Both at the micro and macro levels). The objective of this chapter is to determine a sustainable model for energy consumption. It posits that unsustainable energy consumption will pose more challenges to the future development and advancement of smart grid and communication systems, which may have adverse effects on smart grid analytics for the smart city. An alternative way of thinking about smart grid is explored one which prioritizes dynamics of sustainable and efficient energy consumption.

The chapter puts forward a novel knowledge on Marxian political ecology which centres on sustainability. It argues that a city cannot be smart when its key structures and components are unsustainable. Emphasis is laid on energy resource efficiency, equality, smart and green energy consumption which the chapter argues must be institutionalized in order for a city to lay claims to "Smart".

The rest of the chapter is structured as follows; theoretical framework, Conceptual issues: smart grid, smart city and sustainable energy consumption, smart grid and smart city: analysis of sustainable energy consumption, recommendation, future research directions and conclusion.

THEORETICAL FRAMEWORK

Several theoretical explanations have been provided on studies in smart grid analytics and the smart city in the broad field of sustainable development, reflecting on their key assumptions, analyzing key trends in their trajectories and suggesting future research agenda that might advance the divergent social trends, changes and theoretical threads in the field of inquiry (Castells,2000; Schor,2001; Shove,2003; Hamilton, et al; 2010; Galli, et al; 2011) . These spate of scholarly-reflections, have set up new research agenda that prove useful in understanding original impetus of sustainable development within various research threads. Sustainable energy consumption is one of such useful themes which seek to explore the question of the relation of present and future energy consumption in the smart city, within the smart grid analytics.

While endorsing the general theoretical orientation of Marxian political ecology, its less adoption in the contemporary real-world contexts driven largely by asymmetrical capitalist resource extraction has overly undermined a sustainable and equitable energy consumption. This has serious consequences for Smart grid analytics and the smart city. Thus, a distinctive component of the ecological critique of smart grid analytics and smart city debate is largely their exclusion of environmental concerns - both the understanding of how smart grid provides energy and the institutional core regarding the "clean or green" concerns within the production and consumption of energy. These less lucid concerns form the theoretical basis of the Marxian political ecology.

Ecology is the study of the relationships of organisms with one another and the relationship of organisms to their environment. Ecological Marxism is a neo Marxist theory which studies the primacy of capitalist power and inequality in access and use of natural resources (Speake & Gismondi, 2005:55; Harvey, 2005; Alkon& Agyeman, 2011;Amadi & Igwe,2016). Ecological Marxists draw attention to political and ecological factors in clarifying how material power and resource accumulation (e.g. capital, wealth, military power) mediate human society and nature relations (Biersack, 2006: 3; Bryant, 1998: 80).

Following the assumptions of the ecological Marxism, the chapter presents an overview and critique of the smart grid analytics and the smart city model.

And then sets out in more details the ways in which the field of political ecology could fruitfully engage with the question of unequal access, underutilization and disproportionate energy consumption which challenges the smart city model. Thus, inequality remains a key strand of unsustainable consumption. Juliet Schor (2001) argued that "economic globalization, militarization, corruption, the monopolization of environmental resources, and the legacies of colonialism have meant that the global "South" doesn't consume enough—at least not in terms of basics such as food, clothing, shelter". Schor (2001) observed that "in 1999, per capita GDP in the Less Developed Countries (or "global South") was $3,410 (measured as purchasing power parity). By contrast the Developed Countries ("global North") enjoyed average per capita GDP of $24,430, a gap of about eight times. One third of the population in the global South (1.2 billion) lives on less than $1 per day; 2.8 billion live on less than $2 per day. Together, this 4 billion comprises two-thirds of the world's population".

Elizabeth Shove (2003,p.1) re-echoes an insightful concern and argued that it is evident that lifestyles, particularly in " the West, will have to change if there is to be any chance of averting the long-term consequences of resource depletion, global warming, the loss of biodiversity, the production of waste or the pollution and destruction of valued 'natural' environments".Schor(2001:4)further argued that; "The number of vehicles per person has increased; as has the size and luxuriousness of those vehicles. The culture of excess has yielded $20,000 outdoor grills; $17,000 birthday parties for teen girls at FAO Schwarz; diamond studded bras from Victoria's Secret; a proliferation of Jaguars, Porsches and other luxury cars; status competitions in stone walls; professional quality appliances for people who are never home to cook; designer clothes for six-year olds; and bed sheets costing a thousand dollars apiece".

On this basis, Marxian political ecology provides a theoretical and conceptual model that interrogates both smart grid and smart city studies and offers an alternative to the mainstream debates on equitable and sustainable energy consumption.

CONCEPTUAL ISSUES: SMART GRID, SMART CITY AND SUSTAINABLE ENERGY CONSUMPTION

There is need for conceptual clarity on the issues of smart grid, smart city and sustainable energy consumption by exploring these concepts and offering a way of thinking about how they could be interrelated within the understanding of city life and sustainable development.

Smart grid is the 'third great revolution' of electricity (Mazza,2004) which represents a wide range of 'changes in its production, delivery and use' including concepts in the present and future research agenda linked to the challenges of electricity supply. The wide range of terms associated with smart grid analytics and absence of consensus on scholars on a universal definition suggests divergent terminological and conceptual complexities. This points out that a number of concepts have been used to examine smart grid analytics encompassing complex systems and networks. The smart Grid Energy resource Centre in the United States considered the power grid within the context of optimal control, ecology, human cognition, glassy dynamics, information theory, microphysics of clouds, and many others. Mulgan(1991)had suggested the relevance of control of networks within the new economies of communication. Geoffrey Werner-Allen et al.;(2006, p.1)propose the deploying of a wireless sensor network on an active volcano .They suggest that sensor-network application for volcanic data collection relies on triggered event detection and reliable data retrieval to meet bandwidth and data-quality demands.

Quite apart from specific and proven smart grid technologies in use as explicated, Smart grid remains an evolving field of inquiry involving a set of related technologies which points out the need for conceptual clarity. In a number of studies, smart grid represents an advance on electricity as it aims at energy consumption in relation to a variety of uses in the society. A number of research had recently been written in the field of smart grid, providing its salient remit, reviewing key trends in its content and possible future research prospects (Lichtenberg,2010; Hamilton et al., 2010; Galli et al., 2011). The concept of power line communication, which examines the use of electricity infrastructure for data transmission, is equally gaining scholarly relevance in smart grid analytics (Berger et al.;2013).

Since the 1990s and early 2000s, there have been proliferation of grid programmes in most technologically advanced societies of the world. This includes the Telegestore system developed in Italy in 2005, the Austin, Texas mesh network which dates around 2003, and the Boulder, Colorado smart grid in which was set up in 2008. Other grid programmes in the United states include *IntelliGrid* – a brainchild of the Electric Power Research Institute (EPRI), *Grid 2030* developed by the U.S. electric utility industry and the *Modern Grid Initiative (MGI)* among others. The Pacific Northwest National Laboratory (2007) report shows that wide area networks were revolutionized following the Bonneville Power Administration (BPA) which broadened its smart grid research with prototype sensors simulated for rapid analysis of anomalies in electricity quality over very large geographic areas. This resulted in the foremost operational Wide Area Measurement System (WAMS) in 2000. A number of countries gradually assimilated the model. For instance, China reopened a detailed national WAMS system on completion of its 5-year economic plan in 2012 (Yang,2001).

Scant conceptual exploration on the smart grid model underpins theoretical and conceptual impasse which gives rise to various analytical difficulties in arriving at the extant terminological underpinnings of the concept. For example, institutional exploration of smart grid underscores the corporate dynamics of consumption (Fine

& Leopold,1993), while micro grid examines individual or household network consumption patterns (Brohmann, et al;2013). Gender dynamics of grid consumption have also been given some conceptual attention (Thompson,1996). There are debates on grid integration for renewable energy linking aspects of communication for resource efficiency (Yu,et al;2011).

The complex conceptual exploration provides the original impetus to investigate smart grid which the present chapter seeks to explore. This is on the basis that the question of the relation of smart grid to sustainable energy consumption and the smart city appears not to have been adequately conceptualized. A common thread in the conceptual issues raised on smart grid is the application of digital processing and communications to the power grid, making data flow and information management central to the smart grid. There is a widespread assumption that smart grid analytics have had a robust relevance within the sphere of technological advancement, particularly in the smart city.

A common thread in the conceptual issues raised on smart grid is the application of digital processing and communications to the power grid, making data flow and information management central to the smart grid. There is a widespread assumption that smart grid analytics have had a robust relevance within the sphere of technological advancement, particularly in the smart city.

The transition from "smart machine" (Zuboff,1988) to "smart city" has gained recent scholarly attention. This includes debates on the rise of the "network societies" (Castells,2000), "power and networks" thesis(Huges,1983). Research on networks examines the modern wireless networks which have been an important conceptual exploration of the smart grid studies. This encompasses issues of "network congestion" (Ahmad et al.; 2009). There are studies on technological innovations to salvage congestion including several proposed algorithms. Both the "global city" (Sassen,2005) and the "green or eco-city" architecture aim at sustainable environmental impact to mitigate unsustainable energy, water and food consumption, pollution and waste (Register, 1987; Lehmann,2010). Technological innovations linked to smart city have also been conceptualized as "invention of comfort"(Crowley. 2003; Shove, 2003). This suggests the ease and comfiture of technological advancement.

On its part, the term smart city has been widely used in a variety of scholarships including seminars, journal articles, conferences and in particular among social scientists, engineers, city planners and a wide range of professionals. This suggests a multidisciplinary discussion which accounts for the emergence of various technologies, such as Information communication technology(ICT), Social Networks, Web Applications and Internet Technologies, immersive technologies, virtual and augmented reality, wearable technologies, cloud computing, Data science, Big Data insights etc.

The internet of things(IoTs)have also provided useful insights on the smart city debates. According to the Digital Agenda for Europe (2014), the Smart City concept means smarter urban transport networks, upgraded water supply and waste disposal facilities, and more efficient ways to light and heat buildings. And it also encompasses a more interactive and responsive city administration, safer public spaces and meeting the needs of an ageing population. Smart City is a new paradigm for the integration of Internet Technology in the Urban context.

"The Future of Cities is Smart", and similar mantra have emerged to suggest the plausibility of the smart city paradigm including several partnerships as well as various research organizations worldwide which are currently engaged in Smart City potentials and applications. Advanced Internet Technologies currently provide smart applications for virtually every sphere of human endeavor.

Smart City internet technologies are deployed in contemporary times as the basis of scientific debate for the new architect of Urban Computing. Smart Cities technologies and applications have received growing attention in recent years from various perspectives. The thriving numbers behind their adoption and exploitation in different application contexts have captured the attention of Internet Technology specialists, computer engineering and business researchers that, in the past years, have been trying to decipher the phenomenon of urban computing, its relation to sustainable energy consumption and its implications for new research opportunities that effect innovations in modern living, way of conducting business and developing economy for sustainability.

There are concerns in the smart city schema related to dichotomy arising from globalization. Thomas Barnett (2004) had argued on the penetration of globalization and disconnection of the outside world. David Ronfeldt, links this to "cyberocracy" a concept he deployed to explain the new state as a proliferation of transnational network rooted in a primarily cybernetic vision of the government that rules by the use of information and of society. Ronfeldt (1992) connects this to the concept of the "nexus-state," to explain a new network state entity enabled primarily by technology and networks. There are works which justify technological advancement as a 'modernization paraphernalia' and panacea to ecological challenges popularly known as the ecological modernization school (Spaargaren,1997). The ecological modernization school has been critiqued by the ecological Marxian scholars as less inclusive and rather exploitative. They contend that relying on technology alone might fail to deliver inclusive development (Harvey,2005; Schor,2005).

Despite its robust offshoot there are conceptual reflections on sustainable energy consumption (Barr et al., 2005). This becomes perceptive as the mainstream notion of smart grid suggests a "smart and greener model'(Mazza,2004). This orientation reinforces the conceptual revaluation of sustainable energy consumption.

Energy consumption arguably since the post-industrial revolution era has been overly influenced by the logic of capitalist pattern of consumption (Harvey,2005). It is not enough to dismiss deleterious effects of unsustainable energy consumption in an attempt to forecast the future of smart cities in an increasingly capitalist social formation. Thus, unsustainable energy consumption has serious consequences for smart city. This is discernible from the sustainable development perspective. Conceptualized in the Brundtland Commission Report (1987)as development that meets the needs of the present generation without compromising the ability of the future generation from meeting theirs, Sustainable development has three key development threads namely social, economic and environmental. In this regard, the chapter conceptualizes aspects of the environmental component linked to sustainable energy consumption.

In the last thirty years or so, there has been a proliferation of studies on various aspects of sustainable development. However scant studies have linked sustainability to smart grid analytics and the smart city specifically in the context of energy consumption. This has resulted in disciplinary attention to sustainable energy consumption linked to the smart grid analytics. Sustainable consumption has become central in virtually every field of development inquiry, particularly the dynamics of capitalist energy resource appropriation or extraction and its effects on the wider society. This puts the smart grid analytics at the centre of the sustainability debate resulting in a number of new useful research agenda, which among others point out the need for sustainable energy consumption.

Conceptual advances in this direction have provided seminal arguments. For instance, Schor (2005) offers a useful insight on the effects of disproportionate consumption of the advanced societies notably the United States and the UK, while Daniel Milner (2012) suggests the modest consumption patterns of Norway. Such comparative analysis has been insightful in current debates on global unsustainable energy consumption. This presents a persuasive strand of the implications of unsustainable consumption. On this basis, sustainable energy consumption as argued aims at a conceptual model of smart grid that offers an alternative to the largely Eurocentric paradigm in smart grid analytics.

More inclusive studies on smart grid analytics are needed to address the "global digital divide" which not only affects smart grid and smart city policy scholarship rather undermines the future of sustainable and equitable energy consumption. Thus, the arguments for further studies will build on prior conceptualization of smart grid and smart city.

The sustainability theoretical debates create linkages between the overall energy consumption implications to both to human and non-human species and the ecosystem. Responsible consuming has been an important question in response to sustainability signals (Warde, 1999; Davidson & Hatt, 2005; Dobbyn & Thomas, 2005). This accounts for the changing patterns of energy consumption often ignored in policy and research agenda.

There has been evidence of disparity on the smart grid scholarship largely propagating the needs of the developed societies. For instance, Gontar et al.; (2013) argued that smart city concept is an 'attempt to answer the following problems: urbanization, aging of social infrastructure in developed countries, cutting CO^2 emissions'.

The nexus between smart grid and sustainable consumption model tends to require more inclusive approach encompassing the urbanization dynamics of both the developed and developing societies. Thus, concepts like "smart grid" and "smart city" form part of resurgent debates on urbanization which are part of everyday lives of humanity. Yet, they are also contentious, as several political, social and technological interactions are often made on the basis of specific interests or lifestyles which might be at variance with the tenets of sustainability.

Smart Grid and Smart City: Analysis of Sustainable Energy Consumption

The ecosystem is in a state of flux as it is impacted by a number of factors notably rapid urbanization, population growth, technological advancement and global connectivity. These external factors will have significant impact on smart grid analytics in the future. These suggest that the question of smart and sustainable city has never been adequately resolved.

Analysis should advance in a way which provides analytical and conceptual capacity to explore the interrelatedness and overlapping spheres of practices and processes within both concepts. Thus, while smart grid analytics is fundamental to a smart city, it should be conceptualized within sustainability strand, distinct from the capitalist resource exploitation. Shove (2003) provides such interface between "convergence and conventions of comfort", in the context of "cleanliness and convenience" regarding analysis of a "green consumption".

Another concern is the cost of institutionalizing a smart grid particularly among the poor societies. Recently, the World Economic Forum (2016) reported that a transformational investment of more than $7.6 trillion is needed over the next 25 years (or $300 billion per year) to modernize, expand, and decentralize the electricity infrastructure with technical innovation as key to the transformation.

Although we now live in an almost digital world, Mulgan, (1991)argues 'that the late twentieth century world is covered by a lattice of networks'. Castells (2000) described this as the 'network society' which will develop 'through synergy among relevant theorizing, computational literacy, and sociological imagination dominated by technology' and internet hypertext. Several technological systems still require inclusive and sustainable components.

As recent trends suggest, the rise in smart cities in the 21st century including smart grids for signaling and conditioning energy consumption including meters, sensors and actuators for gauging, point out the increasing need to explore the sustainability linkages. Current trends in global energy use are far from sustainable. Oil demand continues to grow, while experts expect a historic peak in oil production within the next 20 years. Carbon dioxide (CO2) emissions from fossil fuel combustion in 2002 were about 13% above the 1990 levels, whereas a stabilization of the climate would demand a reduction by 50%until 2050 and further reductions thereafter (UNEP, Centre on Sustainable Consumption and Production (CSCP),2005).

There is significant variation on energy consumption at individual levels across countries and regions of the world which has implication for sustainable consumption. Evidence of unsustainable energy consumption includes the rise in pollution, flaring of gas, ecological disruptions etc. For instance, while OECD energy consumption would need to decrease, there is evidence of limited access to energy supply in the low-income societies. The world's richest people, earning over 20,000 US dollars per annum, consume nearly 25 times as much energy per person as the poorest people (UNEP, CSCP, 2005). On the other hand, currently, nearly one third of the world population has no access to electricity and another third has only poor access (UNEP, CSCP, 2005). Reliance on traditional fuels for cooking and heating can have a serious impact on health and the environment. This has different implications and impacts in terms of ecological breakdown, supply voltage, power rating and usage. It is evident that there are different needs for energy consumption serving various social, economic and technological needs. This may yield different results positively or negatively as far as efficient and optimal use of energy is concerned.

Sustainable energy consumption suggests policy relevant dynamics aimed at eco efficiency. This reinforces the notion that sustainable energy consumption is aimed at meeting the energy needs of the present generation without tainting energy use for the future city users. It integrates energy into a single, smart-power analytics to meet energy consumption needs of the city. It equally allows multifarious -integration of energy consumption as an inclusive logic as well as a modality for efficient use. High voltage and complex systems could be made compatible through modalities of efficient and sustainable consumption. This includes stepping the grid down to the poor which forms the thesis of a research agenda that argues that both city and grid are smart to the extent that they are inclusive (gender, class, race), affordable (pro poor), clean and green (pro nature), value driven, renewable, replicable, cost effective (efficient).These are constituents of sustainable grid /sustainable city.

A grid is 'smart' to the extent that it is compatible, inclusive, and responsive and conforms to the tenets of efficient energy consumption. It links equity with energy consumption. This underpins the fact that there have been a number of challenges associated with the task of providing improved quality of energy services in modern cities and urban centres. This largely includes the transformation of the global energy system, creating more viable service experiences.

In particular, this will guide the future of smart grid analytics as it will show how sustainable energy consumption could be deployed in a wide range of application which justifies the classification of a city as smart or otherwise. This encompasses the broad fields of transportation, automotive systems, manufacturing/production, consumer electronics, communications etc.

Irrespective of its robust analytics, the dearth of studies exploring the revolutionary role of the Smart grid within sustainability contexts is less explored. This is a key requirement for the advancement of a future research direction for the smart city. Analyses of technologies for the integration and interconnection of citizens, businesses, government and the natural environment have been superficial. This does not match the actual importance of the proliferation of the smart city scholarship. The specific changes related to unsustainable energy consumption, will shape the future and nature of smart grid analytics over the coming decades. Thus, the changes(high)energy consumption largely industrial and commercial among the high-income societies will taint the sustainability dynamics of the smart city.

Against the background of these trends, a number of alternative and critical perspectives had emerged to interrogate the sustainability assumptions of the smart grid analytics and the smart city (Alkon & Agyeman, 2011). Sustainable smart city transcends deploying technologies and divergent applications, rather involves a more direct question that boarders on the preservation or conservation of energy use to the broader elucidation of the persistent challenges of transition from exclusive to inclusive micro-grids technology access, which identifies individual energy needs with specialized and equitable distribution model among end users.

Bridging rural/urban dichotomy, which creates "grid divide" accounts for a more robust grid. The re-engagement with this debate offers alternative inclusive model particularly among the poor societies of the Third World. Thus, as the developing societies are rapidly urbanizing, sustainable grid becomes a critical component of contemporary urbanization. This largely includes protecting the ecosystem from hazardous effects of energy consumption pointing out the need for grid accountability, reliability, responsible and transparent grid which ensures stability.

Scant research inquiry had examined sustainable smart city including aspects of eco efficiency, city ecological footprints, eco labeling and triple bottom line. Smart city research for sustainable innovation should be deepened in contemporary sustainable development studies as important bloc in contemporary urbanization. Eberle et al.., (2004) identified sustainable consumption as a more ecological but also socially … way of buying and using goods and services.

The challenges of disproportionate consumption among the high, middle and low-income countries create disparity regarding access and equitable energy consumption. For instance, there are concerns on billing systems, smart meters and sustainability challenges. This includes the question regarding accountability, precision, remote control, censorship, variable rate pricing. Thus, grid preference could be given to some sections of the society, country or class of individuals than others. A socially inclusive grid accounts for sustainability.

Control mechanisms are central concern to the future of smart grid for a smart city. Who controls the grid? And what yardsticks make for a viable control mechanism? These include indices that account for meter reading dynamics.

The extent to which smart grid guarantees corporate and individual privacy is less clear. This suggests the overriding influence the grid providers might wield which could be at variance with sustainable production of energy and implicitly underscores supplier and end user disparity.

Equally, complex and divergent security challenges are linked to sustainable grid consumption. This points out the need for modalities to protect and mitigate threats arising from grid related hazards.

SOLUTIONS AND RECOMMENDATIONS

The proliferation of smart grid scholarship and absence of policy and research consensus on the future of smart city and smart grid has been a central concern of the sustainable grid scholarship. This has led researchers to advance knowledge on future trajectories of smart city within the sustainability paradigm.

Smart grid analytics should be designed within natural resource friendly distribution linked to complex but mutually related networks based on cost effective and eco- friendly processing power. Pro poor grid architecture should be an integral part of sustainable grid.

This includes affordable grid, reliable grid-monitoring dynamics such as meters, grid maintenance and management mechanisms. All point out the saliency of smart grid and need for a policy shift in conceptualization from the stand point of grid model to sustainable smart grid. This increasingly results in new modes of inquiry including debates on smart micro-grid architecture, smart city authentication and authorization model, metering and billing, power surge and economic power distribution models, smart city priority scheduling, grid surge and control models etc.

Cognizant of the central role cities play in the overall development of an economy, being the hub of social, technological and economic transactions, including innovation, hospitality, connectivity and creativity, as well as various types of service delivery centers, sustainable city development models becomes a priority area within the smart city studies, global energy efficiency action plan is inevitable. Proponents argue that energy efficiency is building bloc to sustainable future (Blok,2005; Young, 2008).

Technological advancement suggests that several external factors will have significant impact on smart grid analytics in the future. This will have effects on the sustainability component of smart grid which inevitably will impact cities. With the interconnectedness of the city and the economy, government, the people and social interactions, and the ecosystem- a complex interplay of challenges are imminent as increasing changes in consumption pattern and lifestyles of people will affect the level of energy consumption, and point to the need for improved smart grid models to contain rising demands on energy.

'An increase in green energy producers and consumers may lead to more agencies with potentials to provide smart grid analytics'. To achieve these objectives, there are a number of salient initiatives: regulatory dynamics on energy consumption, the institutionalization of green and smart energy, efficient use of resources, and social inclusiveness among others.

Given the importance of Smart Cities, it is mandatory to define and standardize indicators to measure the evolution of cities to the challenge of achieving higher competitive rates within the sustainable cities. This helps to critically explore, identify and align certain indicators as measurable indices to determine how "sustainable" a "smart city" is. The aim is to transcend rhetoric and move into realities and specifics. For instance, are there indicators measuring corruption or terrorism in a smart city? What are the evaluation models for determining the smart nature and variables of an alleged smart city? How has ecological challenge of deleterious resource extraction including capitalist exploitation impacted the natural environment? How has climate issues such as vulnerability of climate change and uncertainties of global warming and natural disasters notably tsunamis, cyclones, hurricanes, earth quakes etc been tailored for possible mitigation and mainstreaming into the smart city model? These and similar concerns are pertinent in smart city and sustainability studies.

Existing studies on the adoption of energy-efficient measures both industrially and for household purposes are examined from a multi-disciplinary conceptual lens. This includes geographers, sociologists, regional and town planning studies, consumer behavior studies as well as market related dynamics. Research and policy discourse on smart electricity metering should be enlarged to check consumption levels and preferences of various end users.

It comprises institutions, practices and forms of tariff, yet all of these practices are less revolutionary in charting a sustainable energy consumption as they may not ameliorate emissions and unequal access. Kannberg et al.;(2003) provide a number of benefits to smart grid as a modernized energy system. This among several others includes reduction in power consumption particularly at the consumer side conceptualized within the *demand side management.* Such amelioration of consumption facilitates including load balancing, mitigates power failures. It is argued that increased efficiency and reliability of the smart grid will save consumers some money and aid in reduction of CO_2 emissions.

We suggest that sustainable energy points out that energy-saving measures should include amelioration of waste, low-cost or no-cost measures to check both capital investment and individual consumption patterns. There is consensus among experts that the implementation of more sustainable consumption behavior requires not only awareness among consumers, but also changed social and economic structures: Consumption is a "socially constructed historically changing process" (Bocock, 1993: 45). Both UNEP (2005) and Centre on Sustainable Consumption and Production (CSCP)(2005) suggest the following measures to sustainable energy consumption;

1. Demand-side energy efficiency (also termed energy end-use efficiency): This most important option relates to technical, organizational and individual measures to reduce the final energy needed to heat/cool our houses, produce goods etc.
2. Co-/tri-generation: Introduction of on-site co- or tri-generation of heat, cold, and power can dramatically improve energy efficiency on the supply side. This option is largely related to the issues of energy generation and distribution. It is, therefore, only further discussed here in relation to its applications at consumers' sites (e.g. in industry or public buildings).
3. Renewable energy: The third option is renewable energy produced and used onsite through biomass or solar thermal collectors etc. as well as that fed into the electricity grids.
4. Limiting energy services: The final option could be to limit the amount of energy services we use (e.g. by capping dwelling floor space) to a level sufficient to cover our energy-related needs.

Similarly, adopting equitable energy access policies have been suggested. IEA (2009b) posits that, "sound statistical data…and a clear description of the (energy services) situation" are the first of the preconditions for successful rural energy access policies.

Again, the understanding of the smart grid analytics within a smart city should not be treated in isolation of the prevailing global asymmetry in which a colossal technological gap exists between societies of the affluent North and the poor South. Thus, how the smart city is framed to meet the demands and needs of the increasingly urbanizing poor countries of Asia, Latin America and Africa is less lucid. These are some of the salient issues of future development concern.

FUTURE RESEARCH DIRECTIONS

The future research agenda is confronted with a number of trends. A human centered grid should drive much of the smart grid analytics. Johnson (2009) had proposed a "human centered information integration for the smart grid".This has not been given adequate attention.

Samir Succar and Ralph Cavanagh (2012) have examined "the promise of smart grid" and came to a conclusion that the smart grid can give us cleaner air, better health, lower electricity bills, and reduced carbon dioxide (CO2) emissions in the atmosphere. Beyond the alleged energy efficiency, future research agenda will be able to examine more critical and in-depth evaluation of these claims. Particularly in the context of "cost effective smart grid". The point has been that the grid analytics have been quite expensive.

Next-generation transmission and distribution infrastructure will be better able to handle possible directions of the valuation of more sustainable producers and consumers. Smart Grid which is easily the next big technological revolution since resurgence of internet has a critical role to play in the future cities.

From the foregoing, the future policy direction of a smart city one which is built on sustainability is imperative. There is need to review and determine a new vision on smart grid analytics and the smart city. The move toward smart city platforms will require sophisticated models and approaches for a sustainable and secure future. There is need for new strategies to maximally leverage on the new technologies in the daily life.

The likely trajectories of sustainable city proposed in this chapter is inclusive and participatory within all facets of city development including communication, transportation, urban spaces etc which are linked to the understanding of the linkages between smart grid analytics for a smart city. The point the chapter makes at intervals is that future research agenda should focus on the process and dynamics of achieving sustainable consensus on the future vision for technology development that could direct access and alignment with divergent environments within the idea of smart city in the neo liberal order.

Ideas linking smart city to neo-liberal resource consumption have been scant. This suggests the saliency of a research agenda that should keep track of resource use and resource conservation. Driven largely by the notion of capitalist exploitation there is increasing need to always interrogate the place of environmental sustainability in the smart city research. This includes issues such as green city, eco efficiency, urban renewal, triple bottom line, eco labeling, ecological footprint and smart city accounting (Amadi & Imohita, 2017). With the notion of the smart grid as a fundamental attribute of a third industrial revolution and neo-industrialization, future research should keep track of the feasibility of such attributes within the core tenets of sustainability blocs namely social, economic and environmental in a distinct manner to explore patterns of equitable and resourceful energy consumption. The basis of this trend is to deploy control mechanism for mitigation of deleterious energy use. Thus, a new research direction that emphasizes process mining, sustainability analysis, inclusive, pro nature, pro poor and pro energy consumption particularly among the poor societies of the global South will provide relevant future trajectories for all stakeholders.

In particular, future research direction should chart a collective cause aimed to understand and deploy sustainability strategies including the strategic role of smart grid analytics at different levels of the smart city, involving different stakeholders both domestic, industrial and end users.

CONCLUSION

The point this chapter has been making and continues to make is that increase in green energy production and consumption may lead to more agencies with capacity to provide future smart grid analytics. However, the concomitant challenge of 'efficient energy consumption' remains inevitable within a capitalist system riddled with inequality. This underscores the saliency of sustainable energy consumption policy response.

This chapter at intervals communicates and disseminates the urgency of sustainable energy consumption policy response. It insists that research agenda in smart grid analytics and sustainable smart city is essential. The aim is for efficiency in energy consumption and to preserve the smart city, promote- smart grid analytics and the user experience with the provision of sustainable services for all stakeholders including businesses, government and end users. There is ample projections on the state-of-the art potentials of Smart City Systems and applications and related innovative technologies. However, the sustainability model driving the smart grid process remains less lucid. This has been contestable worldwide and to a large extent shows how the future of smart grid could be analyzed and in particular deepened within the sustainability frontiers in the advanced societies and how the poor societies could be included in the smart grid analytics.

The chapter partly initiates a dialogue between the smart grid analytics and smart city on the basis that the smart city agenda has not been adequately and sufficiently explored within the broader sustainability context. Furthermore, this sustainability orientation provides a rationale for advancing an inclusive, renewable, and pro poor analytics. It is suggested and desirable to define a smart city within sustainability contexts constituted by resource efficiency. This will help address the question of the relation of capitalist resource consumption, the natural environment and the poor. This model as argued moves smart city beyond the largely technological advancement frameworks that prioritizes capitalist resource exploitation, to an understanding of the natural resource as a set of inclusive and collective sphere for the future of the natural ecology, human and non-human species.

REFERENCES

Ahmad, S., Liu, M., & Wu, Y. (2009). *Congestion games with resource reuse and applications in spectrum sharing*. CoRR, abs/0910.4214

Alkon, A., & Agyeman, J. (2011). *Cultivating food justice: race, class, and sustainability*. Cambridge, MA: The MIT Press.

Amadi, L., & Igwe, P. (2016). Maximizing the Eco Tourism Potentials of the Wetland Regions through Sustainable Environmental Consumption: A Case of the Niger Delta, Nigeria. *The Journal of Social Sciences Research*, *2*(1), 13–22.

Amadi, L., & Imoh-Ita, I. (2017). Intellectual capital and environmental sustainability measurement nexus: a review of the literature. *Int. J. Learning and Intellectual Capital, 14*(2), 154-176.

Barnett, T. (2004). *The Pentagon's New Map*. Putnam Publishing Group.

Barr, S., Gilg, A., & Ford, N. (2005). The household energy gap: Examining the divide between habitual- and purchase related conservation behaviors. *Energy Policy*, *33*(11), 1425–1444. doi:10.1016/j.enpol.2003.12.016

Berger, L., Schwager, A., & Escudero-Garzás, J. (2013). Power Line Communications for Smart Grid Applications. *Journal of Electrical and Computer Engineering*. doi:10.1155/2013/712376

Biersack, A. (2006). Reimagining political ecology: culture/power/history/nature. In A. Biersack & J. B. Greenberg (Eds.), *Reimagining political ecology* (pp. 3–40). Durham, NC: Duke University Press. doi:10.1215/9780822388142-001

Blok, K. (2005). Improving Energy Efficiency by Five Percent and More per Year? *Journal of Industrial Ecology*, *8*(4), 87–99. doi:10.1162/1088198043630478

Bocock, R. (1993). *Consumption*. London: Routledge. doi:10.4324/9780203313114

Bryant, R. (1998). Power, knowledge and political ecology in the third world: A review. *Progress in Physical Geography*, *22*(1), 79–94. doi:10.1177/030913339802200104

Castells, M. (2000). *Toward a Sociology of the Network Society*. American Sociological Association.

Crowley, J. (2003). *The invention of comfort*. Baltimore, MD: Johns Hopkins University Press.

Davidson, D., & Hatt, K. (2005). *Consuming Sustainability Critical Social Analysis of Ecological Change*. Fernwood Publishing.

Digital Agenda for Europe. (2014). *The EU explained: Digital agenda for Europe European Commission European Commission Directorate-General for Communication Citizens information 1049*. Publications Office of the European Union

Dobbyn, J., & Thomas, G. (2005). *Seeing the light: the impact of micro-generation on our use of energy*. Sustainable Consumption Roundtable, London, UK.

Eberle, U., Brohmann, B., & Graulich, K. (2004). Sustainable consumption needs visions. Position Paper. Institute of Applied Ecology, Öko-Institut, Freiburg/Darmstadt.

European Commission. (2005). Doing More with Less: Green Paper on energy efficiency. Brussels: Author.

European Commission. (2010b). *Communication From the Commission to the European Parliament, The European Council, The Council*. Brussels: The European Central Bank, The Economicand Social Committee and the Committee of The Regions.

European Environment Agency. (2005). The European environment — State and outlook 2005. Copenhagen: Author.

European Environment Agency (EEA). (2005). *Household consumption and the environment*. EEA Report No. 11/2005. Copenhagen: Author.

Fine, B., & Leopold, E. (1993). *The world of consumption*. London: Routledge.

Galli, S., Scaglione, A., & Wang, Z. (2011). For the grid and through the grid: The role of power line communications in the smart grid. *Proceedings of the IEEE, 99*(6), 998–1027. doi:10.1109/JPROC.2011.2109670

Gontar, B., Gontar, Z., & Pamuła, A. (2013). Deployment of Smart City Concept in Poland. Selected Aspects *SIsteminiai Tyrimai, 67*(3), 39-51. doi:10.7220/MOSR.1392.1142.2013.67.3

Hamilton, B., Miller, J., & Renz, B. (2010). *Understanding the benefits of smart grid*. Tech. Rep. DOE/NETL-2010/1413. U.S. Department of Energy.

Harvey, D. (2005). *Brief History of Neoliberalism*. New York: Oxford University Press.

Hughes, T. (1983). *Networks of power: Electrification in Western society*. Baltimore, MD: Johns Hopkins University Press.

IEA. (2009b). *Comparative Study on Rural Electrification Policies in Emerging Countries*. Paris.Available at http://www.iea.org/papers/2010/rural_elect.Accessed 18/7/2017

Johnson, P. (2009). *Human centered information integration for the smart grid*. Technical Report CSDL-09-15, University of Hawaii, Honolulu, HI.

Kannberg, L., Kintner-Meyer, C., Chassin, D., Pratt, R., DeSteese, J., Schienbein, L., . . . Warwick, W. (2003). *Grid Wise: The Benefits of a Transformed Energy System*. Pacific Northwest National Laboratory under contract with the United States Department of Energy: 25. arXiv:nlin/0409035.

Lehmann, S. (2010). Green Urbanism. *Formulating a Series of Holistic Principles Surveys and Perspectives Integrating Environment and Society, 3*(2), 1–10.

Lichtenberg, S. (2010). *Smart Grid Data: Must There Be Conflict Between Energy Management and Consumer Privacy?* National Regulatory Research Institute.

Mazza, P. (2004). The Smart Energy Network: Electricity's Third Great Revolution. *Climate Solutions, 1*(2), 1–7.

Milner, D. (2012). *Consumption and Its Consequences*. Cambridge Polity Press.

Mulgan, G. (1991). *Communication and Control*. Cambridge, UK: Polity Press.

Pacific Northwest National Laboratory. (2007). *Gridwise History: How did GridWise start?* Available at http:/gridwise.pnl.gov/foundations/history.stmAccessed 20/9/2017

Register, R. (1987). *Ecocity Berkeley: Building Cities for a Healthy Future*. North Atlantic Books.

Rennings, K., Brohmann, B., Nentwich, J., Schleich, J., Traber, T., & Wüstenhagen, R. (Eds.). (2013). *Sustainable Energy Consumption in Residential Buildings*. Springer. doi:10.1007/978-3-7908-2849-8

Ronfeldt, D. (1992). Cyberocracy is coming. *The Information Society*, *8*(4), 243–296. doi:10.1080/01972243.1992.9960123

Saleh, M., Althaibani, A., Esa, Y., Mhandi, Y., & Mohamed, A. (2015). Impact of clustering microgrids on their stability and resilience during blackouts. *2015 International Conference on Smart Grid and Clean Energy Technologies (ICSGCE)*, 195–200. 10.1109/ICSGCE.2015.7454295

Sassen,S .(2005). The Global City: introducing a Concept. *The Brown Journal of World Affairs*, *11*(2), 27-43.

Schor, J. (2001). *Why Do We Consume So Much?* Clemens Lecture Series 13 Saint John's University.

Schor, J. (2005). Prices and quantities: Unsustainable consumption and the global economy. *Ecological Economics*, *55*(3), 309–320. doi:10.1016/j.ecolecon.2005.07.030

Shove, E. (2003). Converging conventions of comfort, cleanliness and convenience. *Journal of Consumer Policy*, *26*(4), 395–418. doi:10.1023/A:1026362829781

Spaargaren, G. (1997). *The ecological modernisation of production and consumption: in environmental. sociology*. Wageningen, The Netherlands: Landbouw University Wageningen.

Speake, S. (2005). Water:A Human Right. In *Consuming Sustainability Critical Social Analysis of Ecological Change*. Fernwood Publishing.

Succar, S., & Cavanagh, R. (2012). *The Promise of the Smart Grid: Goals, Policies, and Measurement Must Support Sustainability Benefits*. NRDC Issue Brief.

Thompson, C. (1996). Caring consumers: Gendered consumption meanings and the juggling lifestyle. *The Journal of Consumer Research*, *22*(4), 388–407. doi:10.1086/209457

UNEP/Wuppertal Institute Collaborating Centre on Sustainable Consumption and Production (CSCP). (2005) *Sustainable Energy Consumption*. A Background Paper prepared for the European Conference under the Marrakech Process on Sustainable Consumption and Production (SCP), Berlin, Germany.

U.S. Department of Energy. (2003). *"Grid 2030" A National Vision for Electricity's Second 100 Years*. Office of Electric Transmission and Distribution.

U.S. Department of Energy. (2004). *National Electric Delivery Technologies Roadmap*. Office of Electric Transmission and Distribution.

US Energy Independence and Security Act 2007 (EISA-2007)

Warde, A. (1999). Convenient food: Space and timing. *British Food Journal*, *101*(7), 518–527. doi:10.1108/00070709910279018

WCED (World Commission on Environment and Development). (1987). Our Common Future. Oxford, UK: WCED.

Werner-Allen, G., Lorincz, K., & Marcillo, M. (2006). Deploying a Wireless Sensor Network on an Active Volcano. *IEEE Internet Computing*, 1-25.

World Economic Forum. (2016). *Digital Transformation: Digital trends in the automotive industry*. Available at http://reports.weforum.org/digital-transformation/digital-trends-in the automotive-industry.Accessed 9/21/2017

Yang, Q. (2001). *Electrical engineering and its automation*. Beijing Sifang Automation Co. Ltd.

Young, D. (2008). When do energy-efficient appliances generate energy savings? Some evidence from Canada. *Energy Policy*, *36*(1), 34–46. doi:10.1016/j.enpol.2007.09.011

Yu, F., Zhang, F., Xiao, W., & Choudhury, P. (2011). Communication Systems for Grid Integration of Renewable Energy Resources. *IEEE Network*, *25*(5), 22–29. doi:10.1109/MNET.2011.6033032

Zhao, J., Huang, W., Fang, Z., Chen, F., Li, K., & Deng, Y. (Eds.). (2007). *Proceedings of Power Engineering Society General Meeting*. doi:10.1109/PES.2007.385975

Zuboff, S. (1988). *In the Age of smart machines*. New York: Basic Books.

KEY TERMS AND DEFINITIONS

EISA: Energy Independence and Security Act.

ISDN: Integrated services digital network.

ISO: International Organization for Standardization.

LSI and VLSI: Large scale integration and very large scale integration. The integration of large numbers of computers on a single microchip.

MAP: Manufacturing automation protocols.

ONA: Open network architecture.

ONP: Open network provision, set of proposals devised in 1987 by European Commission to allow for open access to the communication infrastructure.

SNA: Systems network architecture.

This research was previously published in Smart Grid Analytics for Sustainability and Urbanization edited by Zbigniew H. Gontar, pages 138-161, copyright year 2018 by Engineering Science Reference (an imprint of IGI Global).

Chapter 8
Sustainability Evaluation of Green Urban Logistics Systems:
Literature Overview and Proposed Framework

Jesus Gonzalez-Feliu
Ecole des Mines de Saint-Etienne, France

ABSTRACT

This paper aims to propose a systemic vision of literature on sustainable urban logistics assessment and evaluation. Although non-extensive, this overview pretends lack of unification in the subject of assessing and evaluating the impacts of green urban logistics systems, and, through the proposal of a general assessment and evaluation framework, the steps done and being in course towards standards on this field. First, an overview of the research in urban logistics is provided, after what the main visions of sustainable development and their derived issues for urban logistics assessment and evaluation are presented. Then, a framework to assess and evaluate green urban logistics systems via scenario comparison is proposed. This framework aims to propose a methodological framework to use and combine existing methods to assess scenarios, and not a "black-box" model of software ready to use. This is done to make synergies between existing methods, and to show that, although they remain at an initial stage, steps on the way of defining standards are made. After that, the main applicability and application issues of the proposed methodological framework are addressed, showing those principles of standard from the literature. Finally, and to conclude, future developments on urban logistics research are proposed.

DOI: 10.4018/978-1-7998-0441-3.ch008

Copyright © 2020, IGI Global. Copying or distributing in print or electronic forms without written permission of IGI Global is prohibited.

INTRODUCTION

In the last years, the importance of urban goods transport has changed the way city planners and transport and logistics practitioners organize their activities and define their priorities for action. Indeed, in various European countries, urban goods transport needs to be taken into account in transport plans for medium and big cities (CERTU-ADEME, 1998; Spinedi, 2008; Macharis and Melo, 2011; Arndt et al., 2013) and the debates on urban logistics are numerous. Indeed, in the last five years, various conferences on urban logistics (or having it as one of the main topics) have been held in different countries (Gonzalez-Feliu, 2016a). Opposing to pioneer events, more related to the scientific community, since 2012 those conferences have a higher implication of practitioners, which shows the interest of the different stakeholders on the subject and the needs of improving urban goods transport (Taniguchi et al., 2001; Macharis and Melo, 2011; Arndt et al., 2013; Gonzalez-Feliu et al., 2014b).

At the same time, those needs contrast with the difficulty of well identifying and quantifying urban goods transport. Indeed, we observe in literature a plethora of methods and procedure to collect and estimate data for assessing the efficiency and the sustainability of urban goods trips. Moreover, the visions and terms used by the different actors do not always coincide, making more difficult the communication and understanding of the different stakeholders.

However, and although several authors highlighted in a recent past a lack of unification in urban logistics planning and management (Macharis and Melo, 2011; Gonzalez-Feliu et al., 2014b), unified frameworks are starting to be observed, and the road towards a standard is maybe started. As a contribution to that path, this chapter aims to provide an overview, as well as a guide to scenario assessment in urban goods transport, mainly for sustainability analyses. The paper is organized as follows. First, an overview of the different times and developments in urban logistics is provided, as well as the main definitions needed for our overview and analysis purposes. Then, the main visions of sustainable development, their derived issues for urban logistics assessment and evaluation and the main indicators used in those assessments and evaluations are presented. After that, a methodological framework to assess and evaluate green urban logistics systems via scenario comparison is proposed. It is aimed at using and combining existing methods to assess scenarios, to make synergies between existing methods, and to show the potential of defining unified paths from existing works by similarity. After that, the main applicability and application issues of the proposed methodological framework are addressed by relating it to existing works and showing those principles of standard from the literature. Finally, and to conclude, future developments on urban logistics research are proposed.

An Overview on Urban Logistics: From Jules Caesar to City Logistics 4.0

The planning and organization of goods transport in urban areas is not new. Indeed, the first written document that deals with the regulation of the transport of goods in a city is attributed to Jules Caesar, in the 1st B.C. (Quak, 2008): the *Legge Iuliana Municipalis*, who regulated urban deliveries by setting up night transit slots. Far from stopping here, the planning and organization of Rome city's food supply for the Empire, which since the reign of Augustus, named a figure, the prefect to the Annona, in charge of Rome city food supply, and dedicated to this purpose a set of logistical areas called Horrea, (Petit, 1974; Virlouvet, 2000, 2006). In the middle ages, guilds of merchants and craftsmen, as well as feudal lords ensured those functions of logistics city and supply (Boone *et al.*, 1990; Britnell, 1995; Kidd and Stumm, 2005). It was only with the industrial revolution and the consequent development of cities that the supply of these urban centers has changed drastically (Williamson, 2002), following the market economy developments and a more liberal and less centralized organization. The organization of transport in general was a private affair, and that period was characterized by small public intervention in terms of urban transport and logistics, for both people and goods.

This trend is highly changed in the 20th century, mainly in the 50's-60's with an important public investment in urban accesses and transport infrastructures, and a public intervention mainly in terms of personal transport, but not in goods transport. Cities are (even nowadays) planned and organized by giving priority to the mobility of people. Nevertheless, economic activities are crucial for the life of the cities, and the increase in the urban population indirectly increases the flow of goods for the supply of the cities. That is why, in the 1970s, with the first problems of congestion and overload road infrastructure and parking in some cities, many scientists and public authorities started wondering about the quantification and qualification of the transport of goods in urban areas (Demetsky, 1974; Watson, 1975). Those works remain mainly related to the context of the cities of the United States.

In Europe, the main actions in terms of urban logistics proposed between 1980 and 1990 were promoted and worn by the public authorities to fight congestion, but were related to punctual interventions or to proposals for regulation to deal with the specific emergencies, in a vision of urban transport where the priority remains the transport of people and not that of freight (Gonzalez-Feliu, 2008). It was only after the 1990s the main concepts of urban logistics are developed, with the French program "Goods in the City" in 1993 (Gonzalez-Feliu, 2016a), the notion of citylogistik in Germany (Ruske, 1994; Kohler, 1997) and The COST program on urban goods (COST 321, 1998) the ECMT seminar on urban goods transport in Europe (ECMT Economic Research Center, 1999).

The first International Conference on City Logistics (Taniguchi and Thompson, 1999) is held in Cairns, Australia, in 1999 and a first definition of city logistics is generalized by Taniguchi *et al.* (2001). Moreover, several coordinated actions (on the initiative of public or private stakeholders) are increasing in Europe, mainly in Germany, in France, in Italy, in the Netherlands, in United Kingdom or in Switzerland, among others (Rosini, 2005; Gonzalez -Feliu, 2008). However, the notion of city/ urban logistics is not considered in the same way for the different disciplines. Anyway, we find nowadays two main definitions of the field related with research that seems widely adopted by most scientific and practice communities.

The first is that of Taniguchi et al. (2001) who use the term city logistics and define it as

the process for totally optimizing the logistics and transport activities by private companies in urban areas while considering the traffic environment, the traffic congestion and energy consumption within the framework of a market economy.

The second, defined by Gonzalez-Feliu et al. (2014b), refers to urban logistics and defines it as

the pluridisciplinary field that aims to understand, study and analyze the different organizations, logistics schemes, stakeholders and planning actions related to the improvement of the different goods transport systems in an urban zone and link them in a synergic way to decrease the main nuisances related to it.

The notion of city/urban logistics, and the main research priorities have evolved in time. Nowadays, with the information era and the time of big data, physical internet and flexibility, a new notion, City Logistics 4.0, is starting to be developed (Taniguchi, 2016). However, in the 10 years, the most popular actions have taken place, and we observe nowadays thousands of specific actions in this, many directly issued from research programs, not all of them remaining still operational. Those solutions are of different nature, but share different components. According to Gonzalez-Feliu (2008, 2016a), the main components of an urban logistics systems are the following:

- Infrastructures, which can be of two main natures:
    - Linear, i.e. roads (reserved lanes, new corridors, accesses), railways, waterways.
    - Nodal, mainly logistics platforms, parking areas or other urban logistics spaces.

- Organization components, that are directly related to the distribution and logistics schemes deployed and the management methods carried out to operate them.
- Regulation, which can be made via (1) restrictive policies, (2) Incentive actions or (3) tarification-based regulations.
- Technologies, which can be related to:
    - Vehicle engine (fuel, gas or electric mainly).
    - Other vehicle-based technologies (double-triple temperature vehicles, elevation accessories or speed/acceleration limitation devices, among others).
    - Information and communication technologies, such as GPS devices, eco-driving support systems, etc.
    - Intelligent Transportation Systems, or ITS, mainly related to communicating vehicles and smart freight distribution systems.
- Communication and valorization campaigns.
- Financing mechanisms, mainly subsidies, tarification for financing purposes, fault-paying schemes or Public-Private Partnerships (PPPs).

Urban logistics actions and solutions will then result from combining those different elements (a solution does not have only one component, but includes various elements of the above list). However, to assess their potential on improving quality of life and their sustainability it is important to use methods that allow comparing those solutions among them, by keeping a coherence and robustness. To do this, it is important to define first what sustainability in urban logistics is and how the issue of evaluating sustainable (or green) urban logistics systems has been made in literature.

Main Sustainability Questions in Urban Logistics

Sustainability is one of the main keywords associated to contemporaneous urban transport, either for people than for goods. However, sustainable transport is often seen from the viewpoint of town planners or public authorities, and sometimes individuals or companies acting in their name or making policy analysis on their demand. However, a city's overall urban transport system is a complex entity where various stakeholders interact. Several authors have studied those stakeholders (van Duin, 2012; Gonzalez-Feliu et al., 2013b; Boudouin et al., 2014). Those stakeholders can be of different nature and have different objectives and stakes. Without entering in the detail of those stakeholders, we aim to highlight that sustainability is not observed, measured and perceived in the same way by the different stakeholders of urban goods transport. For example, if we distinguish between public and private stakeholders[1], public stakeholders will have the following sustainability aims and stakes:

- To revitalize the economic activity of the urban areas, particularly the town centres (with retail shops and various services);
- To master urban sprawl, controlling strategic town planning;
- To diminish congestion issues in the more dense (central) urban areas;
- To diminish the impacts of nuisance due to greenhouse gas and other atmospheric pollutants emissions, and to reduce noise levels;
- To implement convenient goods distribution services and to create social cohesion.

And private ones regard sustainability as follows:

- To provide quality services (commercial and tertiary activities) to the customers;
- To reduce the economic costs related to the last mile management;
- To apply sustainable policies of the firm to urban transport practices.

In this context, even when both categories of stakeholder regarding sustainable development with the classical three spheres vision (economic, environmental and social), the precise objectives, their quantification and the measurements to assess if the objectives are met are different and sometimes specific to a stakeholder and a planning horizon (i.e. operational objectives are not the same as strategic ones, as well as the figures in a firm that will be related to them). But sustainability has nowadays different interpretations. For that reason, we present below the main visions of sustainability in urban logistics as well as our viewpoint regarding it.

Visions of Sustainable Development Related to Urban Logistics

The visions of sustainable development are not always the same, as shown in literature for other fields (where some definitions remain close to environmental-friendly strategies, others are going to the social responsibility priorities, etc.), so it is important to define also which are the visions of sustainability in urban logistics. Traditionally, sustainability is divided in three spheres, and is applied to many fields, including logistics and goods transport (Morana, 2013). In urban logistics, the three spheres can be defined as follows (Morana, 2014; Gonzalez-Feliu and Morana, 2014):

- The economic sphere includes all economics and logistics performance elements, i.e. all issues that show if the urban logistics system is economically viable and logistically efficient. In other words, an urban logistics solution or initiative needs to be economically sustainable to ensure its continuity in the timeline and the human factor is central to the management of any organization.

- The environmental sphere refers to the respect of the environment and the rational use of natural resources, which has traditionally be seen as a constraint in logistics. However, urban logistics has been defined on a mainly environmental perspective. Most works on urban logistics deal with this sphere, and most of them confound sustainable urban logistics with green urban logistics.
- The social sphere is the less studied, and the most complex one. Indeed, the respect of society and the search of a more ethic world is difficult to study and measure. However, many public authorities see urban logistics as an opportunity to improve quality of life, increase the accessibility of urban populations (for example to shopping areas or basic needs) or create and improve employment of sensible populations.

To this classical vision, an extended viewpoint of sustainable development has also been defined in literature (Lindholm, 2010; Patier and Browne, 2010). The three classical spheres of sustainable development can be completed by their interactions, resulting on 7 elements to be considered (Gonzalez-Feliu and Morana, 2014), as shown in Figure 1.

Figure 1. Components of sustainable urban logistics
Adapted from Stivers, 1976.

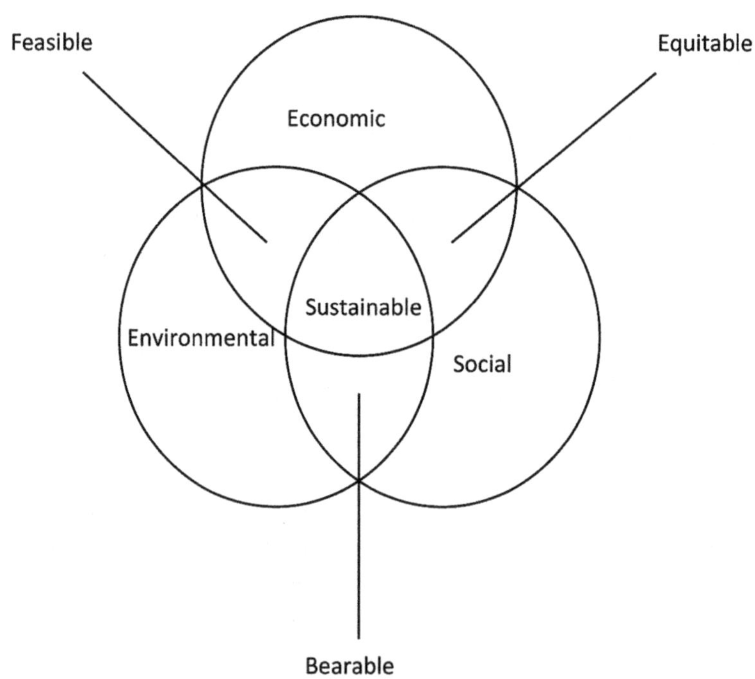

In this extended vision, feasibility is defined as the respect of both economic and environmental performance. Bearableness addresses simultaneously the respect of environment and social equity. Equitability is defined as the joint consideration of economic and social aspects, and is mainly related to equal distribution of benefits or the respect of social constraints but maintain an economic vision. Finally, we could define global sustainability when all six components are included, i.e. not only economic, environmental and social aspects separately but also their interactions, i.e. feasible, bearable and equitable aspects and their combinations.

This vision can be also declined into the relationships between logistics and territory (Masson and Petiot, 2012). The economic sphere is therefore connected to the issue of logistics performance, the environmental one to the respect of the environment and the social one to the equity of the space. Combinations of these three areas (logistics viability, livable spaces and equitable communities) are equivalent to those of Figure 1, and authors discuss the conditions under which sustainable logistics can be defined. The authors state that their approach does not follow the conventional logic of the three spheres that shows an antagonism between logistics performance (economic sphere) and spatial equity (social sphere), and between it and environmental sustainability (environmental sphere). Indeed, according to the authors, logistics performance is not antagonistic to environmental sustainability, as is also emphasized by different authors in logistics which show the trend of logistics performance towards environmental management vision of the overall logistics (Srivastava, 2007; Gonzalez-Feliu and Morana, 2014).

An alternative vision of sustainable development has arisen recently (Macharis and Kin, 2016). Instead of defining sustainability on its components, dealing then with the classic spheres, this new approach proposes to see sustainability on the basis of their capabilities, which are four, according to (Macharis, 2014). This vision is noted as "the four A's of sustainability". This concept is more related to transport and is organised around the following four elements:

- The Awareness is defined as the state or ability to perceive, to feel, or to be conscious of events. The first step to sustain, then to take conscience of the needs to reach it.
- Act and shift can be intended as the reactivity and capacity to operate changes (initially modal splits, but then this notion can be extended to any change) in order to reduce the nuisances imputed to transport, but also to logistics. The second step of sustainability can then be seen as the voluntarism to change the current transport and logistics organizations into cleaner and more socially equitable ones.
- However, those changes will not however be efficient without the notion of Avoidance, seen as the capacity to avoid increasing the nuisances of logistics

and freight transport. Indeed, the third step to sustainability is that of act to avoid, instead of needing to repair later.

- Last but not least, such avoidance needs to be combined with a will of Anticipation. Indeed, the fourth step of sustainability is that of forecasting and identifying the possible nuisances of logistics and freight transport in advance to have the possibility of avoiding them.

To those visions of sustainability in urban logistics, it is important to add the notion of dynamics. Indeed, a system will not be sustainable if it is static, so it needs to be flexible, lean and evolutionary, i.e. in a long-term, global, complex and collaborative systemic vision (Brodhag, 2016). Indeed, many actions are made for short-term actions (like alternate circulation to fight pollution peaks) but need to be complemented with more long-term actions to have a sustainable effect (for example by implementing dynamic land use zones, clean vehicle incentives and change policies, etc.); most actions start to be implemented in a local context, and need to evolve to be transferable and applied in a global context (this is the case of many urban logistics platforms or local policies like night deliveries, with the Manhattan's pilot that inspired many other cities in the world which are nowadays implementing such policies); in the same way, actions at the beginning need to be simple to be implemented and accepted, then they will evolve to take into account the complex system nature of the urban environment; finally, the good balance between the market competition and a stakeholders' collaboration stage is needed to ensure the sustainability of an urban logistics system (a synthetic schema of this vision of Sustainable Development is presented in Figure 2).

Figure 2. Synthetic vision of sustainability in urban logistics
Author's elaboration, extending and aggregating following visions: Stivers, 1976; Macharis, 2014; Brodhag, 2016.

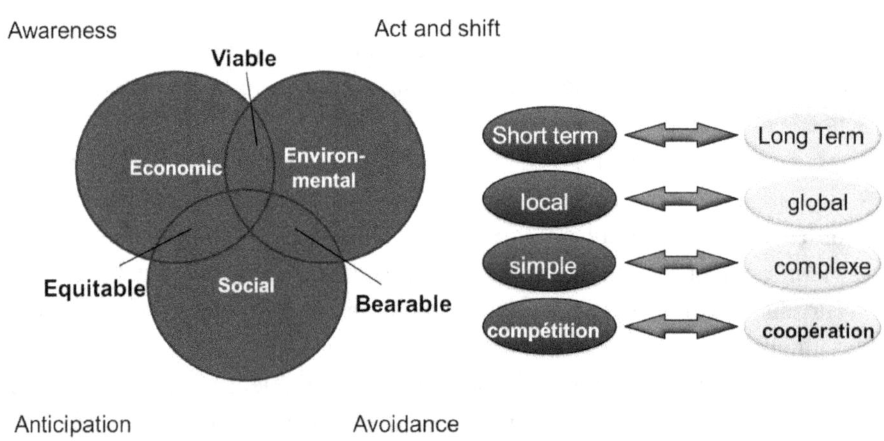

ASSESSING THE SUSTAINABILITY OF URBAN LOGISTICS SYSTEMS: A NON-EXHAUSTIVE OVERVIEW

The assessment of sustainable (or green) urban logistics initiatives is a main issue for various communities of research and practice and has become a priority both in transport (Banister et al., 2008) and logistics (Morana, 2013). In urban logistics, the sustainability assessment of solutions and actions has been the object of a large amount and variety of work, but none of these methods has yet reached to become perceived by the community as generic and transferable. Indeed, even if some authors advocate their method as generic (Filippi et al., 2010; Patier and Browne, 2010), they are little used outside the context of the teams having proposed them (i.e. the work in papers defining them or related to them). In addition, by their variety and heterogeneity, some authors suggest the idea that it seems difficult to propose a general method applicable and transferable to any case. But are these really different works, methods and approaches? Is it so difficult to find common patterns and points of convergence? And, can we still define unified bases for the evaluation of sustainable urban logistics initiatives? To understand this, it is important to make an overview on the literature in this field.

The analysis of the literature on sustainable urban logistics assessment and evaluation is difficult because it does not exist, to the best of our knowledge, detailed syntheses of works dealing with both the identification of indicators of evaluation of sustainable urban logistics and the methodologies to assess them. Although dozens of research works propose assessment indicators and methods, their positioning on the literature is by the method of simulation, and mainly by the type of demand estimation model, field where literature reviews are more numerous (Ambrosini et al, 2008; Anand et al., 2012; Comi et al., 2012; Gonzalez-Feliu and Routhier, 2012; Taniguchi et al., 2012; Cattaruzza et al., 2015). Other approach to present evaluation and assessment methods, is that of start on the basis of the non-existence of standard principle, thus avoiding extensive analyses of literature (Behrends et al., 2008; Patier and Browne, 2010). Finally, some works deal with the identification of indicators, referring to transport or logisitcs literature (Melo and Costa, 2011; Gonzalez-Feliu and Morana, 2014; Morana and Gonzalez-Feliu, 2015) but do not present the assessment methods nor make an analysis of the literature on them. In conclusion, few works refer to the complete methodology in scenario assessment and evaluation, and none of them propose an overview of the literature on those methodologies. Consequently, we propose here an analysis of the literature which is not intended to be exhaustive, but who will try to identify the main lines of thought on the evaluation of sustainable urban logistics (including the definition of indicators, their estimation and the general scenario assessment/evaluation methodology). Given the complexity and diversity of works in this area, we will consider more particularly those proposing the evaluation and analysis of results. The works whose purpose is to present models without a real analysis of results or an assessment objective are not considered here.

The evaluation of sustainable urban logistics was one of the first subjects to be identified in the theme. Indeed, most early works on the evaluation of urban logistics came out of the First International Conference on City Logistics (Mizutani, 1999; Ooishi and Taniguchi, 1999; Takahashi and Hyodo, 1999; Visser et al., 1999; Whiteing, 1999) and the main city/urban logistics refer directly to sustainability (Taniguchi et al., 2001; Gonzalez-Feliu et al., 2014b). The first works on urban logistics evaluation mainly propose methods of simulation or analysis of scenarios using several sets of indicators, mainly logistical (distance-based) or environmental (mainly of the greenhouse gas emissions or pollutants, directly related to distances travelled by the vehicles); those works focus more on modeling aspects than on evaluation indicators or on the analysis and discussion of the results.

Taniguchi and van der Heijden (2000) propose as a methodology for evaluating urban logistics initiatives on the basis of a dynamic traffic simulation combined with probabilistic vehicle routing optimization. The authors propose three main indicators: the total travelled distance, the cost of the entire route set and the total greenhouse gas emissions of the transportation system. The proposed method is tested on a theoretical and restricted case (a small grid network) and is not applied on realistic data. This method was extended in Taniguchi et al. (2003) and Tamagawa et al. (2010) by focusing on the refinement of mathematical models; the indicators are similar to those proposed in Taniguchi and van der Heijden (2000). Similar approaches can be found in the literature (as for example Hosoya et al., 2003; Crainic et al, 2004, 2012; Qureshi and Hanaoka, 2006); they are mainly based on the proposal and refinement of models rather than in impact assessment techniques. They propose either four-step models or vehicle routing approaches to estimate the transport flows. Main indicators of this category of methods are traveled distances, monetary costs, greenhouse gas and NOx emissions.

Moreover, Segalou et al. (2004) provide an estimation of the environmental impacts of urban logistics, under its extended vision (i.e. including IEM, ECM and UMM flows). Authors make a quantification of the flow from various sources of data (mainly derived by quantitative surveys and/or professional registers). The main proposed indicators are direct emissions, mainly greenhouse gases as well as several other pollutants (primarily NOx, SOx, CO and HC). Other assessments on the basis of surveys or measures are also proposed for specific systems, such as the e-commerce deliveries (Esser and Kurte, 2006) or Urban Consolidation Centers, or UCCs (Morganti, 2011; Vaghi and Percoco, 2011; Tozzi et al., 2014; Andriankaja et al., 2015b). Those evaluations are environmental (mainly direct emissions from transport but the last cited work proposes a first life cycle analysis approach to include indirect emissions).

Behrends et al. (2008) address the issue of the sustainability of the transport of goods in urban areas from the point of view of transport companies to define a global indicator of sustainability on the basis of a relationship between the level of emissions of CO_2 and GDP (gross domestic product), the intensity of transport (for example, the activity of transport to GDP), the intensity of traffic (traffic level) and technical capacity (ratio between emissions of CO_2 and the intensity of the traffic). The view above is echoed by Lindholm (2010) to illustrate the point of view of local authorities.

Previous methods were primarily related to distances covered and thus the operational performance of urban logistics. Also, they do not take into account investment costs or their depreciation. Related to that issue, van Duin et al. (2008) are the first to present, to our knowledge, a scientific paper about cost analysis benefits (CBA) to evaluate the suitability of urban logistics. The authors simulate several scenarios of development of UCCs (Urban Consolidation Centers) in the medium term through vehicle routing optimization of touring and assess their monetary costs with CBA. After them, only few authors have recourse to the CBA to assess the feasibility of the projects of urban logistics, and in public initiatives this type of approaches remain also little deployed. However, some examples of application are seen in Gonzalez-Feliu et al. (2013a, 2014c), De Langhe (2014) and Gonzalez-Feliu (2016b).

Other works address the definition of indicators. To the best of our knowledge, the first work proposing a set of indicators to evaluate urban logistics is that of Melo and Costa (2007), then extended in Melo (2010) and Melo and Costa (2011). Authors define a set of indicators define, mainly economic and related to the transport of goods. The aim is to evaluate the performance of urban goods transport and to this purpose authors adapt the main indicators of interurban transport. However, the list is extensive (more than 50 indicators are proposed) and no method to priorize or combine them (for example into a multi-criteria method) are proposed. Henriot et al. (2008) and Patier and Browne (2010) propose a methodology to evaluate logistics projects, defining a core of 30 indicators, grouped into five categories (economic, environmental, societal, regulatory and ergonomic), then a secondary indicators set of more than other 30. However, the authors point out the impossibility of comparing evaluations between them when the indicators used in these assessments are not the same, but the methodology allows to compare assessment methodologies and robustness. Although the authors insist on the "standard" nature of this methodology, it has been proven difficult to transfer and apply outside of the group of researchers having developed it (Massari et al., 2013). A third group of works deal with the construction of sustainable dashboards, i.e. sets of indicators in a limited number but without combining them into a unique value, to support decision making in tactical and operational urban logistics management (Gonzalez-Feliu and Morana, 2014; Morana et al., 2014; Morana and Gonzalez-Feliu, 2015). Two different methods to construct the dashboards are proposed, both on a multi-stakeholder and group decision viewpoint: the first is based on a group of experts (Gonzalez-Feliu and Morana, 2014; Morana et al., 2014) whereas the second follows a socio-constructive learning approach (Morana and Gonzalez-Feliu, 2015).

Finally, a set of work that takes account of the of multi-stakeholder and multi-criteria nature of urban logistics is proposed by Macharis et al. (2014), by adapting the well-known MAMCA (Multi-Actor Multi-Criteria Analysis) framework (Macharis et al. 2010). This work allows to set priorities on the criteria taken into account and their importance that can guide the search for a consensus, but the definition of these criteria and their quantification (or qualification) remains a given entry for those methods.

From all those works, we can define the list of criteria and indicators for assessing and evaluate the sustainability of urban goods transport (see Table 1). Nine criteria have been identified as the most recurrent in literature. From those criteria, 23 indicators are identified; eight of them are directly related to goods transport routes, and other six can be estimated from routes giving complementary information (mainly costs or employment rates related to routes).

From this non-exhaustive overview, we can conclude that the main approaches of sustainable urban logistics assessment and evaluation give much importance to environmental aspects, and economic and social issues remain secondary. This is, according to Morana and Gonzalez-Feliu (2015) a typical example of the gap between public authorities' stakes and those of private companies (authors show the differences of priorities between operational managers and strategic figures of an enterprise, and relate them also to those of public stakeholders). Generally, the only economic (or efficiency) aspects are related to distances covered and loading rates, and little financial evaluation approaches are found in literature.

We also observe that there is not currently standardized method or unified for the assessment of patterns and sustainable urban logistics solutions. In addition, although these ratings are used to compare the alternative to two different times, approaches like before-after assessment are not systematic (which contributes to this lack of standard). For those reasons, and thinking it is important to be able to compare evaluations and give standard frameworks to support decision making in urban logistics, we propose a general framework to guide evaluation. Indeed, rather than proposing a unique method that needs to be strictly followed and would result difficult to deploy, we propose to structure the evaluation and give a unified vision of distance estimation in order to be able to compare evaluations but at the

Table 1. Main criteria and indicators in literature to evaluate sustainable urban logistics

Sphere	Criteria	Main Indicators	Type of Actor
Economic	Transport effectiveness	• Loading rates (%) • Loading paths (%) • Travelled distances (km)	Public and private
		• Weight.distance (Ton.km)	Private
	Quality	• Service rates (%)	
		• On time service rates (%)	
	Financial viability	• Net margin (€, $, %) • Internal Return Rate (%)	Mainly private
Environmental	Energy consumption and climate change	• GHG emission rates - transport (kg CO_2 eq.) • GHG emission rates - construction and end-of-life phases (kg CO_2 eq.) • Energy consumption - transport (MJ eq.) • Energy consumption - construction and end-of-life phases (MJ eq.)	Public, private
	Direct pollutants	• Air pollution rates - transport (kg SO_x, NO_x, VOC, PM10)	Public, private
	Indirect pollutants	• Fossil depletion potential • Metal depletion potential • Air pollution rates - construction and end-of-life phases (kg SOx, NOx, VOC, PM10)	Public, private
Social	Enterprise social/societal impacts	• Customer satisfaction ratio (%) • Average customer satisfaction rates (1-5 or 1-10) • Employee satisfaction ratio (%) • Absenteism ratio (%)	Private
	Impacts on employment	• Employment creation/conversion rates (%)	Public, private
	Congestion	• Congestion rates of running vehicles (km. CEU) • Congestion rates of parking vehicles (h.CEU)	Public

Sources: Author's elaboration as a conclusion of the proposed overview.

same time giving stakeholders enough freedom and flexibility to deploy evaluation methods that they control and understand. However, it is important to note that urban goods transport is not only related to last mile distribution but other flows are related to urban goods movement, so in the estimation of routes those components of urban goods transport need to be estimated properly. It will then be important to distinguish those flows. This is why we present in newt section the main categories of urban goods flows.

Main Categories of Urban Goods Transport Flows

One of the main reasons (but not the only one) of this difficulty of standardization in evaluation and assessment methods is the difficulty of defining a common scope in what are the urban goods transport flows. In literature, we find different visions of urban goods transport (Woudsma, 2001; Taniguchi et al., 2001; Russo and Comi, 2006; Gonzalez-Feliu et al., 2010). In this paper, we aim to be positioned in the most global viewpoint, as on Gonzalez-Feliu et al. (2014d), who distinguish three categories of flows.

The most known group of flows is that of inter-establishments movements (IEM, Ségalou et al., 2004), that include all transport flows between two economic activities (B2B flows). This is the most common definition of urban goods transport (Woudsma, 2001), but it only represents between 45 and 55% of the total road occupancy issues of urban transport related to goods (Ségalou et al., 2004). According to Ségalou et al. (2004), within those flows we can distinguish different modes of management (shipper' own account, receiver's own account and third party transport). We can then define subcategories of flows related to the activity of the receivers (i.e. the destination of goods) as shown in Table 3.

The category that contributes the most to urban goods transport nuisances (more than 50% of road occupancy rates according to Ségalou et al., 2004) is however that of end-consumer movements (ECM, Russo and Comi, 2006), i.e. the movements of goods from the selling places or warehouses to the final consumer. If IEM concern B2B flows related to intermediary or final products, ECM represent B2C flows. Those flows include of course shopping trips (motorized or not), home deliveries, out-of- home deliveries and other related transport flows to deliver directly or indirectly the final consumer.

Finally, a category often associated with residual and less important flows (Patier, 2002) is that of urban management movements (UMM, Gonzalez-Feliu et al., 2014d). That category is the most difficult to characterize since it is the most heterogeneous and although it represents between 8 and 10% of road occupancy rates (Ségalou et al., 2004), some of the flows included in this category are, in distances, very small with respect to those of other categories. The main categories of UMM flows are: waste collection, construction materials transport and logistics, network maintenance, household and enterprise moving flows and postal services (only mail).

To better understand the contribution of such flows to traffic and congestion, it is important to consider the road occupancy issues not only for vehicles in movement but also for those stopped. However, this second indicator is difficult to estimate, and although some works attempt to deal with them (Gonzalez-Feliu et al., 2014d; Cattaruzza et al., 2015), they do not directly relate both road occupancy issues.

Nevertheless, we can examine those issurs in Cattaruzza et al. (2015) with the assessment of both vehicle flows and number stopped vehicles in the urban area of Lyon, then converted into road occupancy issues.

Traffic road occupancy rates reflect the occupancy of infrastructures by vehicles in movement, and are expressed in km × equivalent vehicle (generally car-equivalent units or CEU). In this assessment, authors used conventional equivalence rates (i.e. one private car or light vehicle = 1 CEU, one light goods vehicle = 1.5 CEU, one small single truck = 2 CEU, one big single truck = 2.5 CEU and one semi-articulated vehicle 3 CEU). To estimate traffic road occupancy rates the total traveled distances are estimated by type of vehicle then converted it to traffic road occupancy rates using the conversion rates shown above (Cattazurra et al., 2015). Park road occupancy rates are estimated in a similar way, but since parked vehicles are not travelling distances, other estimations are needed. Cattaruzza et al. (2015) estimate the number of vehicles parked (related to the number of trips), then estimate the average total parking time per type of vehicle. Finally, they apply the same conversion ratios to obtain comparable values.

We observe in Table 2 that traffic road occupancy rates of IEM and ECM are almost equivalent in the city of Lyon (ECM rates are lower than on Ségalou et al., 2004, and IEM a little higher). This is due to the fact that Lyon is a city with an efficient public transport system which and a multi-polar retailing structure can have an impact on motorized shopping trips, i.e. resulting on an increase of non-motorized trips, which will decrease ECM since those flows are mainly related to motorized shopping trips (Gonzalez-Feliu et al., 2010). UMM are less numerous but are far from being negligible (about 8% of traffic road occupancy). When dealing with parking vehicles, we observe that those who consume the most part of the road are personal cars for shopping trips, with 64% of the road occupancy rates. We recall that in this assessment, vehicles not dealing with goods transport are not considered, but they are however impacting goods transport since they are occupying the urban

Table 2. Road occupancy rates per week in the city of Lyon

Type of Flows	Traffic Road Occupancy	Park Road Occupancy	
		Whole Urban Area	Inner City
IEM	45,7%	28,2%	47,5%
ECM	46,1%	64,0%	33,7%
UMM	8,2%	7,8%	18,8%
Total	100,0%	100,0%	100,0%

Cattaruzza et al., 2015.

space and many times they limit goods transport operations (Gonzalez-Feliu et al., 2013a). When dealing with the inner city (and not the entire urban area) the impact of IEM increases, as well as that of UMM, due to the fact that the space is lower and the circulation constraints higher, mainly in city centers.

From those synthetic results, we can observe that each component of urban goods movement has a non-negligible contribution to the overall impacts (in terms of trips, traveled distances or road occupancy rates), so it is important to consider all in a general framework regarding sustainability evaluation. Moreover, actions on one of the components (i.e. structural changes that increase e-commerce for a sector) can have a direct impact on all three components (in this example, a decrease of sales in classical retailing channels that decrease their distribution logistics flows, an increase in e-commerce distribution flows, both intermediary, i.e. B2B; and end-consumer, i.e. B2C, and finally an increase of waste that in some generalized contexts can modify the waste collection in some areas of the city). So it is important to propose to take into account all flows, even if not all evaluations and assessments will need an accurate estimation of all of them.

A General Framework for Evaluation and Assessment of Green Urban Logistics Initiatives

After reviewing the main indicators to assess sustainable urban goods transport systems and categories of transport flows, it is important to state on the use of assessment methods. The aim of this chapter is not to make a comparison of models in terms of efficiency, since models have been conceived for different uses and with different aims, and this type of comparisons lead, in our opinion, towards a competition or towards statements on which model is better or preferable to others. Since a standard or dominant model cannot be defined, we prefer to conceptualize here a framework showing how the various models can be used or combined in a synergic way to better answer to city planners' needs in terms of sustainable urban goods transport assessment. To do this, we propose the following framework. Given a goal for a planner (i.e. the object to assess and the aims of this assessment), the proposed framework guides the corresponding scenario assessment. The first action to make in scenario assessment is to define the type of scenarios that will be used in the assessment. We distinguish from literature four main types of scenarios: Business as usual forecasts, realistic alternatives or contrasted alternatives.

Once the type of scenario(s) to test is defined, it is important to identify the type of analysis that will be carried out. Five main different types of analyses are identified in the literature:

1. Before-after analyses aim to study new systems or organizations, by comparing an initial situation where the new system is not still implemented to a final situation where this system is already operating. They are usually applied to ex-post evaluation (Leonardi et al., 2012). However, the same principles can be applied to ex-ante evaluation. In those cases, an initial situation is constructed then one or more scenarios are defined. In general, only one scenario is tested (Nuzzolo et al., 2014). The comparison is then made between one simulated scenario and the initial situation.
2. Another possibility is that of comparing alternatives among them, without defining an initial situation (Gonzalez-Feliu et al., 2010; Macharis et al., 2014). This type of analysis is often made when an initial situation cannot be defined, or when the aim is to find the most suitable alternative from a set of solutions, independently of the initial situation (i.e. the goal is to choose an alternative so the change the initial situation, so this reference scenario is not required in such analyses.
3. Incremental analyses can be defined as scenario comparisons when, from an initial situation, each scenario is made by changing some parameters increasingly (Durand and Gonzalez-Feliu, 2012). For example, economic analyses of systems when demand is increasing, cost-benefit analyses (van Duin et al., 2008; Gonzalez-Feliu et al., 2013a), deployment assessment analyses or market developments are main applications of this type of analysis.
4. Combination-based analyses (Gonzalez-Feliu et al., 2012c) are those in which first, several alternatives are compared to an initial situation, then, combinations of those alternatives are compared. They are usually developed to test the impacts of single actions or functionalities then to identify scale effects or synergies among alternative solutions and actions.
5. Finally, backcasting analyses (Lopez-Ruiz and Crozet, 2010; Gonzalez-Feliu et al., 2012a) are mainly used in long term issue assessment. First, several future configurations are tested to find a set of suitable scenarios then examine, by coming back to present, how those scenarios can be met, at what cost and in which time. The most suitable scenario in terms of evolution can be then selected.

In all cases, a comparison between at least two situations is made. And even when more than one scenario are proposed, each of them is compared to a reference or to another one. In this case, we can relate each type of analysis to the comparison of two situations, as synthesised in Figure 3.

Figure 3. Chart of the proposed unified before-after analysis procedure for both ex-ante and ex-post assessment and evaluation
Author's elaboration.

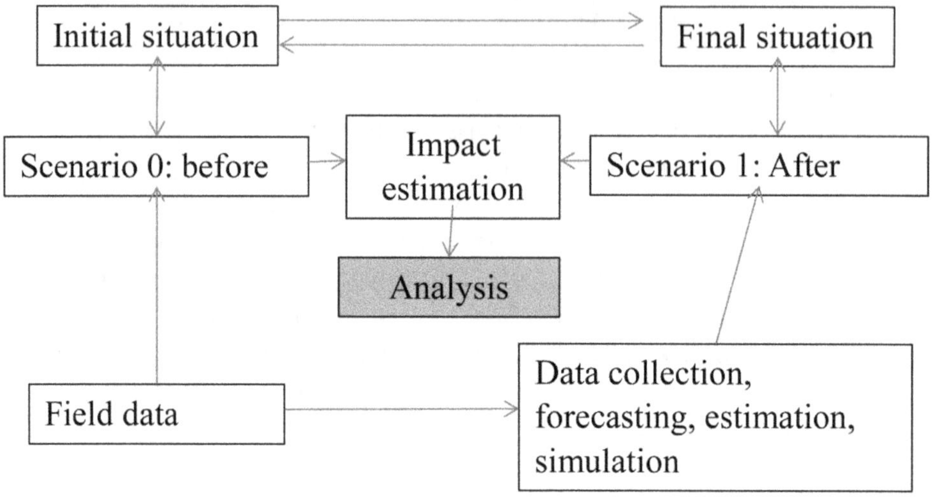

Chart and Structure of the Proposed Methodology

After that, the scenarios can be defined and the assessment carried out. To do this, we propose a generic framework organized in the following steps:

1. **Initial Situation (Input Data):** The definition of an initial situation, when applicable, is made from current data. In general, it is important to define before the simulation the input data needs, in order to collect and produce the information necessary to the different assessments. To this, it is important to define the variable of the flow estimation models (step 3) taking into account the data available and the different data production modalities.
2. **Scenarios to Assess:** Analogously to initial situation definition, the different scenarios need to be characterised. That means in general that several assumptions and previous estimations need to be realized (mainly related to the variables of the flow estimation models and the parameters of the impact estimation methods).
3. **Estimation of Flows (for Each Scenario):** Once the scenarios are characterized, a flow estimation needs to be carried out. We think that the estimation of flows need to be as complete as possible, even if its accuracy is not always guaranteed, but it will give a global idea of the importance of considered transport systems with respects to the overall urban goods transport. To do this, the main stages of the flow estimation can be divided as follows:

a. **Demand Generation:** For each zone (or establishment, or household group, when applicable), a number of trips is assigned. This is done respectively for each subcategory of inter-establishments, end-consumers and urban management movements. The generation function is in general related to demographic variables (related to the population), to socio-economic variables (related to economic activities) and eventually to geographic variables (for end-consumers and urban management flows, the distribution of populations in individual and collective habitat zones can have different influences on goods transport generation patterns). Those models are in general category class approaches mainly estimated by a constant or an econometrics approach.

b. **Route Construction:** All flows can be represented by routes. To do this, it is important first to define the origin of the route (a depot for a delivery route, a household or a workplace for a shopping trip chain and a depot or a technical location for several urban management flows) the different destinations. To do this, different approaches can be used: catchment area models, discrete choice models, O-D categorization approaches, etc. Then, routes can be defined either via empirical, statistical or vehicle routing optimization approaches.

c. **Distance and Time Calculation:** From the results of route estimation distances (directly) and times (indirectly, via average travel distances in the different and streets of the urban area) can be estimated. Also, the number of vehicles and eventually, the number of drivers, can be estimated. Those basic indicators are needed for the further step.

4. **Estimation of Impacts (for Each Scenario):**

a. At economic level, several indicators can be estimated, like net margins, operational costs, global costs (by translating investment costs to each route), etc.

b. At environmental level, two main approaches can be envisaged: direct emission estimation (CO_2, NO_x, SO_x, VOC, PM10, …) or live cycle analysis that include those emissions (for all construction, use and end-of-life phases of vehicles, infrastructures and buildings) but also several indirect impacts like metal depletion, soil and water pollution, etc.

c. At social level, congestion can be a good indicator. Note only that in urban goods transport both running and stopped vehicles contribute to congestion. Congestion by running vehicles is expressed in km.CEU (car equivalent unit) and congestion by stopped vehicles in h.CEU. However, since distances can be converted into times, a suitable congestion indicator can be the one that combines both congestions and is expressed in a homogeneous unit (in this case, only h.CEU would be applicable).

5. **Scenario Assessment and Analysis:** Taking into account the type of scenarios and analysis to carry out, the scenarios are compared on the basis of the identified indicators and decision support and practice implications identified. Several methods, like cost-benefit analyses, classical multi-criteria stakeholders, decision distance methods or multi-criteria multi-actor analyses can be used in this phase when several criteria and/or stakeholders are considered.

Application Issues of the Proposed Methodology

Although the proposed methodology is enounced here in a conceptual form, it derives from the statement of current research developments and their derived practical issues. Indeed, although the evaluation methods are various, some trends in the deployment of assessment and evaluation frameworks are observed. Indeed, one of the conclusions of the 3rd VREF conference of urban freight held in Goteborg in October 2016 regarding data production is that although methods are various and heterogeneous they have several common points.

If we extend this to all the assessment and evaluation framework, we can state that the definition of the initial situation, which is necessary to every before-after assessment, is crucial for the final result of the assessment or evaluation, as well as for explaining and communicating results and conclusions. This need is common to all evaluations but the initial situations are nowadays estimated in different ways and this heterogeneity makes sometimes difficult a possible comparison among evaluation or assessment frameworks. However, since to define initial situations it is important to estimate the urban goods transport flows, the stage of freight trip generation (FTG) seems crucial in all assessment frameworks (Holguin-Veras, 2016). Moreover, and although in the last 20 years a standard in urban goods modelling has not been found as universally accepted, a methodological framework for FTG is nowadays widely extended.

This framework, first proposed by Holguin-Veras et al. (2011) then refined in Holguin-Veras et al. (2013) and Sanchez-Diaz et al. (2015a, 2016), has been applied to different contexts (cities, countries and economic activities). According to those authors, FTG can be made using three simple functional logics: a constant trip rate by establishment, a proportional number of trips per establishment (linear function without constant) or a linear function of type $y=a+b.x$ (linear function with constant). The basic idea is to estimate, for both production and attraction trips (i.e. trips at origin and destination) the three models then compare them to the same data sample using an error measure, for instance the root mean square error, RMSE. The US remain the country with the highest number of applications of this methodology (Holguin-Veras et al., 2011, 2013; Lawson et al., 2015; Jaller et al., 2015; Sanchez-Diaz et al., 2015a,), but other countries have applications of the same framework, like Sweden (Sanchez-Diaz, 2016), Portugal (Alho and de Abreu e Silva, 2015), Brazil (Silva and Waisman, 2007; Oliveira, 2014), The Netherlands (González-Calderón et al., 2016) or France (Gonzalez-Feliu et al., 2014a, 2016). The generalization of that framework is an evolution of the classical constant trip generation (Eriksson, 1997; Ministerie VROM, 1999; Aubert and Routhier, 1999; Holguin-Veras et al., 2012) and shows that a unification of frameworks is being actually carried out.

The second step is that of route construction. However, and since the unification in generation is recent, this step is far from reaching a standard. Four-step models (D'Este, 2001) remain very popular although not always adapted to urban goods transport (Gentile and Vigo, 2013). However, and although some authors claim new paradigms not generating O-D matrices (Bonnafous et al., 2014; Gonzalez-Feliu et al., 2014a), in practice the influence of personal transport make that planners need O-D based approaches (Gonzalez-Feliu et al., 2014d). So, and starting from freight trip generation, a trip distribution (i.e. the assignment of the different destinations from each origin, in number of trips, deliveries or freight quantity, as well and its intensity) is needed. Methods are various but remain strongly connected to those of inter-urban freight transport or urban personal transport (those methods are introduced in Ortuzar and Willumsen, 2001). The two main methods used are gravity models (Gonzalez-Feliu, 2014) or entropy-based approaches (Muñuzuri et al., 2009, 2011; Sanchez-Diaz et al., 2015b).

Once O-D matrices are estimated, routes can be constructed. To do this, different methods are found in literature. We find three main schools of thought in urban logistics route construction. The first is that of vehicle routing optimization (Toth and Vigo, 2001), which aims to adapt vehicle routing problem (VRP) algorithms to the questions of urban logistics. Although many works deal with vehicle routing in urban logistics (for overviews on the main issues related to city logistics and VRP, see Taniguchi et al., 2012, 2014; Mancini, 2013; Cattaruzza et al., 2015), most works deal with the traditional "problem solving" vision of operations research (OR), i.e. the search of an optimal or near-optimal solution with respect to a given objective function (mainly obtained by deduction or proposed with few or no statistical analyses of the field of application). The proposed algorithms on this vision are constructed on a computer science perspective, and do not always follow an application goal. Opposing to this main vision of OR, a small set of works follow the "solution probleming" complementary vision of Ackoff (1977) who states that an optimization problem then the consequent methods to solve it and obtained solutions need *in fine* to be examined with respect to the context they aim to be applied, then be eventually reviewed and modified to approach this reality. Although not all authors position their work under this banner, when examining their methodological framework, we can consider that the following group proposes methods under this vision: Sonntag (1985); Quak and De Koster, 2009; Deflorio et al., 2012; Muñuzuri et al., 2012; Gonzalez-Feliu et al., (2015); Salanova-Grau and Gonzalez-Feliu (2015).

The second is that of empirical methods, who aim to estimate travelled distances from current practices (Vaghi and Percoco, 2011; Bonnafous et al., 2014; Gonzalez-Feliu et al., 2014a; Aditjandra et al., 2016). Those methods do not build the entire route but estimate their distances from category class models which define the main characteristics of routes with respect to established typologies. An alternative to category classification based route generation is that of discrete choice models (Hunt and Stefan, 2007; Crocco et al., 2010; Russo and Comi, 2010; Nuzzolo and Comi, 2014). Route characteristics are defined empirically but the assignment of a route to an establishment is not made by empirical based category assignment but by discrete choice models like the well-known logit model. The third, less used in urban logistics than the two first but well-known in regional transport, is that of analytical models (Daganzo, 1997), which can give a valid approximation when the aim is to generate sets of routes to estimate initial situations or global impacts of the overall transportation sector. Examples of those models in urban logistics and/or their applications are found in Andriankaja et al. (2015a); Combes (2016); Faure et al. (2016).

The methods proposed above are related to IEM flows. Remain to address ECM and UMM. For ECM, the most important flows are shopping trips, which can be estimated via classical four-steps (Ortuzar and Willumsen, 2001), although specific frameworks for shopping trip generation can be observed in literature (Vickerman and Barmby, 1985; Gonzalez-Feliu et al., 2010, 2012b). In general, shopping trip generation follows a more standard viewpoint, with models developed by linear regression approaches related to several variables, such as population, employment of retailers or motorization rates, among others. Cubukcu (2001) proposed a general methodological framework for shopping trip generation, relating shopping trip rates to three categories of variables: demographic, socio-economic and technologic. Concerning trip distribution and shopping round construction, classical models are used (trip distribution is made using gravity or entropy-based models). However, in addition to modal choice some authors preconize to develop retailer choice models (Nuzzolo and Comi, 2014).

UMM flows remain less studied and only empirical models allow to give estimates of the whole category of flows (Gonzalez-Feliu et al., 2014d). However, for waste collection flows, construction trips and service/maintenance flows, when estimated separately, vehicle routing approaches can be used (Toth and Vigo, 2002).

Once routes are estimated, travelled distances and times can be calculated. In general, distance calculation is directly made from route construction, and travel times can be quickly deduced when average speeds are known. For example, after

Table 3. Mean speeds for urban goods transport flows in the City of Bilbao, delivery stops are not considered in speed estimation

	Part of the Duration (Travel Time)	Part of The distance	Speed (Harmonic Mean)	CO_2 Emissions per Truck
Residential	58%	29%	7 km/h	1 047 g/km
Main road	22%	25%	16 km/h	813 g/km
Motorway	20%	46%	33 km/h	761 g/km

Pluvinet et al., 2012.

analysing an important set of GPS-based data on vehicle paths, Pluvinet et al. (2012) give a general average speed for the city of Bilbao, according to the type of street (see Table 3). Those results are close to those obtained in French urban goods surveys (Ambrosini et al., 2010) as well as to those of other works in the world on urban goods transport GPS data (a comparison of methods is given in Gonzalez-Feliu et al., 2013c).

From distances and times, a transport cost can be estimated. Although there are many methods to estimate it (as for example considering an operational cost, the variable cost, the marginal cost, or making a complete cost-benefit analysis, among others), in a first time it can be interesting to estimate the operational cost of routes, i.e. considering all variable costs (fuel and driving/stopping time) related to routes, then adding fix costs related to the transport operations (vehicle acquisition/rental costs, insurances, etc.), to obtain a homogeneous and comparable indicator. For example, Gonzalez-Feliu et al. (2013b) estimate this cost to estimate the suitability of collaborative urban freight transport in Lyon (France) from CNR (2012) data, adapting the method proposed in Generalitat de Catalunya (2009). The resulting cost function is the following:

$$C_{ij} = 0.35d_{ij} + 34.52t_{ij}$$

where d_{ij} is the average travel distance and t_{ij} the average travel time between i and j.

Another direct indicator obtained from traveled distances (and eventually stop times) is the road occupancy index, mainly related to occupancy rates of the space by running vehicles. Several authors state on the possibility of estimating also occupancy rates by stopped vehicles (Bonnafous et al., 2014; Cattaruzza et al., 2015), but for the moment the first indicator is related to traveled distances and the second to stop times, and for the moment they are generated separately. However, since travel times can be directly estimated from distances (if the average speed of urban goods traffic is known), a possibility of combining both indicators seems to be to convert all in time, before applying ponderation rates. Indeed, road occupancy rates (for running vehicles) are obtained by pondering travelled distances by a coefficient related to the vehicle size to express those rates in km.CEU (car-equivalent unit, estimated as presented above).

We present in Table 4 the percentage impacts of the overall urban goods transport flows related to road occupancy, for the city of Lyon. Those results are the extension of those presented in Cattaruzza et al. (2015) to precise the impacts inside each category of flows:

We can highlight from those values the importance of motorized shopping trips (about 60% of the total travelled distances, lowered to 40% of road occupancy issues because they are made by personal car, and deliveries, either to establishments or households, are in general made by light vehicles or heavy trucks, which have higher impacts on road occupancy). Retailer deliveries, the most studied segment in urban goods transport, represent only 10% of road occupancy rates and it is

Table 4. Importance of the different categories of urban goods transport flows, in percentage of trips and road occupancy rates

Categories of Urban Goods Transport Flows	Subcategories of Urban Goods Transport Flows	Percentage of Trips	Distances (%)	Road Occupancy (%)
Inter-establishment movements (IEM)	Last mile transport to retailers, offices and services establishments	16 %	6 %	10 %
	Intermediary distribution logistics	11 %	9 %	17 %
	Urban industry and agriculture	5 %	3 %	5 %
	Maintenance and other services at destination	29 %	7 %	8 %
	Total	***61 %***	***25 %***	***40 %***
End-consumer movements (ECM)	Motorized shopping trips	27 %	63 %	41 %
	Home deliveries	5 %	5 %	6 %
	Out-of-home deliveries	3 %	1 %	2 %
	Total	***35 %***	***69 %***	***49 %***
Urban management movements (UMM)	Waste collection and management	-	1 %	2 %
	Construction (infrastructure and building) and network maintenance	2 %	3 %	6 %
	Postal distribution and moving logistics	1 %	1 %	2 %
	Total	***3 %***	***5 %***	***10 %***
Other trips	***Trips not entering in precedent categories***	***1 %***	***1 %***	***1 %***

Extended from Cattaruzza et al., 2015.

intermediary distribution (wholesalers, warehouses, etc.) which has a higher impact on road occupancy, mainly due to the number and size of trucks deployed to those purposes. Urban management flows remain less important but construction and network maintenance has a similar impact than home deliveries or urban industry and agriculture (about 5-6%), although travelled distances, vehicle uses and sizes of transport management are different for each category.

From those direct indicators, more specific measures can be defined. Without entering in detail, we propose, as an example, a dashboard canvas of indicators extracted from Gonzalez-Feliu and Morana (2014) that shows a possible set of indicators to use for sustainable evaluation and assessment of green urban logistics systems (see Table 5).

This proposal is an example of use for urban logistics dashboards, and the final composition of each dashboard will be related to the precise application field, i.e. the number and nature of the indicators, as well as the way they are calculated and reported, will be specific to each application, but since most of them are issued from a unified estimation of flows, it will be relatively easy to re-calculate them on other applications if a comparison is required. So, even when producing different indicators for different evaluations, if flows are estimated in a comparable way, it will: be possible to reproduce comparable indicators, surpassing in this way the limits of frameworks like Henriot et al. (2008) and Patier and Browne (2010), or at least trending into a more collaborative and shared vision of evaluation that takes into account the different stakeholders of urban logistics and their related needs (Morana and Gonzalez-Feliu, 2015).

Table 5. An example of indicators set for urban logistics evaluation

Category	Nature	Indicator	Stakeholder
Awareness	Economic	Loading ratio	Public or private
	Economic	Warehouse fill rate	Private
	Service quality	Service rate	Private
Act and shift	Environmental	Saving in number of trucks used	Public or private
Avoidance	Environmental	Greenhouse gas emissions	Public or private
	Environmental	Emissions of NOx	Public or private
	Environmental	Emissions of PM10	Public or private
Anticipation	Economic	Operational costs	Private
	Social	Rate of jobs to be converted	Public

Adapted from Gonzalez-Feliu and Morana, 2014.

CONCLUSION

This chapter is a first step towards a systemic vision of scenario assessment for evaluating and assessing the sustainability of green urban logistics systems. After defining the main stakes of urban goods transport planning for both public and private actors, this paper addressed the sustainability issues of urban logistics and proposed a set performance indicators used in literature to assess urban goods transport sustainability. Moreover, this paper presented the main categories of flows, showing the contribution of each category to congestion by running vehicles. Furthermore, the paper proposed a systemic vision of urban goods transport estimation, not focusing only in classical retailer deliveries but in the most global and inclusive vision of flows. Finally, the meta-structure of a framework for sustainability scenario assessment in urban goods transport planning was proposed, not to choose the "best" approaches but to use a synergic combination of methods and procedures presented above.

We saw in this paper that the vision of sustainability depends on the stakeholder, and is not unique. However, several indicators seem to be used by most stakeholders, and we observe dominances in the definition of evaluation frameworks.

Those dominances are not observed when estimating goods transport flows. Various families of approaches are deployed and no standard seems to be found, but we observe that most approaches start by generating flows and matching the origin of goods to the possible destinations, then constructing routes and trip chains, and finally estimating distances and other indicators. This metastructure is observed in most families of method for all three categories of urban goods transport flows. This statement allows us to conclude on the need of unifying the methodological path (i.e. the modelling metastructure) and not necessarily the precise mathematical framework.

It is with this vision that we propose a general framework for scenario assessment related to urban goods transport sustainability. This framework seems corresponding to most approaches and can also be used to combine those methods and models in a synergic way. Which is important is to define the more suitable scenario analysis method (including type of scenarios and family of analysis) for the given aims and objectives, as well as to construct the scenarios in a coherent way, by taking into account the available data and the data needs of the assessment methods.

In conclusion, although urban goods scenario assessment is often seen as a complex jungle, it seems that common methodological structures can be found and help researchers and practitioners to have common reference guidelines towards collaboration, consensus and common understanding for a more suitable urban goods transport overall system.

Remains then to address the main future research and practice directions. As seen in this chapter, a "unique" method, procedure of software does not seem to be the best solution to propose unified assessments and evaluations, and as the literature shows there standards can be found in methodological frameworks. Since FTG has already reached a stage of maturity that is leading to a standard, the first research direction seems to be in completing this framework with a distribution and route construction framework. Finding both analytical models for quick and simple assessments and VRP-based algorithms following the "solution probleming" vision of Ackoff (1977), i.e., contextualized and in direct relation with the reality they aim to represent, seems to be crucial to propose unified frameworks from flow estimation. In this sense, the framework will deal with the general methodology (as the four-step model in personal transport) and not with the proposal of a "black-box" model or a unique software. Although some steps are done on this path, the work is far to be unified and important opportunities, both in research and practice (to deal with the applicability and real application of those models) will appear in the next years.

A second direction is in the definition of indicators and their estimation methods. The generalization of economic analysis methods seems crucial to provide practitioners with correct tools to evaluate and assess their urban logistics systems and to provide quantitative and justified values that show the potential and limits of those systems. On another side, the environmental evaluation needs to take into account indirect effects, and the development of life cycle analysis approaches seems to go into a unification of practices. Finally, social and spatial evaluations, as for example based on accessibility, remain little used but are crucial to assess the acceptability and the social impacts which become important in many public decisions. The quantification of those issues seems to be one of the next research challenges to address.

To those two directions, it is important to add the standardization of data production and the search of a good compromise between the resources needed to produce data and the quality of the estimates obtained with this data. Finally, the collaboration among researchers and practitioners seems to be one of the most important keys to reach unified frameworks and make them be deployed in practice.

ACKNOWLEDGMENT

Part of this research was made on the context of the French National Research Agency project ANR VBD 2013-001 ANNONA. Moreover, the present document synthesises the author's vision of urban goods evaluation proposed in his memory of Habilitation to direct PhD's, directed by Dr. Corinne Blanquart, director of research at IFSTTAR and director of the SPLOTT laboratory. Author aims also to thank

Professor José Holguin-Veras, Dr. François Combes, Dr. Christian Ambrosini and Dr. Christian Brodhag for their enriching comments and suggestions regarding this general vision of urban goods assessment and evaluation.

REFERENCES

Ackoff, R. L. (1977). Optimization+ objectivity= optout. *European Journal of Operational Research*, *1*(1), 1–7.

Aditjandra, P. T., Galatioto, F., Bell, M. C., & Zunder, T. H. (2016). Evaluating the impacts of urban freight traffic: Application of micro-simulation at a large establishment. *European Journal of Transport & Infrastructure Research*, *16*(1).

Alho, A., & de Abreu e Silva, J. (2014). Freight-Trip Generation Model: Predicting Urban Freight Weekly Parking Demand from Retail Establishment Characteristics. *Transportation Research Record: Journal of the Transportation Research Board*, (2411), 45-54.

Ambrosini, C., Patier, D., & Routhier, J. L. (2010). Urban freight establishment and tour based surveys for policy oriented modelling. *Procedia: Social and Behavioral Sciences*, *2*(3), 6013–6026. doi:10.1016/j.sbspro.2010.04.015

Ambrosini, C., Routhier, J. L., Sonntag, H., & Meimbresse, B. (2008). Urban freight modelling: a review. In E. Taniguchi & R. G. Thompson (Eds.), *Innovations in City Logistics* (pp. 197–211). New York: Nova Science Publishers.

Anand, N., Quak, H., van Duin, R., & Tavasszy, L. (2012). City logistics modeling efforts: Trends and gaps-A review. *Procedia: Social and Behavioral Sciences*, *39*, 101–115. doi:10.1016/j.sbspro.2012.03.094

Andriankaja, D., Gondran, N., & Gonzalez-Feliu, J. (2015a). Assessing the environmental impacts of different IPSS deployment scenarios for the light commercial vehicle industry. *Procedia CIRP*, *30*, 281–286. doi:10.1016/j.procir.2015.02.159

Andriankaja, D., Gondran, N., & Gonzalez-Feliu, J. (2015b). Evaluation environnementale dun centre de distribution urbaine par une approche dACV. *Logistique & Management*, *23*(1), 51–69. doi:10.1080/12507970.2015.11665672

Arndt, W. H., Beckmann, K. J., Gies, J., & Gonzalez-Feliu, J. (Eds.). (2013), Städtischer Wirtschaftsverkehr - Commercial/Goods Transportation in Urban Areas - Transports Commerciaux/Marchandises en Ville. Dokumentation der Internationalen Konferenz 2012 in Berlin, Deutsches Institut für Urbanistik, Difu Impulse 2013/3, Berlin, Germany.

Aubert, P. L., & Routhier, J. L. (1999), Freturb: Un modèle de simulation des transports de merchandises en ville. In *Transport Modes and Systems: Proc., 8th World Conf. on Transport Research* (*Vol. 1*). Pergamon, Amsterdam, 531-544.

Banister, D., Hickman, R., & Stead, D. (2008). Looking over the horizon: visioning and backcasting. In A. Perrels, V. Himanen, & M. Lee-Gosselin (Eds.), *Building blocks for sustainable transport - Dealing with Land Use, Environment, Health, Safety, Security, Congestion and Equity*. Amsterdam: Springer.

Behrends, S., Lindholm, M., & Woxenius, J. (2008). The Impact of Urban Freight Transport: A Definition of Sustainability from an Actors Perspective. *Transportation Planning and Technology*, *31*(6), 693–713. doi:10.1080/03081060802493247

Bonnafous, A., Gonzalez-Feliu, J., & Routhier, J. L. (2014). An alternative UGM paradigm to OD matrices: the FRETURB model. *Selected proceedings of the 13th World conference in Transport Research, WCTR 2013*, Rio de Janeiro, Brazil.

Boone, J. L., Myers, J. E., & Redman, C. L. (1990). Archeological and historical approaches to complex societies: The Islamic states of Medieval Morocco. *American Anthropologist*, *92*(3), 630–646. doi:10.1525/aa.1990.92.3.02a00050

Boudouin, D., Morel, C., & Gardrat, M. (2014). Supply chains and urban logistics platforms. In J. Gonzalez-Feliu, F. Semet, & J. L. Routhier (Eds.), *Sustainable urban logistics: Concepts, methods and information systems* (pp. 1–20). Heidelberg: Springer. doi:10.1007/978-3-642-31788-0_1

Britnell, R. H. (1995). *A Commercialising economy: England 1086 to C. 1300*. Manchester: Manchester University Press.

Brodhag, C. (2016, November 7). Le développement durable, enjeux climat et biodiversité. Lecture Notes and presentation. Master PPIEV 1b. Introduction à l'économie verte. Oran.

Cattaruzza, D., Absi, N., Feillet, D., & Gonzalez-Feliu, J. (2015). *Vehicle Routing Problems for City Logistics*. EURO Journal of Transportation and Logistics. Doi:10.100713676-014-0074-0

CERTU-ADEME. (1998). *Plans de déplacements urbains. Prise en compte des marchandises, guide méthodologique*. Lyon: CERTU.

CNR. (2010). *Evolution du prix du gazole et incidence sur le prix de revient. Situation au 30 juillet 2012*. Comité national routier, Paris, France, available at: www.cnr.fr/content/download/357/3457/version/15/file/Note%20gazole%2001%20ao%C3%BBt%202012.pdf

Combes, F. (2016, June 1-4). A theoretical analysis of the cost structure of urban logistics. *Proceedings of the 6th International Conference in Information Systems, Logistics and Supply Chain*, Bordeaux, France.

Comi, A., Delle Site, P., Filippi, F., & Nuzzolo, A. (2012). Urban freight transport demand modelling: A state of the art. *European Transport/Trasporti Europei, 51*(7), 1-17.

COST. 321 (1998). Urban Goods Transport, Final report of the action. European Commission Publications, Bruxelles,

Crainic, T. G., Mancini, S., Perboli, G., & Tadei, R. (2012). Impact of generalized travel costs on satellite location in the two-echelon vehicle routing problem. *Procedia: Social and Behavioral Sciences, 39*, 195–204. doi:10.1016/j.sbspro.2012.03.101

Crainic, T. G., Ricciardi, N., & Storchi, G. (2004). Advanced freight transportation systems for congested urban areas. *Transportation Research Part C, Emerging Technologies, 12*(2), 119–137. doi:10.1016/j.trc.2004.07.002

Crocco, F., De Marco, S., Iaquinta, P., & Mongelli, D. W. (2010). Freight transport in urban areas: An integrated system of models to simulate freight demand and passengers demand for purchase trips. *International Journal of Mathematical models and methods in applied sciences, 4*(4), 295-273.

Cubukcu, K. M. (2001). Factors affecting shopping trip generation rates in metropolitan areas. *Studies in Regional and Urban Planning, 9*, 51–68.

D'Este, G. (2000). Urban freight movement modeling. In Hensher, D., & Button, K. (Eds.), Handbook of Transport Modelling (pp. 539-552). Amsterdam: Pergamon.

Daganzo, C. F. (1997). *Fundamentals of transportation and traffic operations*. Emerald Group Publishing Limited.

De Langhe, K. (2014). Analysing the Role of Rail in Urban Freight Distribution. *Next Generation Supply Chains: Trends and Opportunities, 18*, 223.

Deflorio, F. P., Gonzalez-Feliu, J., Perboli, G., & Tadei, R. (2012). The influence of time windows on the costs of urban freight distribution services in city logistics applications. *European Journal of Transport and Infrastructure Research, 12*(3), 256–274.

Demetsky, M. J. (1974). Measurement of Urban Commodity Movements. *Transportation Research Record, 496*, 57–67.

Durand, B., & Gonzalez-Feliu, J. (2012). Impacts of proximity deliveries on e-grocery trips. *Supply Chain Forum: An International Journal, 13*(1), 10-19.

ECMT Economic Research Centre (Ed.), *Freight Transport and the City. Round Table 109*. Paris, France: ECMT publications.

Eriksson, J. R. (1997). *Urban freight Transport forecasting – an empirical approach. Urban Transportation and the Environment II* (pp. 359–369). Ashurst: Computational Mechanism Publications.

Esser, K., & Kurte, J. (2006). B2C e-commerce: impact on transport in urban areas. In E. Taniguchi & R. G. Thompson (Eds.), *Recent Advances in City Logistics*. Amsterdam: Elsevier.

Faure, L., Burlat, P., & Marquès, G. (2016). Evaluate the viability of Urban Consolidation Centre with regards to urban morphology. *Transportation Research Procedia, 12*, 348–356. doi:10.1016/j.trpro.2016.02.071

Filippi, F., Nuzzolo, A., Comi, A., & Delle Site, P. (2010). Ex-ante assessment of urban freight transport policies. *Procedia: Social and Behavioral Sciences, 2*(3), 6332–6342. doi:10.1016/j.sbspro.2010.04.042

Gentile, G. and Vigo, D. (2013). Movement generation and trip distribution for freight demand modelling applied to city logistics. *European Transport / Trasporti Europei*, 54 (6), 1-27.

González-Calderón, C., Holguín-Veras, J., & Tavasszy, L. (2016, October 17-19). Freight (trip) generation modeling in the Netherlands. *Proceedings of the Third VREF conference on urban freight: Plan for the future – sharing urban space*, Gothenburg, Sweden.

Gonzalez-Feliu, J. (2008). *Models and Methods for the City Logistics – The Two-echelon Capacitated Vehicle Routing Problem* [PhD. Thesis]. Politecnico di Torino, Turin.

Gonzalez-Feliu, J. (2014, August 21-22). Modelling urban goods transport demand: determinants, variables, methods and applications. *Proceedings of the Congreso Internacional, Industria y Organizaciones – Logística para la Competitividad*, Bogotà, Colombia.

Gonzalez-Feliu, J. (2016a). *La logistique urbaine durable. Monographie d'Habilitation à Diriger des Recherches*. Université Paris-Est, Champs-sur-Marne.

Gonzalez-Feliu, J. (2016b, June 1-4). Viability and potential demand capitation of urban freight tramways via demand-supply modelling and cost benefit analysis. *Proceedings of the 6th International Conference in Informations Systems, Logistics and Supply Chain*, Bordeaux, France, Kedge Business School, Bordeaux.

Gonzalez-Feliu, J., Ambrosini, C., Henriot, F., & Routhier, J. L. (2012a). La logistique urbaine à l'épreuve du facteur 4. *Recherche Transports Sécurité*, *28*, 88–100.

Gonzalez-Feliu, J., Ambrosini, C., Pluvinet, P., Toilier, F., & Routhier, J. L. (2012b). A simulation framework for evaluating the impacts of urban goods transport in terms of road occupancy. *Journal of Computational Science*, *6*(4), 206–215. doi:10.1016/j.jocs.2012.04.003

Gonzalez-Feliu, J., Basck, P., & Morganti, E. (2013a). Urban logistics solutions and financing mechanisms: a scenario assessment analysis. *European Transport/Trasporti Europei*, *54*(11), 1-16.

Gonzalez-Feliu, J., Cedillo-Campos, M.G., Garcia-Alcaraz, J.L. (2014a). An emission model as an alternative to O-D matrix in urban goods transport modelling. *Dyna, Journal of the Facultad de Minas, Universidad Nacional de Colombia – sede Medellin*, *81*(187), 249-256.

Gonzalez-Feliu, J., Durand, B., & Andriankaja, D. (2012c). Urban logistics and e-grocery: have new B2C trends a positive impact on the environment? In P. Golinska & C.A. Romano (Eds.), Environmental issues in supply chain management - new trends and applications (pp. 251-264). Springer.

Gonzalez-Feliu, J., & Morana, J. (2014). Assessing urban logistics pooling sustainability via a hierarchic dashboard from a group decision perspective. In C. Macharis, S. Melo, J. Woxenius, & T. Van Lier (Eds.), *Sustainable Logistics* (pp. 113–135). doi:10.1108/S2044-994120140000006004

Gonzalez-Feliu, J., Morana, J., Salanova Grau, J. M., & Ma, T. Y. (2013b). Design and scenario assessment for collaborative logistics and freight transport systems. *International Journal of Transport Economics*, *40*(2), 207–240.

Gonzalez-Feliu, J., Muñuzuri, J., Cedillo-Campos, M. G., Ambrosini, C., Taniguchi, E., & Chiabaut, N. (2015). Contraintes de livraison et politiques daccès au centre-ville: À la recherche du véhicule optimal urbain par une simulation de scénarios. *Logistique & Management*, *23*(2), 31–45. doi:10.1080/12507970.2015.11673822

Gonzalez-Feliu, J., Pluvinet, P., Serouge, M., & Gardrat, M. (2013c). Urban Freight Analysis Based on GPS Data. In Y. H. Hsuesh (Ed.), *Global Positioning Systems: Signal Structure, Applications and Sources of Error and Biases* (pp. 73–93). New York: Nova Science Publishers.

Gonzalez-Feliu, J., & Routhier, J. L. (2012). Modelling urban goods movement: How to be oriented with so many approaches? *Procedia: Social and Behavioral Sciences*, *39*, 89–100. doi:10.1016/j.sbspro.2012.03.093

Gonzalez-Feliu, J., Sanchez-Diaz, I., & Ambrosini, C. (2016, January 10-14). Aggregation level, variability and linear hypotheses for urban delivery generation models, *Proceedings of the 95th Transportation Research Board Annual Symposium Compendium of Proceedings*, Washington.

Gonzalez-Feliu, J., Semet, F., & Routhier, J. L. (Eds.). (2014b). *Sustainable urban logistics: Concepts, methods and information systems*. Heidelberg: Springer. doi:10.1007/978-3-642-31788-0

Gonzalez-Feliu, J., Taniguchi, E., & Faivre d'Arcier, B. (2014c). Financing Urban Logistics Projects. From Public Utility to Public–Private Partnerships. In J. Gonzalez-Feliu, F. Semet, & J. L. Routhier (Eds.), *Sustainable urban logistics: concepts, methods and information systems* (pp. 245–264). Heidelberg: Springer. doi:10.1007/978-3-642-31788-0_13

Gonzalez-Feliu, J., Toilier, F., Ambrosini, C., & Routhier, J. L. (2014d). Estimated data production for urban goods transport diagnosis. In J. Gonzalez-Feliu, F. Semet, & J. L. Routhier (Eds.), *Sustainable urban logistics: concepts, methods and information systems* (pp. 113–144). Heidelberg: Springer. doi:10.1007/978-3-642-31788-0_7

Gonzalez-Feliu, J., Toilier, F., & Routhier, J. L. (2010b). End consumer movement generation in French medium urban areas. *Procedia: Social and Behavioral Sciences*, *2*(3), 6189–6204. doi:10.1016/j.sbspro.2010.04.030

Holguin-Veras, J. (2016, October 17-19). Applications of Data from Establishment Surveys. *Proceedings of the Third VREF conference on urban freight: Plan for the future – sharing urban space*, Gothenburg, Sweden.

Holguín-Veras, J., Jaller, M., Destro, L., Ban, X., Lawson, C., & Levinson, H. S. (2011). Freight Generation, Freight Trip Generation, and the Perils of Using Constant Trip Rates. *Transportation Research Record*, *2224*, 68–81. doi:10.3141/2224-09

Holguín-Veras, J., Jaller, M., Sánchez-Díaz, I., Wojtowicz, J., Campbell, S., Levinson, H., & Tavasszy, L. (2012). *NCHRP Report 739/NCFRP Report 19: freight trip generation and land use*. Washington, DC: Transportation Research Board of the National Academies.

Holguin-Veras, J., Sánchez-Díaz, I., Lawson, C., Jaller, M., Campbell, S., Levinson, H. S., & Shin, H. S. (2013). Transferability of Freight Trip Generation Models. *Transportation Research Record*, *2379*, 1–8. doi:10.3141/2379-01

Hosoya, R., Sano, K., Ieda, H., Kato, H., & Fukuda, A. (2003). Evaluation of logistics policies in the Tokyo metropolitan area using a micro-simulation model for urban goods movement. *Journal of the Eastern Asia Society for Transportation Studies*, *5*, 3097–3110.

Hunt, J. D., & Stefan, K. J. (2007). Tour-based microsimulation of urban commercial movements. *Transportation Research Part B: Methodological*, *41*(9), 981–1013. doi:10.1016/j.trb.2007.04.009

Jaller, M., Sánchez-Díaz, I., & Holguín-Veras, J. (2015). Identifying Freight Intermediaries: Implications for Modeling of Freight Trip Generation. *Transportation Research Record: Journal of the Transportation Research Board*, *2478*, 48–56. doi:10.3141/2478-06

Kidd, J., & Stumm, M. (2005). From pre-Medieval to post-Modern times: Logistics routes and their modalities have not changed much. *Management Decision*, *43*(9), 1249–1261. doi:10.1108/00251740510626308

Kohler, U. (1997, October 21-24). An innovating concept for city-logistics. In Mobility for everyone. *Proceedings of the 4th World Congress on Intelligent Transport Systems*.

Lawson, C., Holguín-Veras, J., Sánchez-Díaz, I., Jaller, M., Campbell, S., & Powers, E. (2012). Estimated generation of freight trips based on land use. *Transportation Research Record: Journal of the Transportation Research Board*, *2269*, 65–72. doi:10.3141/2269-08

Leonardi, J., Browne, M., & Allen, J. (2012). Before-after assessment of a logistics trial with clean urban freight vehicles: A case study in London. *Procedia: Social and Behavioral Sciences*, *39*, 146–157. doi:10.1016/j.sbspro.2012.03.097

Lindholm, M. (2010). A sustainable perspective on urban freight transport: Factors affecting local authorities in the planning procedures. *Procedia: Social and Behavioral Sciences*, *2*(3), 6205–6216. doi:10.1016/j.sbspro.2010.04.031

Lopez-Ruiz, H., & Crozet, Y. (2010). Sustainable Transport in France: Is a 75% Reduction in Carbon Dioxide Emissions Attainable? *Transportation Research Record*, *2163*, 124–132. doi:10.3141/2163-14

Macharis, C. (2014, April 28). Innovative solutions for sustainable logistics (invited speech). Presentation at Logistics Day 2014 – Cluster for Logistics, Luxembourg.

Macharis, C., De Witte, A., & Turcksin, L. (2010). The Multi-Actor Multi-Criteria Analysis (MAMCA) application in the Flemish long-term decision making process on mobility and logistics. *Transport Policy*, *17*(5), 303–311. doi:10.1016/j.tranpol.2010.02.004

Macharis, C., & Kin, B. (2016). The 4 A's of Sustainable City Distribution: Innovative Solutions and Challenges ahead. *International Journal of Sustainable Transportation*. doi:/ doi:10.1080/15568318.2016.1196404

Macharis, C., & Melo, S. (Eds.). (2011). *City distribution and urban freight transport: multiple perspectives*. Edward Elgar Publishing. doi:10.4337/9780857932754

Macharis, C., Milan, L., & Verlinde, S. (2014). A stakeholder-based multicriteria evaluation framework for city distribution. *Research in Transportation Business & Management*, *11*, 75–84. doi:10.1016/j.rtbm.2014.06.004

Mancini, S. (2013). Multi-echelon distribution systems in city logistics. *European Transport/Trasporti Europei*, *54*(2), 1- 24.

Massari, F., Monier, V., Serouge, M., Gonzalez-Feliu, J., Le Van, E., Chèze, C., & Morana, J. (2013). La collecte et le transport des produits usages et des déchets dans une optique de logistique inverse (Synthèse/Extended Abstract). Etat des connaissances et propositions méthodologiques. RECORD, Villeurbanne. (English and French versions).

Masson, S., & Petiot, R. (2012). Attractivité territoriale, infrastructures logistiques et développement durable. *Les Cahiers Scientifiques du Transport*, *61*, 63–90.

Melo, S. (2010). Evaluation of urban goods distribution initiatives towards mobility and sustainability: indicators, stakeholders and assessment tools [PhD. Thesis]. Porto University.

Melo, S., & Costa, A. (2007). Effects of collaborative systems on urban goods distribution. *Proceedings of the 11th World Conference on Transport Research, Berkeley*.

Melo, S., & Costa, A. (2011). Definition of a set of indicators to evaluate the performance of urban goods distribution initiatives. In C. Macharis & S. Melo (Eds.), *City Distribution and Urban Freight Transport: Multiple Perspectives* (pp. 120–149). Northampton: Edward Elgar. doi:10.4337/9780857932754.00013

Ministerie, V. R. O. M. (1996). *Handboek vrachtverkeer in gemeenten*. Ede: CROW.

Mizutani, T. (1999). Measures to enhance the efficiency of urban freight transportation in Hiroshima city. *Proceedings of the First International Conference on City Logistics*, Cairns, Queensland, Australia.

Morana, J. (2013). *Sustainable supply chain management*. John Wiley & Sons. doi:10.1002/9781118604069

Morana, J. (2014). Sustainable supply chain management in urban logistics. In J. Gonzalez-Feliu, F. Semet, & J. L. Routhier (Eds.), *Sustainable urban logistics: Concepts, methods and information systems* (pp. 21–35). Springer Berlin Heidelberg. doi:10.1007/978-3-642-31788-0_2

Morana, J., & Gonzalez-Feliu, J. (2015). A sustainable urban logistics dashboard from the perspective of a group of operational managers. *Management Research Review*, *38*(10), 1068–1085. doi:10.1108/MRR-11-2014-0260

Morana, J., Gonzalez-Feliu, J., & Semet, F. (2014). Urban Consolidation and Logistics Pooling. In J. Gonzalez-Feliu, F. Semet, & J. L. Routhier (Eds.), *Sustainable Urban Logistics: Concepts, Methods and Information Systems* (pp. 187–210). Heidelberg: Springer. doi:10.1007/978-3-642-31788-0_10

Morganti, E. (2011). Urban food planning, city logistics and sustainability: the role of the wholesale produce market. The cases of Parma and Bologna food hubs [Thèse de doctorat]. Université de Bologne.

Muñuzuri, J., Cortés, P., Onieva, L., & Guadix, J. (2009). Modeling freight delivery flows: Missing link of urban transport analysis. *Journal of Urban Planning and Development*, *135*(3), 91–99. doi:10.1061/(ASCE)UP.1943-5444.0000011

Muñuzuri, J., Cortés, P., Onieva, L., & Guadix, J. (2011). Estimation of daily vehicle flows for urban freight deliveries. *Journal of Urban Planning and Development*, *138*(1), 43–52. doi:10.1061/(ASCE)UP.1943-5444.0000099

Muñuzuri, J., Grosso, R., Cortés, P., & Guadix, J. (2013). Estimating the extra costs imposed on delivery vehicles using access time windows in a city. *Computers, Environment and Urban Systems*, *41*, 262–275. doi:10.1016/j.compenvurbsys.2012.05.005

Nuzzolo, A., & Comi, A. (2014). Urban freight demand forecasting: A mixed quantity/delivery/vehicle-based model. *Transportation Research Part E, Logistics and Transportation Review*, *65*, 84–98. doi:10.1016/j.tre.2013.12.014

Nuzzolo, A., Comi, A., & Rosati, L. (2014). City logistics long-term planning: Simulation of shopping mobility and goods restocking and related support systems. *International Journal of Urban Sciences*, *18*(2), 201–217. doi:10.1080/12265934.2014.928601

Oliveira, L. K. (2014). Diagnosis of loading and unloading spaces for urban freight distribution: A case study in Belo Horizonte. *Journal of Transport Literature*, *8*(1), 178–209. doi:10.1590/S2238-10312014000100009

Ooishi, R., & Taniguchi, E. (1999). Effects and profitability of constructing the new underground freight transport system. *Proceedings of the First International Conference on City Logistics*, Cairns, Queensland, Australia.

Ortúzar, J. de D., & Willumsen, L. G. (2001). *Modelling transport*. Chichester: John Wiley & Sons.

Patier, D. (2002). *La logistique dans la ville*. Paris: Celse.

Patier, D., & Browne, M. (2010). A methodology for the evaluation of urban logistics innovations. *Procedia: Social and Behavioral Sciences*, *2*(3), 6229–6241. doi:10.1016/j.sbspro.2010.04.033

Petit, P. (1974). *Histoire générale de l'Empire Romain* (Vol. 1). Éditions du Seuil.

Pluvinet, P., Gonzalez-Feliu, J., & Ambrosini, C. (2012a). GPS data analysis for understanding urban goods movement. *Procedia: Social and Behavioral Sciences*, *39*, 450–462. doi:10.1016/j.sbspro.2012.03.121

Quak, H. (2008). *Sustainability of urban freight transport: Retail distribution and local regulations in cities (No. EPS-2008-124-LIS). Erasmus Research Institute of Management*. ERIM.

Quak, H. J., & de Koster, M. R. B. (2009). Delivering goods in urban areas: How to deal with urban policy restrictions and the environment. *Transportation Science*, *43*(2), 211–227. doi:10.1287/trsc.1080.0235

Qureshi, A. G., & Hanaoka, S. (2006, August). Analysis of the effects of a cooperative delivery system in Bangkok. In E. Taniguchi & R. G. Thompson (Eds.), *Recent Advances in City Logistics*. Amsterdam: Elsevier. doi:10.1016/B978-008044799-5/50107-8

Rosini, R. (Ed.). (2005). *City Ports Project. Interim report. Quaderni del servizio Pianificazione dei Trasporti e Logistica n. 7*. Bologna, Italy: Regione Emilia Romagna.

Ruske, W. (1994). City logistics-solutions for urban commercial transport by cooperative operations management. Proceedings of the *OECD Seminar on Advanced Road Transport Technologies*, Omiya, Japan.

Russo, F., & Comi, A. (2006). Demand model for city logistics: a state of the art and a proposed integrated system. In E. Taniguchi & R. G. Thompson (Eds.), *Recent Advances for City Logistics* (pp. 91–105). Amsterdam: Elsevier.

Russo, F., & Comi, A. (2010). A modeling system to simulate goods movements at an urban scale. *Transportation*, *37*(6), 987–1009. doi:10.100711116-010-9276-y

Salanova-Grau, J. M., & Gonzalez-Feliu, J. (2015). Comparing fast VRP algorithms for collaborative urban freight transport systems: A solution probleming analysis. *The IUP Journal of Supply Chain Management*, *12*(2), 7–23.

Sanchez-Díaz, I. (2016). Modeling urban freight generation: A study of commercial establishments freight needs. *Transportation Research Part A, Policy and Practice*. doi:10.1016/j.tra.2016.06.035

Sanchez-Diaz, I., Gonzalez-Feliu, J., & Ambrosini, C. (2016, June 1-4). Assessing the implications of aggregating data by activity-based categories for urban freight trip generation modeling. *Proceedings of the 6th International Conference in Informations Systems, Logistics and Supply Chain*, Bordeaux, France.

Sánchez-Díaz, I., Holguín-Veras, J., & Ban, X. J. (2015b). A time-dependent freight tour synthesis model. *Transportation Research Part B: Methodological*, *78*, 144–168.

Sanchez-Díaz, I., Holguin-Veras, J., & Wang, X. (2015a). An exploratory analysis of spatial effects on freight trip attraction. *Transportation*.

Ségalou, E., Ambrosini, C., & Routhier, J. L. (2004). The environmental assessment of urban goods movement. In E. Taniguchi & R. G. Thompson (Eds.), *Logistics Systems for Sustainable Cities* (pp. 207–220). Amsterdam: Elsevier. doi:10.1108/9780080473222-015

Silva, M. R., & Waisman, J. (2007) Urban Freight: an exploratory study about freight truck generation in bars and restaurants. *Proceedings of the 16º Brazilian Congress of Transport and Traffic*, Maceio, Brazil (in Portuguese).

Sonntag, H. (1985). *A Computer Model of Urban Commercial Traffic . Transport, Policy and Decision Making*, 3(2).

Spinedi, M. (Ed.). (2008). Logistica urbana: dagli aspetti teorici alle applicazioni pratiche. Esperienze italiane e straniere a confronto. *Proceedings of the Emilia Romagna Region, City Logistics Expo*, Bologna, Italy.

Srivastava, S. (2007). Green supply-chain management: A state-of-the-art literature review. *International Journal of Management Reviews*, *9*(1), 53–80. doi:10.1111/j.1468-2370.2007.00202.x

Stivers, R. L. (1976). *The sustainable society: ethics and economic growth*. Philadelphia, PA: Westminster Press.

Takahashi, Y., & Hyodo, T. (1999). A simulation study on the effect of Physical Distribution Facilities in the Tokyo Metropolitan Region. *Proceedings of the First International Conference on City Logistics*, Cairns, Queensland, Australia.

Tamagawa, D., Taniguchi, E., & Yamada, T. (2010). Evaluating city logistics measures using a multi-agent model. *Procedia: Social and Behavioral Sciences*, *2*(3), 6002–6012. doi:10.1016/j.sbspro.2010.04.014

Taniguchi, E. (2016, October 17-19). Planning the future – sharing urban space for city logistics. *Proceedings of the Third VREF conference on urban freight: Plan for the future – sharing urban space*, Gothenburg, Sweden.

Taniguchi, E., & Thompson, R. G. (1999). *City Logistics I*. Kyoto: Institute for City Logistics.

Taniguchi, E., Thompson, R. G., & Yamada, T. (2003). Predicting the effects of city logistics schemes. *Transport Reviews*, *23*(4), 489–515. doi:10.1080/0144164021016399 9

Taniguchi, E., Thompson, R. G., & Yamada, T. (2012). Emerging techniques for enhancing the practical application of city logistics models. *Procedia: Social and Behavioral Sciences*, *39*, 3–18. doi:10.1016/j.sbspro.2012.03.087

Taniguchi, E., Thompson, R. G., & Yamada, T. (2014). Recent trends and innovations in modelling city logistics. *Procedia: Social and Behavioral Sciences*, *125*, 4–14.

Taniguchi, E., Thomson, R. G., Yamada, T., & Van Duin, R. (2001). *City Logistics - Network modelling and Intelligent Transport System*. Amsterdam, The Netherlands: Elsevier. doi:10.1108/9780585473840

Taniguchi, E., & Van Der Heijden, R. E. (2000). An evaluation methodology for city logistics. *Transport Reviews*, *20*(1), 65–90. doi:10.1080/014416400295347

Toth, P., & Vigo, D. (2002). *The vehicle routing problem.* SIAM Monographs on Discrete Mathematics and Applications.

Tozzi, M., Corazza, M. V., & Musso, A. (2014). Urban goods movements in a sensitive context: The case of Parma. *Research in Transportation Business & Management, 11*, 134–141. doi:10.1016/j.rtbm.2014.03.003

Vaghi, C., & Percoco, M. (2011). City logistics in Italy: success factors and environmental performance. In C. Macharis & S. Melo (Eds.), *City Distribution and Urban Freight Transport: Multiple Perspectives* (pp. 151–175). Northampton: Edward Elgar. doi:10.4337/9780857932754.00015

Van Duin, J. H. R. (2012). *Logistics concept development in multi-actor environments* [PhD. Thesis]. Delft University of Technology.

Van Duin, R., Quak, H., & Muñuzuri, J. (2008). *Revival of cost benefit analysis for evaluating the city distribution centre concept? Innovations in city logistics* (pp. 97–114). New York: Nova Science.

Vickerman, R. W., & Barmby, T. A. (1985). Household Trip Generation Choice: Alternative Empirical Approaches. *Transportation Research Part B: Methodological, 19*(6), 471–479. doi:10.1016/0191-2615(85)90042-6

Virlouvet, C. (2000). Les denrées alimentaires dans les archives des Sulpicii de Pouzzoles. *Cahiers du Centre Gustave Glotz, 11*(1), 131–149. doi:10.3406/ccgg.2000.1526

Virlouvet, C. (2006). Encore à propos des" Horrea galbana" de rome: Entrepôts ou ergastules? *Cahiers du Centre Gustave Glotz, 17*(1), 23–60. doi:10.3406/ccgg.2006.898

Visser, J., Van Binsbergen, A., & Nemoto, T. (1999). Urban freight transport policy and planning. In *City logistics I* (pp. 39-69).

Watson, P. L. (1975). *Urban Goods Movement: A Disaggregate Approach, C.D.* Lexington: Heath.

Whiteing, A. E. (1999). Logistics in City Centres: A Planning and Policy Dilemma. In D. Waters (Ed.), Global logistics and distribution planning: strategies for management (pp. 280-290). Boca Raton: CRC Press.

Williamson, J. G. (2002). *Coping with city growth during the British industrial revolution.* Cambridge University Press.

Woudsma, C. (2001). Understanding the movement of goods, not people: Issues, evidence and potential. *Urban Studies (Edinburgh, Scotland)*, *38*(13), 2439–2455. doi:10.1080/00420980120094605

ENDNOTE

1 Public stakeholders are often associated to regulating authorities and their technical services, and are in general in charge of transport plans and urban policy. Private stakeholders can be of three main natures: (1) shippers and receivers, that need the goods, (2) transporters that transport the goods and (3) auxiliary functions like logistics service providers (3PL, 4PL, 5PL), real state stakeholders or other transport auxiliary stakeholders.

This research was previously published in Green Initiatives for Business Sustainability and Value Creation edited by Arun Kumar Paul, Dipak Kumar Bhattacharyya, and Sandip Anand, pages 103-134, copyright year 2018 by Business Science Reference (an imprint of IGI Global).

Chapter 9
Operationalising Sustainability within Smart Cities:
Towards an Online Sustainability Indicator Tool

Conor Mark Dowling
Future Analytics, Ireland

Stephen M Purcell
Future Analytics, Ireland

Stephen Diarmuid Walsh
Future Analytics, Ireland

William M Hynes
Future Analytics, Ireland

Mary Lee Rhodes
Trinity College, Ireland

ABSTRACT

Sustainable decision making for strategic planning requires support frameworks which can analyse and interpret the multiple interactions underpinning complex urban environments. Such frameworks are often based on economic measures, consequently there is a need for a decision support framework designed to aid work towards a more holistic view of sustainable urban planning projects. As an initial step towards this framework, the authors aim to devise sustainable energy indicators designed to be flexible enough to address the vast extent of urban sustainability processes. In addition, we examine how these indicators need be designed to reflect the broad range of data sources that reflect the three pillars of sustainability, namely: environmental; economic; and social elements. The indicators have been developed specifically for the European urban environment and can be used both internally, for identification of "hot spots", and externally, for sustainability reporting and stakeholder engagement.

DOI: 10.4018/978-1-7998-0441-3.ch009

Copyright © 2020, IGI Global. Copying or distributing in print or electronic forms without written permission of IGI Global is prohibited.

INTRODUCTION

Developing urban sustainability performance indicators will be critical for cities aiming to reach the targets set at the 2015 Paris Climate Conference (COP-21). With the signing of the COP-21 conference agreement world leaders have pledged to keep global temperatures "well below" 2.0°C above pre-industrial times and "endeavour to limit" them even more, to 1.5°C. This will involve limiting the amount of greenhouse gases emitted by human activity to the same levels that trees, soil and oceans can absorb naturally. Each country's contribution to cutting emissions will be reviewed every five years so that objectives may be scaled to the challenge. This paper focuses on the advanced economies of Europe due to the substantial contribution they have to global CO_2 emissions.

It has been argued that traditional measures of urban quality of life and economic performance are inadequate as they fail to ensure a composite analysis of the interdependent systems underpinning urban society. The crossover and interdependence between city systems at a regional and European level is not currently reflected in urban decision-making. This deficiency risks the resilience of urban environments and increases the susceptibility of these centres to the consequences of climate change. It also impedes the ability of these urban centres change and adapt their constituent processes in a sustainable manner. This issue is especially pronounced when it comes to the assessment, measurement and remediation of an urban areas' practices in relation to the consumption of energy.

Past attempts to answer the rudimentary question of whether a city is successfully becoming more sustainable cannot be adequately determined due in part to divergence on which indicators should be employed. Drawing from this, the question arises: How should these indicators be selected and then how should they be aggregated? The basis of this paper is that, for successful assessment of urban sustainability, the required first step is to put forward a set of multi-dimensional indicators. The use of a multi-criterion framework to make the 'measurement of sustainability' operational is discussed here by means of an assessment of key stakeholder perspectives including potential end users, academics, and industry leaders.

The concept of sustainable development is based on the observation that a city's economy, environment and wellbeing can no longer be separated. Newman and Jennings (2008, p. 264) define sustainable urban ecosystems as 'systems which are ethical, effective, self-regulating, resilient, self-renewing, and flexible'. Developing from early natural ecosystems our modern urban systems have become more complex with many interrelated structural layers. We adopt the definition of sustainable development put forward by the World Commission on Environment and Development (WCED); "development that meets the needs of the present without compromising the ability of future generations to meet their own needs," (Brundtland & Mansour,

1987). In terms of operationalising sustainable development, this research aims to enhance the urban environment by minimising energy consumption through green building monitoring technologies and reducing greenhouse gas emissions by providing clear analysis of the urban environment. The manner in which this goal will be achieved is linked to the development of a bespoke decision support tool for city managers and citizens with sufficient functionality to equip those stakeholders with the knowledge needed to alter the patterns of energy consumption in the urban environment. The indicators discussed herein are a vital first component in that process. Refined through the use of stakeholder consultation, the indicators presented herein adhere broadly with the three pillars of sustainability.

The Bellagio Principles (Hardi & Zdan, 1997), which served as guidelines for indicator design, express the need for 'indicators' and 'standardised measurements' of sustainability: When considering indicators for sustainable development, the issue can appear complicated by the lack of any firm foundation on which to base their development. Haughton and Hunter (1994) argue that the concepts of building for the future, resource equity and environmental impact must underpin the process of sustainable development, such that the principles of inter-generational equity and trans-frontier responsibility are at the forefront of sustainable development policy. These concepts continue to be widely debated and most recently world leaders discussed how to address sustainability issues at the 2015 United Nations Climate Change Conference in Paris.

For our purposes, the fundamental outcome of this agreement is the need for sustainability analysis tools for the built environment, particularly relating to the management and quantification of energy use. Critically these outcomes will impact on the societal and economic activity of urban communities (Anttiroiko, 2012). Indicators may be used to retrospectively monitor the long-term sustainability trends of a built urban system. The aggregated information these indicators can provide will help in the development of short-term projections. Through this paper, we put forward 31 sustainable urban indicators (SUIs) with the objective of developing a quantitative assessment index to provide a coherent multidimensional view of the urban environment. These SUIs form the basis of an online interactive planning tool which will aid planners in the analysis of and decision making for sustainable urban environments.

The impetus for the creation of the SUIs is increasingly being reflected in the drive to create low carbon energy districts in the European urban environment. The planned creation of such districts is being driven both by European Union policy and national governments. This will be further enhanced by the approach of the binding deadline for the Energy Efficiency Directive (Directive. 2012/27/EU of The European Parliament and of The Council) and the resultant cost arising from the purchase of Emission Trading Scheme credits in order to rectify exceedance of

Greenhouse Gases (GHG) targets. The "Stability Programme Update" published by the Irish Department of Finance (2016, p28) frames the thinking which is beginning to emerge in governance circles around the risk posed by noncompliance to binding European targets associated with emissions reductions:

There are fiscal risks associated with a legally binding EU Effort Sharing Decision on climate change covering the 2013-2020 period. Ireland is obliged to achieve a 20 per cent Greenhouse Gas emissions reduction (compared to 2005 levels) in certain sectors. Current EPA projections estimate that Ireland will not achieve this reduction and failure to comply may incur costs of hundreds of millions of Euro through the purchase of carbon credits until such time as the target is complied with. Similarly, further new costs may arise in the context of a new EU climate and energy framework for the period 2020-2030, which will set new emissions reduction targets.

The pressures identified above therefore cast a new light on the need and utility of SUIs which have the capacity to inform the development of low carbon energy solutions for the retrofit of existing urban environments and creation of new ones.

There are many examples of 'checklists' for indicators available in the literature (Whate et al., 2006) and these can provide a useful tool for reviewing indicators as they are developed, highlighting any obsolescence which may mean the indicator ultimately needs to be abandoned or subsumed into a more encompassing indicator. A sound theoretical framework is the starting point in constructing composite indicators. The framework put forward clearly defines the phenomenon to be measured and its sub-components; the indicators account for and may be applied to all major systems and activities relevant to developing energy-efficient cities. The indicators will aid in the creation of a 'Smart Sustainable Urban Environment', integrated with urban planning tools including architectural master planning and detailed energy optimisation and environmental analysis. Transparency was a key aim in constructing credible indicators.

Once formulated into a decision support tool SUIs can accurately reflect the impact of energy related interventions in the urban environment. This will be a valuable tool for urban planners: To date, initiatives associated with this discipline have focused on retrofitting the urban environment in order to facilitate low carbon mobility, as evidenced by Sustainable Urban Mobility Plans (Wefering, 2013) or through static initiatives focused on the quantification of carbon emissions through Sustainable Energy Action Plans (SEAPs). The availability of an interactive decisions support tool focused specifically on modelling energy interventions to the urban environment give planners additional scope to sustainably develop new urban areas and retrofit existing ones. The potential to model energy interventions on a city or sub city scale open a new area for urban planners to make meaningful contributions to the ‘energy transition’.

Conceptual Indicator Framework

It is important to note that indicators are not intended to accomplish a required change, but rather they act as catalysts where change is necessary, providing an ‘early warning system’, highlighting areas of concern and thereby enabling decision-makers to initiate the necessary remedial measures:

Indicators quantify change, identify processes and provide a framework for setting targets and monitoring performance (Crabtree & Bayfield, 1998, p. 1).

Indicators for the quantification of sustainable development should provide a continual assessment of the overall sustainability of a system and the indicators themselves will require constant review and updating over time, as changes occur; implementing indicators is a dynamic process. In providing a means for monitoring progress towards sustainability, indicators are also an important communication tool: “Communication is the main function of indicators: they should enable or promote information exchange regarding the issue they address,” (Smeets & Weterings, 1999, p. 35). There are often complex issues and intricate processes underlying indicator work and while it is important to maintain a sufficient level of detail and transparency in the process, so that data can be tracked and decisions justified, there remains a need to achieve a certain level of simplicity in the end result. Indicators must be meaningful and useable by all and not limited to the ‘experts’. Thus, a balance between quantity and depth of analytical indicators is needed. Public consultation and stakeholder participation throughout the indicator development process can play a significant role. An indicator should measure what those concerned are interested in and must provide meaningful information, enabling action to be taken.

A conceptual framework allows for the coherent and consistent selection of indicators. This is particularly important given that any indicator selection process is value laden, for example stakeholder opinion may differ over the weight given to different criteria for a good indicator; assuming a trade-off between cost and complexity; the objectives chosen; the baseline and benchmark data, etc. Thus, having an explicit framework allows a more transparent, responsive and robust process for indicator selection. Waldron and Williams describe five broad categories of frameworks (Waldron & Williams, 2002):

1. **Domain-Based**: Addressing a variety of performance issues to include social, economic and environmental but not necessarily linking with specific management goals
2. **Goal-Based**: To identify indicators that respond directly to sustainability goals but do not address interrelationships
3. **Sectoral**: These respond to the function of a specific management group, and thus are useful in assessing management response to specific issues
4. **Issue-Based**: Often provide a short-term response to address the 'issue of the day'; longer term sustainability implications may be overlooked
5. **Causal Frameworks**: These assess the existing conditions, stresses and responses but within-domain interactions are overlooked

Within these frameworks indicator selection can vary greatly. Kelly and Baker (2002) put forward the following criteria: Indicators should be adapted to suit local needs, consist of both objective and subjective data and make links between different issues. In addition, indicators should be based on;

- **Validity**: Acceptable, believable; related to a higher order theme, related to theory
- **Reliability**: Can be used over time, and across space but only within the region
- **Data Availability and Accessibility**: Qualitative and quantitative
- **Methodology**: Timely, uses existing data, considers scale; considers capacity to collect and use data) and relevance to objectives; beneficial to investors/ taxpayers; links to other indicators.

For our own SUI weighting, it will be possible for the end users to establish the weight of each indicator in order to adapt it to the context in which the tool will be deployed.

While the choice of indicators must be guided by the theoretical framework for the composite, the data selection process can be quite subjective as there may be no single definitive set of indicators (Isaacs et al., 2013). A lack of relevant data may also limit the development of sound composite indicators. Given a scarcity of internationally comparable quantitative hard data, composite indicators may include qualitative soft data from surveys or policy reviews (Nardo et al., 2005). Proxy measures can be used when the desired data are unavailable or when cross-city comparability is limited. To the extent that data permits, the accuracy of proxy measures should be checked through correlation and sensitivity analysis. In designing a framework, close attention should be paid to whether the indicator in question is dependent on size-related factors. To have an objective comparison across small and large cities, scaling of variables by an appropriate size measure, e.g. population or populated land area, is required.

The quality and accuracy of composite indicators should evolve in parallel with improvements in data collection and indicator development. The current trend towards constructing composite indicators of city performance in a range of policy areas may provide further impetus to improving data collection, identifying new data sources and enhancing the comparability of statistics. Poor data will produce poor results. From a pragmatic point of view, however, compromises need to be made when constructing a composite: What is essential is the transparency of any compromises.

Methodology

In line with the rationale advanced in Keirstead (2007), before the drafting process commenced, the goal and scope of the indicators was explored in order to provide a robust foundation upon which the indicators could be based. The goal of the indicators is to reflect changes arising from interventions in the simulated built environment in a manner which is accessible to the citizen, decision maker as well as professional practitioner. It was envisaged that a number of the SUIs would be applicable across the spectrum of spatial scales outlined above, depending on the availability of accurate data and technical feasibility for its representation within a geospatial scene. Due to the expected interrelationship between those domains, the framework for the indicator suite would be advanced as a comprehensive and responsive feature to reflect the impact of modelled energy interventions on urban structures. Consideration was given to the need to incorporate an acknowledgement of temporal as well as seasonal influences on the use of energy. As a result, a seasonal and temporal attribute was built in to the indicator drafting process in order to accurately demonstrate the energy dynamics of the chosen intervention at a specified spatial scale where the data permits. The goal of the devised SUIs is to facilitate the creation of a solid evidence base upon which decisions relating to interventions in the urban environment can be assessed and validated against the principles of sustainable development.

"*The Handbook on Constructing Composite Indicators: Methodology and User Guide*" (OECD, 2005) and the "Bellagio Principles" (Hardi and Zdan, 1997) were used in terms of guiding the definition and the structure of the SUIs. The 10 step process outlined in the handbook detailing the formation of composite indicators informed our own methodology for the selection of the SUIs. In addition to the content of the handbook, existing indices were reviewed and the structure described below was devised for the SUIs in order to transparently outline the constituent elements and rationale for each SUI. The structure as seen in Table 2 is set out as follows:

- **Indicator***:* Refers to the name of the indicator
- **Definition***:* Describes the remit and focus of the indicator
- **Unit of Measurement***:* Refers to the measurable attribute of the indicator
- **Scale***:* Refers to the spatial applicability of the indicator
- **Data Source**: Outlines

The methodological approach for the development of the SUIs consisted of three stages:

Step 1: Indicator Development

The first step involved the compilation of an expansive list of sustainable urban indicators, which relate to energy in the urban environment. This list was drawn together from an expansive review of pre-existing indices and industry standards. Prior to the commencement of the indicator drafting process, a literature review was carried out with the aim of establishing a clear understanding of best practice in indicator development, structuring and challenges therein. The material of the literature review comprises peer-reviewed research articles published from 1990s to 2015 in international peer-reviewed scientific journals, internationally recognized sustainable indicator sets (BREEAM, LEED and Green Star) currently being used in the US, UK and Australia focused on criteria, indicators and sustainability assessment systems linked directly or indirectly with urban energy consumption. Research articles approaching sustainability measurement were gathered up in June 2015 with a literature review in the ScienceDirect and Wiley online library electronic databases by using search words for titles, abstracts and keywords. In addition, once the SUIs had been determined the literature was again reviewed to ensure alignment with best practice approaches. The collation of existing indices, projects and energy management standards applied by key stakeholders active in the urban environment was of vital importance to the delivery of this research. Given the prevalence of such standards a best practice approach was used to adopt and adjust existing indicators where possible.

In order to develop indicators which, align with the requirements of the end users and stakeholders as well as the core objectives, 'domains' which reflect the priorities of both were considered. The domains which were decided upon were emissions reduction, cost reduction, and increased energy efficiency.

The rationale for dividing the indicator development process arose from the need to establish a comprehensive assessment of the existing academic work and professional practice in the field of sustainable energy research, to comprehensively collate existing energy focused indicators and to assess that body of indicators against the needs of the end user communities in both Italy and Ireland. By combining this with existing indices relating to energy and sustainable communities, the indicators could then be assessed with reference to their relevance to the identified domains of cost reduction, emissions reduction and increased energy efficiency.

In the first instance, energy indicators which relate to the domains outlined above were developed from a range of existing indices where possible. At the initial phase of the compilation process, the indicators based on existing indices were cross referenced in order to determine if they arose in more than one indicator list. Niche or uncommon indicators amongst those selected were evaluated on their merits and capacities to contribute to the overall index. Niche indicators found to be of merit were retained while those whose focus did not align with the indicator objectives were discharged. The indicators arising from the stakeholder consultation were categorised under the following headings:

- Demographic dynamics
- Accessibility to services and green areas
- Urban mobility and transport
- Health and safety
- Awareness and behaviours

The indicators put forward were carefully chosen to reflect the environmental issues and measure the sustainability performance of the applied area effectively. Through the literature review, 186 indicators were collated. The indicators were then filtered through the framework set out below. Using the criteria set the initial set of indicators were reduced to 84 SUIs.

SUI Selection Consideration:

1. Alignment with Existing Standards & Indices: Indicators need to reflect local policy objectives, but also needs to be consistent with other international indicator programs to allow comparisons across countries
2. Stakeholder Input: Refers to the solicited preferences of members of the end user community, in terms of indicators best suited to reflect the stated objectives
3. Technical Feasibility: Refers to the feasibility of technically representing a SUI in the decision support tool.

4. Data Availability: Refers to the availability of data to effectively support the operation of the SUI in the decision support tool.
5. Clear consistent focus: linkages with the core objectives of carbon reduction, energy cost reduction, and increased energy efficiency
6. Sensitivity and specificity to the underlying energy phenomenon: The indicator should respond relatively quickly and noticeably to changes
7. Be statistically sound: Indicator measurement needs to be methodologically sound and fit for the purpose to which it is being applied,
8. Be intelligible: It should be sufficiently simple to be interpreted in practice,
9. Be consistent over time: The usefulness of each indicator is related directly to the ability to track trends over time

Step 2: Stakeholder Consultation

The second methodological step consisted of a focused consultation exercise involving leading planning experts and representatives of the end user community. The consultation process took place from August to September 2015.

An analysis of existing energy indicators was undertaken after all energy-focused indicators from existing indices had been compiled into a single, comprehensive inventory. An iterative consultation approach was employed in order to select the indicators from the expanded list referred to above. By utilising end user expertise and priorities, and while being cognisant of the need to imbue the key SUI list with the requisite flexibility to be applicable in multiple contexts, a shortlist of preferred indicators was developed. As the shortlisting process progressed, considerations relating to data availability and consultation with the key stakeholders were examined. The technical feasibility of operationalising the individual SUIs was also considered and the outputs of this exercise facilitated the advancement of certain indicators and the omission of others.

Consultation with stakeholders resulted in the development of indicators which reflected their interests with reference to their experiences of, and interactions with, the urban environment cases in Ireland and Italy. The aim of these consultations was to enable experts from various backgrounds and who are familiar with local conditions, environmental needs and policy priorities to voice their opinions. The consultees included academics, local council employees, and energy providers and we worked towards agreement on a consensus view of the relative importance of the indicators based on their experience and judgment.

The high-level SUI domains developed during this consultation phase reflect the needs of decision makers, urban environment professionals as well as citizens. The domains aided in the cataloguing of the impacts of interventions on the urban environment with the aim of generating savings, reducing carbon and other emissions into the atmosphere and enhancing the overall energy efficiency of the urban environment.

Consultation Phase One

The refinement process of the initial 186 SUIs was undertaken in conjunction with a selection of members of the urban environment professions, energy agencies, academics and national energy research organisations:

TrinityHaus (Trinity College Dublin) is an innovation centre which focuses on energy efficient buildings and eco districts. TrinityHaus evaluated the SUIs in terms of their robustness and academic quality with reference to existing indices and manners of quantifying energy consumption and generation in the urban environment.

Integrated Environmental Solutions evaluated and commented on the compatibility of the Sustainable Urban Indicators with the development of the cloud-based tool. Comment was also sought relating to the relevance of the indicators to professionals interacting with and assessing the impacts of energy related interventions in the built environment.

Dundalk Institute of Technology was consulted with relevance to the applicability of the SUIs to the case study area of Dundalk and the availability of data to support each of the indicators should they be applied to any building in the case area.

D'Appolonia S.p.A. an Italian energy engineering consultancy was requested to examine the indicators and their suitability for application to the Italian case context. This consultation also aimed to solicit information in relation to their own end user and stakeholder requirements.

The outputs of the first consultation phase were employed to refine the manner in which each indicator was structured and presented subsequently. This included the manner in which the indicators were categorised in initial sub domains. Where indicators overlapped to a substantial degree or were found not to be relevant, they were amalgamated or removed.

Consultation Phase Two

The next stage of the consultation process consisted of a refining of the indicator list by harnessing the insights of the end user community. Specific feedback was required in order to refine the indicators which would guide the use of an operational tool in the context of monitoring the impacts of energy interventions in the built environment. In order to facilitate meetings and clear communication initial consultation was made with possible Irish end users:

Sustainable Energy Authority of Ireland (SEAI) was established as Ireland's national energy authority under the Sustainable Energy Act 2002. SEAI's mission is to play a leading role in transforming Ireland into a society based on sustainable energy structures, technologies and practices. Individuals from the Energy Policy & Modelling Group, and the Energy Policy Statistical Support Unit were consulted.

Eirgrid is a state-owned company that manages and operates the transmission grid across the island of Ireland, the Public Planning department body was consulted.

CODEMA the Dublin Energy Agency is committed to working with Dublin's local authorities on improving the energy efficiency in Dublin in order to reduce the city's CO2 emissions and achieve the "20-20-20" targets. The energy planning department consulted have experience managing energy efficiency and improvement programmes with Dublin City Council.

Irish Planning Institute (IPI) is the professional body representing the majority of professional planners engaged in physical, spatial and environmental planning in Ireland and Irish planners practicing overseas. The indicator list was referred to the Institute's Policy and Research Group. The purpose of the Group is to enhance the Institute's research capacity and that of the profession as a whole through independent and joint research, this is to be reinforced by and inform policy submissions and the Charter-ship and CPD programmes.

The consultation undertaken in phase two validated the selection of the remaining 50 SUI's from the perspectives of the energy experts who commented on them. They key messages arising from this consultation involved the number of indicators that were present on the list, the availability of data to support their application and the manner in which energy interventions are modelled in the decision support tool. Individual feedback on the indicators, where provided, facilitated the refinement of the definition, scale and unit of measurement, which would be applied within the decision support tool.

Consultation Phase Three

The phase three-consultation stage was focused on the further refinement of the SUI list down to the final 31 SUIs. This phase utilised a host of European end users including Integrated Environmental Solutions Ltd, ESRI Zurich, Dundalk Institute of Technology, Ospedali Galliera, and Louth County Council. Using a workshop format held in Zurich on the 14th and 15th of September 2015. The purpose of this phase was to refine the expanded list of key indicators down to an optimal number for application and eventual integration in the decision support tool. The SUIs were discussed in detail amongst the end users in terms of the utility of each and the manner in which they would be integrated into a cloud based decision support tool.

While our core aim was to devise SUIs which focus on the domains of cost reduction, energy efficiency and reducing carbon dioxide emissions it was also important to characterise the city as an overall integrated system and develop the SUIs to be combined with a "sub-systems of a system" approach to the urban environment. The indicators were effective in the manner in which they considered the components of the built environment and as such, they are attuned to the interrelated nature of the urban environment. The manner in which one indicator may be utilised to correlate the impact of interventions is reflective of the interconnected nature of urban environment systems in relation to the provision and consumption of energy. The characterisation of the city and its energy network in the manner outlined above assisted in the development of indicators which align with the constituent components of the urban environment under the three objectives specified above.

Step 3: Link with Case Data

The third methodological step consisted of the integration of the SUIs into an online visualisation and decision support tool for use in the urban environment. This involved the development of data for case sites of Dundalk, Ireland and Genoa, Italy. These sites were chosen based on access to information provided by the local authorities in both cases which is required to test the SUIs and the decision support tool. The two sites also benefit from the contrasting climates in each location which will impact energy use. The local authorities provided us with data sets which included mapping, building type and material information which we would not have had otherwise. The sites will play a large role in the demonstration and operation of the sustainability tool.

Through the consultation process indicators selected during the initial literature and consultation phases were weighted by key stakeholders including academics, local government officials, energy providers and potential end user with regards to how each indicator pertained to their individual needs. The further review of the indicators selected in relation to supporting data available ensured the final SUIs would be usable across different urban environments and our own case studies in Dundalk and Genoa so selected as two very different cities with very different climates, which could be compared with one another.

The scope of the SUIs relates to two key areas, the first is their feasible area of spatial applicability and the second related to the manner in which the interventions are captured and represented as a unit of measurement with an associated range of effectiveness in line with the goal stated above. The spatial applicability of the indicators are as follows:

- **City Level***:* Indicators at this scale provide a high-level assessment of the performance of urban environments in respect of energy consumption, generation and potential for efficiency focused enhancements.
- **District Level**: Indicators at this scale provide a more focused and detailed assessment of energy generation and consumption at the sub city scale. This focus is valuable as urban environments are composed of sub city districts which are distinctive in terms of their energy characteristics. Developing metrics at this spatial scale requires cognisance of the fact that the definition of a "district" or sub city unit varies from country to country.
- **Building Level**: Indicators at this scale are geared toward the direct assessment of energy related interventions on a building or small cluster of buildings.

The spatial scales outlined above are not mutually exclusive in terms of indicators as some indicators may be applicable across each spatial boundary. In addition to the spatial scale of the SUIs and their applicability, their scope is required to cover the elements of the built environment which have a relationship to energy.

A system of default weightings can be applied to the SUIs in order to afford the end user to customise their selection. The operationalisation of the indicators in subsequent development work requires that default weightings be applied in advance of the final iteration. Assigning a default weighting is congruent with the majority of composite indices, as stated in OECD, 2008. It is recommended that, in order to enhance the utility of the decision making, end users be allowed to establish the weight of each indicator in order to adapt to the context in which the tool will be deployed.

As part of our further research into the area a data catalogue was also developed in a manner which recognised the need to be applicable across multiple urban contexts. A key requirement, in this regard, was the need for commonality and standardisation of data characteristics, associated attributes and accepted terminology in order to present a catalogue that represents most urban contexts within the EU while at the same time providing an extensive data coverage best suited to determine the efficiencies and potential solutions in energy generation, transmission and consumption. Accordingly, eight categories of data were identified as being an appropriate data classification that corresponds to these requirements. Data corresponding to these categories were presented in a manner that captured, where possible, the key characteristics, attributes, units, sources and reference standards. In addition, where possible each data type was delineated according to their spatial scale of applicability based on micro, meso and macro scales (see Table 1).

Table 1. Sustainable urban indicator list

No	Indicator	Definition	Unit of Measurement	Scale	Data Source
1.	**Consumption of renewables**	Renewable energy consumption as a % of total energy consumption in the urban area.	kWh divided by the total kWh consumed in the city over what period of time	City Level	EU SDI theme: Climate change and energy. EU SDI Level 1
2.	**GHG emissions per sector (residential, industry, use type)**	The per capita emissions (Tonnes of Emissions eq. per individual).	C02/CH4/N2O/O3/CFC's emissions per unit GDP (g/€) (GHG emissions in tonnes per capita)	City Level	EU SDI sub-theme: Climate change. EU SDI sub-theme: Climate change Level 2
3.	**Renewable plants (managed by public/private authority) kWh produced**	Renewable electricity produced by LA/PA (kWh). (managed by public/ private authority)	kWh per annum	City Level	EU SDI level 3 Electricity generation from renewable Urban Ecosystem Europe indicator, theme local to global: energy and climate change
4.	**Energy consumption of public buildings**	The energy consumption of municipal buildings per sq. m measure.	Average kWh	City Level	Urban Ecosystem Europe indicator, theme local to global: energy and climate change
5.	**Energy consumption of public buildings II**	% of public buildings that have been audited in terms of their energy usage (i.e. building energy rating/efficiency assessment)	% of public buildings audited	City Level	EU SDI sub-theme: Resource use and waste. EU SDI level 2
6.	**Brownfield versus greenfield development**	Ratio of new developments on brownfield to new developments on Greenfields.	m2	City/ District Level	TISSUE CORE 1
7.	**Roof Space available for PV**	Area of roof space suitable for PV.	m2	Building Level	Determined by Consultation with end users and industry leaders
8.	**Roof Space Available for Solar Thermal**	Area of roof space suitable for Solar Thermal.	m2	Building Level	Determined by Consultation with end users and industry leaders

continues on following page

Table 1. Continued

No	Indicator	Definition	Unit of Measurement	Scale	Data Source
9.	**Final energy consumption, by sector**	The total electricity consumption per capita (kWh/individual) Trends in electricity consumption.	Energy consumption per capita (GJ/inhabitant)	City Level	EU SDI sub-theme: Consumption patterns. EU SDI sub-theme: Consumption patterns level 3
10.	**Energy dependency**	The percentage of total electricity consumption per capita generated elsewhere and imported.	% of total kWh use	City Level	EU SDI sub-theme: Climate change. EU SDI sub-theme: Climate change Level 2
11.	**Total waste generated per year per capita**	Municipal waste per capita	kg/inhabitant	City Level	DCC indicator
12.	**Building stock energy efficiency**	New buildings and renovations assessed in terms of environmental sustainability. % of sustainably classified buildings (both new and renovated).	Energy consumption of residential buildings (MJ/m2) by building type residential, public, commercial as % of total stock	City Level	Determined by Consultation with end users and industry leaders
13.	**Residential Stock Energy Efficiency**	Assessment of Energy Efficiency of Housing Stock based on unit type and year of construction.	Energy consumption of structure (MJ/m2) with regard to adjusted environmental norms	City Level	Determined by Consultation with end users and industry leaders
14.	**Commercial Stock Energy Efficiency**	Assessment of Energy Efficiency of Commercial Stock based on unit type and year of construction.	Energy consumption of structure (MJ/m2) with regard to adjusted environmental norms	City Level	Determined by Consultation with end users and industry leaders
15.	**Industrial Stock Energy Efficiency**	Assessment of Energy Efficiency of Industrial Stock based on unit type and year of construction.	Energy consumption of structure (MJ/m2) with regard to adjusted environmental norms	City Level	Determined by Consultation with end users and industry leaders
16.	**Leisure Stock Energy Efficiency**	Assessment of Leisure Stock Energy Efficiency based on unit type and year of construction.	Energy consumption of structure (MJ/m2) with regard to adjusted environmental norms	City Level	Determined by Consultation with end users and industry leaders

continues on following page

Table 1. Continued

No	Indicator	Definition	Unit of Measurement	Scale	Data Source
17.	**Public Stock Energy Efficiency (Health) I**	Assessment of Public (or private) healthcare facility energy efficiency	Energy consumption of structure (MJ/m2) with regard to adjusted environmental norms	City Level	Determined by Consultation with end users and industry leaders
18.	**Public Stock Energy Efficiency (Education) II**	Assessment of public (or private) educational facility energy efficiency	Energy consumption of structure (MJ/m2) with regard to adjusted environmental norms	City Level	Determined by Consultation with end users and industry leaders
19.	**Special Heritage Status**	Building with heritage designation attached.	Y/N-Specification	City/ District Level	Determined by Consultation with end users and industry leaders
20.	**Electricity Exported to Grid**	Amount of surplus energy exported to the national grid.	Kwh	Building Level	Determined through literature review
21.	**Residential population density**	Total resident population per km 2 of built up area.	Population per km2	City Level	TISSUE CORE 1
22.	**Percentage of buildings assessed under a Building Energy Rating Scheme**	This indicator defines the percentage of buildings classified among A+ and D.	% of total	City Level	Determined through literature review
23.	**Heat Demand Density**	Heat density demand informs viability of District Heating.	Kwh/m2	Sub City Level	Determined by Consultation with end users and industry leaders
24.	**Occupancy**	Percentage of buildings which are vacant.	% of the vacant buildings in relation to the total buildings	City/ District Level	Determined by Consultation with end users and industry leaders
25.	**Smart Metering**	Number of Buildings with Smart Meters installed.	Number of buildings equipped	City/ District Level	Determined by Consultation with end users and industry leaders

continues on following page

Table 1. Continued

No	Indicator	Definition	Unit of Measurement	Scale	Data Source
26.	**Capital Expenditure on Energy Infrastructure**	Amount expended on the management, upkeep and development of energy infrastructure.	€ per head of population per annum	City Level	Determined by Consultation with end users and industry leaders
27.	**Total employment rate**	% of working aged population employed in the locality.	City unemployment rate and % of people in full time employment	City Level	EU SDI sub-theme: Employment, level 2
28.	**Retrofitted Elements (Generation)**	Presence of specific interventions designed to generate energy (micro) at building level.	Y/N-Specification	Building Level	Determined through literature review
29.	**Retrofitted Elements (Insulation)**	Presence of specific interventions designed to reduce heating loss.	Y/N-Specification	Building Level	Determined through literature review
30.	**Retrofitted Elements (Glazing)**	Presence of specific interventions designed to reduce heating loss.	Y/N-Specification	Building Level	Determined through literature review
31.	**Retrofitted Elements (Heating)**	Presence of specific interventions designed to efficiently generate heat.	Y/N-Specification	Building Level	Determined through literature review

Limitations

The suite of indicators that form the starting basis for the SUI selection is based on the indicators put forward in the academic and research literature. The SUIs developed in this paper may therefore be constricted by the available lists. Furthermore, the SUI selection process is limited to those which specifically address our overview goals of a reduction in energy costs, energy related CO_2 emissions reduction, and enhancement of urban energy efficiency. A broader more inclusive tool or a supporting tool which would work in conjunction with the SUIs developed here may be required to address other areas of sustainable development.

The time effectiveness of the use of the selected SUIs might still require demonstration, especially around existing methodologies. The consequences of this led to the use of a reduced set of metrics supported by readily available data. The same issues were encountered during the development of this paper as a substantial number of indicators were found to be of theoretical and academic relevance to the SUI yet were not implemented due to the constraints imposed by data availability. The operationalisation of the indicators in the decision support tool and the development of the wider apparatus should be able to compensate for any detected or undetected issues with the indicators theoretical framework.

Morse (2003) questioned the quest for quantitative indicators, highlighting how qualitative measurements may provide better understandings of sustainability, given its socially constructed nature and the need for perceptions to change in order to achieve more sustainable behaviour. Considering the potential for indicators to mislead, or be manipulated and 'cherry-picked' to show what is desired, Meadows (1998) warns that, as an integral part of the decision-making process, indicators can be a 'dangerous tool'. The desire of stakeholders and decision-makers, for simplicity, comparability and interpretability of indicators may inadvertently result in over-aggregation and/or over-simplification of complex relationships. Consequently, this may mislead or result in false representation (DSCWG, 2001). Standardisation, incorporating the alignment of the SUIs with existing industry standards (e.g. ISO standards) may be needed. During the development of the SUIs, it was determined during consultation that the incorporation of qualitative indicators would present a technical challenge as well as a complication of transferability in terms of normalising such indicators across a variety of socio-environmental, cultural and economic contexts.

The proposed SUI model is transparent, user-friendly, communicative and adjustable to local contexts. This last feature (flexibility) is undoubtedly a great advantage, however it raises doubts about another intended principle of the model: comparativeness. Should the weightings of the SUIs be adjusted to specific needs of particular stakeholders, (or indeed the contrasting needs of the two case studies) their direct comparability is reduced. For an accurate and fair comparison, the weightings in this case would need to be maintained at the same level in both cases.

CONCLUSION

Sustainable development is becoming an increasingly important issue on the global stage. To respond to many sustainability challenges that we are facing, we must be able to measure progress towards sustainable urban development. The generic development framework for SUIs proposed in this paper could be used by local governments as a tool for assessing the level of sustainability of urban systems. The proposed indicators have been developed specifically for a European context but could also be applicable to other areas where the weightings could be tailored for alternative needs. While the framework aims to capture most key issues for the urban environment, it should be emphasised that it would be impossible and indeed impractical to try to translate all the issues into indicators of sustainable development. Nevertheless, the framework is comprehensive, and given its scope, it may be more suitable for large-scale city systems which can afford the time and resources for its implementation.

Having completed a comprehensive assessment of literature relating to the formation of indicators it was concluded that the indicator suite be a synthesis of composite indicators which lend themselves to intra-national comparison and indicators in tandem with expert evaluation. This will support decision making in respect of energy based on the accurate reflection of changes in the manner in which a specific, building, district or city manages its relationship with energy. The latter indicators examine metrics arising from the modelling of interventions to the structure or group of structures.

Through our initial compilation process the indicators obtained were found across different relevant domains. These indicators were taken from a combination of sources, including existing indices relating to the built environment as well as from a study of industry standards in the management of energy. The structure, scope, framework and weighting of the indicators have been developed in line with academic practice relating to the formation of composite indicators. Once compiled, the indicators were advanced for consideration at the consultation phase of the methodology.

The SUIs successfully facilitate the integration of the energy characteristics of buildings, in terms of consumption and generation, with other components of a city's infrastructure and governance. This set of SUIs may be used to aid decision support within the urban environment. In order for urban decision making to be transferable across a number of urban contexts, the selection of these SUIs will be normalised in order to recognise differences in climate, culture, lifestyles, governance and building typology in the settings in which they will be employed. The core aim of these SUIs is to enable end users to identify potential energy efficiency measures, which can be implemented within urban areas, yielding quantifiable reductions in GHGs and energy costs, and indicating the manner in which urban energy efficiency can be enhanced through modelled interventions. The data underpinning these decisions can be added by users or approximated through the use of simulated benchmark values.

The innovative character of the described procedure is based on the central procedural feature: The broad and systematic participation of different consultation groups during the process helps to strengthen the local identity by providing the means for all citizens to better identify with their community and its development. Each community has specific individual factors therefore the weighting for a particular city requires local user knowledge but also requires oversight from industry experts who can use their experience to aid with local efforts. In order to achieve the development of a functional set of sustainability indicators that can be used to compare different communities it is important to connect an indicator system of common goals with a more detailed program that shows the specific subjects of the local urban system.

REFERENCES

Anttiroiko, A. V. (2012). Urban Planning 2.0. *International Journal of E-Planning Research, 1*(1), 16–30. doi:10.4018/ijepr.2012010103

Arksey, H., & OMalley, L. (2005). Scoping studies: Towards a methodological framework. *International Journal of Social Research Methodology, 8*(1), 19–32. doi:10.1080/1364557032000119616

Bell, S., & Morse, S. (1999). *Measuring the immeasurable. The Theory and Use of Sustainability Indicators in Development*. London: Earthscan.

BrundtlandG.H.MansourK. (1987The World Commission on Environment and Development (WCED).

Crabtree, B., & Bayfield, N. (1998). Developing sustainability indicators for mountain ecosystems: A study of the Cairngorms, Scotland. *Journal of Environmental Management, 52*(1), 1–14. doi:10.1006/jema.1997.0159

DSCWG (Douglas Shire Community Working Group). (2001). Douglas Shire Sustainable Futures Draft Strategy. Douglas Shire Council: Mossman.

Hardi, P., & Zdan, T. (1997). Principles in Practice; Assessing Sustainable Development. Retrieved from https://www.iisd.org/pdf/bellagio.pdf

Haughton, G., & Hunter, C. (1994). *Sustainable cities, regional policy and development series 7*. London and Bristol, Pennsylvania: Regional Studies Association.

Isaacs, J. P., Blackwood, D. J., Gilmour, D., & Falconer, R. E. (2013). Real-Time Visual Simulation of Urban Sustainability. *International Journal of E-Planning Research*, *2*(1), 20–42. doi:10.4018/ijepr.2013010102

Jennings, I., & Newman, P. (2008). *Cities as Sustainable Ecosystems*. Principles and Practices.

Keirstead, J. (2007). Selecting sustainability indicators for urban energy systems. *Proceedings of the International Conference on Whole Life Urban Sustainability and its Assessment.*

Kelly, G., & Baker, B. L. (2002). *An evaluative framework and performance measures for the sustainable regions programme*. CSIRO Sustainable Ecosystems.

Meadows, D.H. (1998). Indicators and information systems for sustainable development.

Mitcham, C. (1995). The concept of sustainable development: Its origins and ambivalence. *Technology in Society*, *17*(3), 311–326. doi:10.1016/0160-791X(95)00008-F

Morse, S. (2003). For better or for worse, till the human development index do us part. *Ecological Economics*, *45*(2), 281–296. doi:10.1016/S0921-8009(03)00085-5

Nardo, M., Saisana, M., Saltelli, A., Tarantola, S., Hoffman, A., & Giovannini, E. (2005). *Handbook on constructing composite indicators*. OECD Publishing.

OECD. (2003), Environmental Indicators Development, measurement and use. Retrieved from http://www.oecd.org/env/indicators-modelling-outlooks/24993546.pdf

Pearce, D.W., Markandya, A., & Barbier, E. (1989). *Blueprint for a green economy* (Vol. 1). Earthscan.

Roderick, M. L., Hobbins, M. T., & Farquhar, G. D. (2009). Pan evaporation trends and the terrestrial water balance. II. Energy balance and interpretation. *Geography Compass*, *3*(2), 761–780. doi:10.1111/j.1749-8198.2008.00214.x

Smeets, E., & Weterings, R. (1999). *Environmental indicators: Typology and overview* (p. 19). Copenhagen: European Environment Agency.

Tang, T. S. K. (2011), Building Journal; Sustainable Systems Integrated Model; Modelling Techniques for Low-Carbon Cities. Retrieved from http://www.building.com.hk/forum/2011_0719modeling.pdf

United Nations. (2012) World Urbanization Prospects: The 2011 Revision. New York: United Nations.

United Nations. (2015). Framework Convention on Climate Change. *Adoption of the Paris Agreement.* Retrieved from http://unfccc.int/resource/docs/2015/cop21/eng/l09.pdf

United Nations Framework Convention on Climate Change. Retrieved from http://unfccc.int/meetings/paris_nov_2015/meeting/8926.php

Waldron, D., & Williams, P. W. (2002). Steps towards sustainability monitoring: the case of the resort municipality of Whistler. In Sustainable tourism: A global perspective (pp. 180-194).

WCED, U., (1987). *Our common future. World Commission on Environment and Development.* Oxford University Press.

Wefering, F., Rupprecht, S., Bührmann, S., & Böhler-Baedeker, S. (2013). Guidelines. Developing and Implementing a Sustainable Urban Mobility Plan. *Proceedings of Workshop* (p. 117).

Whate, V., McCrum, G., Blackstock, K. L., & Scott, A. (2006). The Macaulay Institute: Indicators and Sustainable Tourism: Literature Review. Retrieved from http://www.macaulay.ac.uk/ruralsustainability/LiteratureReview.pdf

Worden, K., Bullough, W. A., & Haywood, J. (Eds.). (2003). *Smart technologies.* World Scientific. doi:10.1142/4832

World Commission on Environment and Development. (1987). Our Common Future. Retrieved from http://www.un-documents.net/our-common-future.pdf

This research was previously published in the International Journal of E-Planning Research (IJEPR), 6(4); edited by Carlos Nunes Silva, pages 1-17, copyright year 2017 by IGI Publishing (an imprint of IGI Global).

Section 3

Sustainable Urban Development and Planning

Chapter 10
Developing a Sustainable Eco-City in Pre-Olympic Tokyo:
Potential of New Methods and Their Limits in an Urban Era

Mariko Ikeda
University of Tsukuba, Japan

ABSTRACT

This chapter intends to elucidate the emergence of sustainable urban development in Tokyo in light of the upcoming 2020 Olympics by exploring various administrative and commercial practices, such as urban development plan with rooftop and wall greening or river projects in the inner city. The research methods involved a review of the empirical literature, an analysis of existing statistical data, and a detailed examination of case-specific data collected in a field survey. This chapter concludes that since Tokyo Metropolitan Government encourages urban greening projects as a solution of urban heat island from 2000s, utilization of "green spaces" in the landscape design of commercial and office facilities is gaining attention. This chapter concludes that various practices for sustainable urban development in Tokyo, which faces a restructuring process in light of the 2020 Summer Olympics, exist and that some of these could be further developed by the private sector.

DOI: 10.4018/978-1-7998-0441-3.ch010

Copyright © 2020, IGI Global. Copying or distributing in print or electronic forms without written permission of IGI Global is prohibited.

INTRODUCTION

Environmental Problems in the Urban Era: Overinvestment, Overpopulation, Overshoot

The concepts of "city" and "nature" are perceived as intrinsically contradictory, and authors who stand for the latter have criticized cities for their negligence of and dominance over the environment. Moreover, on a global scale, the concentration of populations in urban areas is growing. It is estimated that in 2030 more than 60% of the world's population will live in cities, and 27% will reside in cities with at least 1 million inhabitants (UN Habitat, 2016). The agglomeration of gross domestic product (GDP) in cities escalates accordingly and 80% of the world's total GDP is concentrated in urban areas. Paris, for instance, houses only 16% of the French population but produces about 27% of France's GDP (UN Habitat, 2016, p. 27). This condition is known as the 'overdevelopment, overpopulation, overshoot' issue, suggesting the excessive burden to the natural environment caused by urbanization (Butler, 2015). The United Nations has therefore adopted a new 2030 Agenda for Sustainable Development, which presents 17 goals that replace the previous Millennium Development Goals (United Nations, 2016, pp. 38-39). This agenda includes several new elements, such as making cities and human settlements inclusive, safe, resilient, and sustainable. Furthermore, it demands the development of low-carbon societies and green technology or the replacement of conventional energy resources with renewable energy, especially in cities, which represent more than 70% of the global energy demand.

Environmental Problems in the Urban Era: Heat Island

Japan's capital Tokyo is, together with its surrounding areas, the largest megacity in the world with approximately 38.1 million inhabitants, which is more than 10 million people more than the population of Delhi, the second-largest city (United Nations, 2016). It is also the city with the largest GDP worldwide. Tokyo's population increase began during the period of Japan's rapid economic growth (1955–1973) with an influx of people from rural regions into suburban areas, which were constructed around the inner-city area; this trend of urbanization is still increasing (Figure 1). The extreme concentration of the population in the metropolitan area has led to an increase of housing, offices and commercial facilities and to a decrease of the natural environment, causing an environmental problem characteristic for large cities, the heat island effect.This phenomenon, also called 'city climate', occurs rather in the city's center than in its outskirts and is mainly caused by solar heat during the day and by waste heat from human economic activities. This artificial heat disposal is said to be equal to 10% of the sun's radiation and to have a stronger effect with increasing city size (Sakakibara, 2001). Observations of the temperature difference between the center of a city and its suburbs show that the difference during the daytime is small but tends to increase during the night because the temperature in the city center does not fall (Maeda, Kondo, Akikawa, Kaga, & Inoue, 2008; Masumoto, 2007; Mizuno, 1993). The temperature is not only dependent on such a parallel distance from a city center, but it is also said to be inversely proportional related to the altitude of the observation spot, and it is supposed that cold air in the atmosphere over the city traps the urban air in a dome.[1] For example, the urban air dome so-called temperature inversion layer above New York is situated at a height of about 310 meters (Bornstein, 1968), in the case of Japan's Nagoya it is at a height of approximately 60 meters (Aikawa, Hiraki, & Eiho, 2006). If one assumes that one floor of a building has a height of 4 meters, this would correspond to a building with 77 floors in the case of New York and to a building with 15 floors in the case of Nagoya. In other words, this phenomenon is not only related to physical factors such as anthropogenic heat but also strongly connected to social factors such as the increase in artificial building structures (Maeda et al., 2008). For example, in the city center, there exist only intensive land use such as high-rise buildings with offices and apartments, which can sustain the high land prices. This leads to the weakening of natural winds [mechanical effects], reducing humidity, generating a complex reflection of shortwave and longwave radiation between buildings, generating thermal effects that further intensify the heat island phenomenon (Kondo, 2009; Oke, 1979). In addition, this urban air dome contains air pollutants, which stem from exhaust fumes of cars and from tall smoke sources, such as thermal power stations (Uno, Wakamatsu, & Ueda, 1988). It is inferred that such an urban climate phenomenon will further change through the expansion of the suburban area with increasing population concentration in the metropolitan area (horizontal expansion) and the vertical expansion of the inner city. In other words, regarding environmental concerns in this urban era of overpopulation, case study in Tokyo, in which approximately 25% of Japan's total population are living, could show a best challenging practice.

Olympics and Paralympics: Urban Competitiveness Towards Sustainable Eco-City

In September 2013, Tokyo was selected as host city for the 2020 Olympic and Paralympic Games (from here on Olympics). During the candidacy, the bid committee for the Olympics pledged in its campaign to give priority to the environment. As reason behind this can be seen that the Organizing Committee for the Olympic Games has pushed forward actions of environmental consideration since 1994.

Environmental protection was added to the Olympic Charter and the promotion of sustainable development is pointed out as mission and role (Saito, 2014). The year 2020, when the Tokyo Olympics will be held, is the target year for the reduction of the world's greenhouse gas emissions, and it is also the target year for the Aichi Targets formulated at the Convention on Biological Diversity. Therefore, parallel to the improvement of the environmental infrastructure with the latest environmental technologies the recovery of natural water and green spaces is also expected (Saito, 2014).

In addition, environmental improvement in the urban era may run into difficulties, in Tokyo in particular, where it often is difficult to reach a consensus regarding redevelopment due to its subdivided groundrights. In reorganizing the city and structure for a national event such as the Olympics, the awareness of universal standards regarding global city environments, there may be instances where certain considerations are applied for the first time to a city's environment. For example, in the reports of the Japanese Ministry of the Environment, in which all of Japan's environmental administrations are summarized, natural spaces are identified as a major factor in the competition between cities.

Figure 1. Rate of net migration for the three major metropolitan areas in Japan from 1954 until 2017
Source: (e-Stat, 2017)

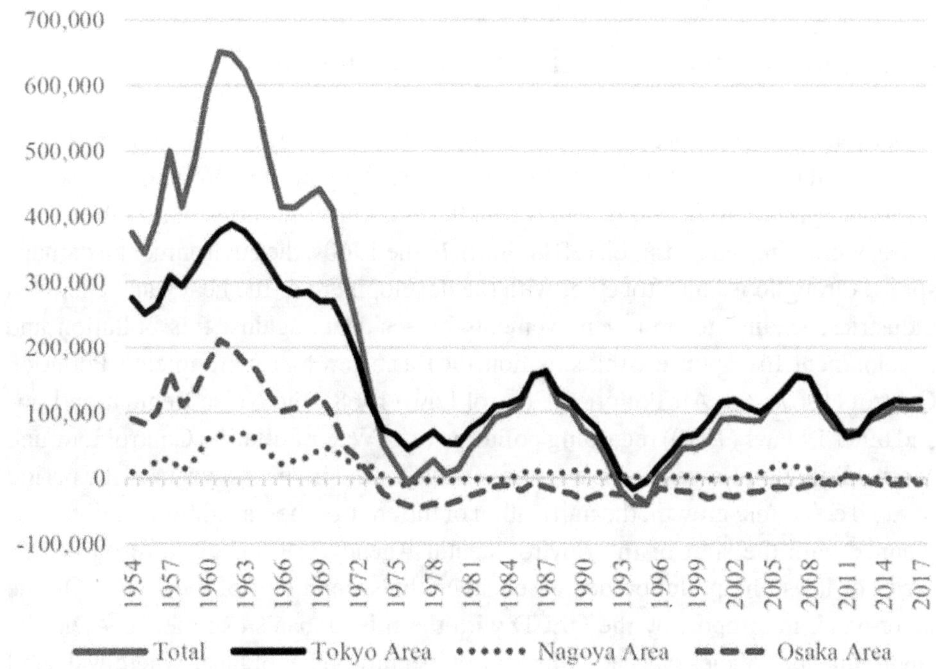

Accordingly, the problem of Tokyo's low green space ratio is addressed. Based on the assumption that the green space ratio of the inner cities of Stockholm, Geneva, and Berlin is 100%, London and Paris would have a ratio of 62.5%, New York would be at 50%, and Tokyo would be just over 25% (Ministry of the Environment, 2014). In this way, in the time before the Olympics, environmental problems are reconsidered together with the ongoing large-scale restructuring of inner-city areas, such as Ginza, Nihonbashi or Shibuya.

Research Purpose and Method

This chapter intends to elucidate the emergence of sustainable urban development in Tokyo in light of the upcoming 2020 Olympics by exploring various administrative and commercial practices in the inner city. The research methods involve a review of existing literature and published documents, an analysis of statistical data and qualitative interviews with authorities, and a detailed examination of case-specific data collected in a field survey. The study will first examine the historical change of environmental policies in Japan, followed by an overview of the environmental policies in Tokyo regarding their relation to the Olympics.

URBAN ENVIRONMENTAL POLICIES FOR A SUSTAINABLE ECO-CITY IN HOST CITIES OF MEGA EVENTS

Overview of Environmental Policies in Japan

Japan's environmental problems became apparent during the period of rapid economic growth that began in 1955. There were mainly seven pollution problems caused by the industry, namely air pollution, water pollution, soil pollution, ground subsidence, noise, vibration, and bad smell (effluvium). In the 1960s, the environmental damage spread throughout Japan together with the development of the heavy and chemical industries, leading to protest movements by residents against this pollution and development. In response to this situation, the Basic Law for Environmental Pollution Control (1967), the Air Pollution Control Law (1968), the Noise Regulation Law, and other 14 laws (1970) regarding pollution (e.g. Water Pollution Control Law and Marine Pollution Prevention Law) were enacted. And in the last years of the period of rapid economic growth, the unification of the environmental administrations was planned with the start of the Environmental Agency (1971).[2] As a result of this series of laws, the pollution decreased since the second half of the 1970s. On the other hand, in a report by the OECD with the title 'Japan's experience – Did the environmental policies succeed?', the environmental policies of Japan where evaluated as inclined to the 'prevention of pollution' (Takeuchi & Lee, 1988). Therefore, the 'creation of a comfortable environment' (or environmental amenities) became the key concept during the 1980s, and in the Regional Environment Management Plan the viewpoint of environmental creation, including landscape preservation, was introduced. In addition, the relocation of production facilities to other countries due to an increased oil price, trade frictions, the strong Yen, and rising labor costs lead to a fading consciousness of pollution, and a response to environmental problems on a global scale became necessary from the mid-1980s (Lee, Harashima, Lee, & Morita, 1995). Furthermore, taking the amendment of the Urban Green Space Conservation Act (1994) as an example, each municipality became able to carry out greening measures in the 1990s.

For such environmentally friendly projects, examples from Europe, in particular from the U.K. and Germany, have always provided points of reference. For example, the establishment of communal areas, such as green spaces or public squares, since the 1970s; or the establishment of allotment gardens (Japan uses the German word 'Kleingarten') and biotopes since the 1990s. Especially the Kleingarten represents a place for city dwellers to enjoy gardening at the weekend and is modeled after Germany's green tourism. In fact, these gardens were mainly developed in the countryside under the name of promoting green tourism after the enactment of the Rural Regions Leisure Law in 1994, a time when the revitalization of rural areas was expected/desired. In addition, taking the biotope, a German invention, and its ideas about preservation, regeneration, and creation as model, public facilities at urban factory sites and schools were constructed and the conservation of agricultural areas was pushed forward. However, because the biotope is a concept opposed to agricultural productivity, it has not spread to other areas outside the rural regions promoting green tourism. On the other hand, a site-specific spread can be found in urban areas, depending on land development design and a limitation of land prices, and rather than a solution to environmental problems, such as the heat island effect and air pollution, the biotope's functions to preserve a green landscape and as a place for recreation and relaxation are emphasized.

Environmental Policies in the Tokyo Metropolis

The history of environmental policies in the Tokyo Metropolis is, depending on the point of view, old. The previously explained pollution problems were at first regional and not national and therefore the 'Regulations for the control of Tokyo's factory suburbs' were enacted in Tokyo in 1949, which were countermeasures at a very early stage even from a national perspective (Lee et al., 1995). Furthermore, a garden city plan was drawn up before the Second World War, and occasionally, town planning and an all-over environmental development occurred.[3]

On the other hand, some turning points exist when looking at the environmental policies in Tokyo after the war. One of them is the hosting of the mega event Tokyo Olympics. Mega events promote improvements of traffic and electricity infrastructure and the redevelopment in a city, and in the case of Tokyo, it greatly affects the city's environment. Therefore, I will survey the aspects of the 1964 Tokyo Olympics and the planned 2020 Tokyo Olympics from the viewpoint of the city environment.

Tokyo Olympics 1964: Developmentalism in the Postwar Era and Its Environmental Load

The population growth in Japan during the economic recovery after the Second World War and during the period of rapid economic growth modified the city environment rapidly. Beginning with the conversion of forest and agricultural land into residential land, backed by the developing industrialization through the formation of industrial complexes along the Pacific belt zone, the improvement of major roads and harbors and their connection by the construction of expressways was needed. During this time, the 55th International Olympic Committee general meeting was held in 1959, where Tokyo was selected as host city for the 1964 Olympics. Postwar Tokyo with its revitalizing economy devised an urban planning for the hosting of this large event five years later and underwent rapid changes. The urban planning of this period is summarized according to Echizawa (2014) in Table 1.

The characteristic of this urban planning was that land that was requisitioned after the war by the American military was returned to Tokyo metropolis under the nominal cause of the holding of a national event with high publicity called Olympics. For example, Yoyogi in Shibuya Ward where the Olympic village was developed, was used until World War II as Yoyogi parade ground and was requisitioned after the war by the United States for use as residential area "Washington Heights" for

Table 1. Developmentalism for Tokyo Olympics in 1964

Development	Urban greening	Infrastructure
Re-use of Olympic village as an urban park	✔	
Construction of Olympic village		✔
Deregulation of City Planning Act (parks and greens) and construction of international accommodation		✔
Roadway and highway infrastructure		✔
Development of rail infrastructure		✔
Construction of the National Stadium		✔

Source: (Echizawa, 2014)

Americans and used until the early 1960s. The Japanese cabinet decided that the same grounds should be turned into a forest park after the end of the Olympic Games, therefore enabling the return of the land and the securing of a forest park which is in use as forested green space until today (Ozaki, 2002). Furthermore, based on the Capital Region Improvement Law in 1956 the First Metropolitan Area Improvement Plan (1958) followed a pre-war plan to establish a green belt of 5 to 10 kilometers with an area of 108,000 hectares around the city area of Tokyo (23 wards), which was intended to supply greenery, secure land for public use, and preserve superior agricultural land for the provision of fresh food (Ministry of Environment, 2014; Ministry of Land, Infrastructure, Transport and Tourism, 2011). This plan was completely disapproved of in the Second Metropolitan Area Improvement Plan and eventually not realized. However, the establishment of a green belt between the core city and the satellite cities had the aim to reduce the urban sprawl and had used the plans for Greater London as reference (Takeuchi & Ishikawa, 2008; Maruyama & Nakagawa, 2004). In this way, the Olympics were taken as opportunity to plan the conversion of large tracts of military-used land that dotted Tokyo into green spaces of which parts are still functioning as rare forest resources at the present.

However, the Olympics also had negative outcomes that resulted from the improvements of the infrastructure that had suddenly become necessary. Especially the construction of the Metropolitan Expressway and the Tokaido Shinkansen were of the highest priority for the country and for Tokyo and improvements in a short time were promoted as procedure. Construction of the former was already demanded before the decision of holding the Olympic Games by the Capital Committee on Construction since 1953 (Table 2), and in the same year the Olympics were decided the Metropolitan Expressway Public Corporation Law was approved (1959 Legislative Act No. 133), making the construction of the expressway an even more urgent matter (Horie, 1996; Furukawa, 2013). Therefore, to avoid time-consuming land purchases, the expressway was built by using the space above state-owned roads, rivers, moats, and waterways (Figure 2). This is also clear from the land use ratio of the expressway, with river land having a ratio of 9.1% (4.9 km) in the planning stage of 1953, increasing to 41.8% (62.5 km) in 1958, and in the process making 35% (71 km) of the used land obtainable without the need to purchase it (Table 2). Among them are planned expressways located between Olympic facilities (National stadiums, Olympic Village) and Haneda Airport such as the Metropolitan Expressway Route 1 Haneda Line (Chuo Ward Honcho 3-Chome – Haneda Airport) and the Route 4 Shinjuku Line (Chuo Ward Nihonbashi Kokucho – Shibuya Ward Yoyogi Hatsudai) with a total of 31.3 kilometers which have been designated as Olympic-related roads and therefore have been partly constructed as bridges above and along canals (Figure 2).

Table 2. Rate of properties for Urban Express Highway

Planning agent	Length of Extended Express Highway	River	Public road	Other properties		Total
				Total	Civil properties	
Committee for capital construction	49km	9.10%	7.70%	83.20%	70.20%	100%
Tokyo Metropolitan Government	62.5km	41.80%	34.40%	23.80%	10.90%	100%
Urban planning	71km	35.00%	37.70%	27.30%	14.40%	100%

Source: (Ishihara & Takahashi, 2015)

During the era of high economic growth, the relation between human society and the water environment in Tokyo changed remarkably. First, in Japan with its high risk of natural disasters, in the city area mainly flood damage occurred caused by large-scale typhoons, and at the same time, chronic water shortages were caused by rapid urbanization, demanding immediate water management measures. Furthermore, Tokyo has the topographic characteristic that it can be divided into an eastern and a western part. In the lowlands in the eastern part, irrigation canals became unnecessary with the transformation of the agricultural structure. The city's moats as focus of the waterside remnants were filled with earth from construction sites during the reconstruction of war damages and as already stated used for the construction of the Metropolitan Expressway. In the western Yamanote (uptown) area, rivers without headspring were filled up and the construction of sewage and drainage canals was furthered (Tsuboi, 2003). Additionally, especially from the late 1950s until the late 1960s, the water pollution of small and medium-sized rivers in Tokyo increased due to the inflow of domestic and industrial wastewater, leading to a very low amenity value for the city's inhabitants. Therefore, it was demanded, and requested by the city's residents as well, that small and medium-sized rivers were either to be used as sewage system or covered up (Ishibashi & Takahashi, 2015). For example, the Shibuya River, which is the origin of the place name Shibuya, the world-famous tourist spot, and which is one of the small rivers that flows southwards through the city's center, was covered up in 1961 according to the 36th report of Tokyo and remains in this state until today (Figure 3) (Nakamura & Oki, 2009). For that reason, the number of Tokyo's rivers and waterways has declined after the Meiji era and the water amenity of this period has further decreased with the vanishing of the rivers from the city's surface (Arai 1996).

Figure 2. Metropolitan Expressway in Tokyo, Chiyoda Ward
Source: (Photo taken by the author, June 23, 2018)

Tokyo Olympics 2020: Towards Sustainable Development Replacing Developmentalism

Among the activities of the Japanese government to vitalize its Visit Japan Campaign, which was started in 2003, the wish to host the Olympic Games emerged and promotional activities commenced in 2011 and in September 2013, Tokyo was selected as host city of the 2020 Olympic Games. In the same year, because of the great relaxation of visa requirements for the Southeast Asian countries Thailand and Malaysia (July 2013), a low exchange rate of the Japanese Yen, the expansion of airline routes, and countermeasures to harmful rumors, the number of foreign visitors to Japan reached more than 10 million people, making the importance of the Japanese tourism industry apparent (Hompo & Yagasaki, 2015). With Tokyo's experience as host city of the Olympics, it can make full use of the athletic facilities constructed in 1964 and the infrastructure improvements of that time to keep the environmental impact low and host sustainable Olympics, merits with which the city tried to appeal (Saito, 2014). In fact, in the evaluation reports for each of the candidate cities for the 2020 Olympics—Madrid (Spain), Istanbul (Turkey), and Tokyo—Tokyo received recognition in this respect with a special mention (International Olympic Committee, 2013). The awareness of reducing the environmental impact in hosting the Olympics is set forth in the Olympic Charter of 1994 and contains that it shall not be a big event with environmental sacrifices led by development, since those

Figure 3. Riverside landscape in Tokyo, Shibuya Ward
Source: (Photo taken by the author, May 2, 2017)

evaluation standards were specified in the selection points (Ministry of Environment, 2014), protection of the environment and reduction of waste, usage of low-carbon energy, and improvement of sustainability have been consciously followed for every event (Table 3). Especially during the construction of the facilities for the 2012 Summer Olympics in London, evaluation systems for the building environment (e.g. BREEAM) were introduced and ISO standards for event management were created, which were highly appreciated when they became standards for the subsequent Olympics (Ministry of Environment, 2014).

Under such circumstances Tokyo was selected as host for the 2020 Olympics and Paralympics and concerns were directed at environmental maintenance as well as an increased awareness of sustainable development. The concrete environmental policies of the Japanese Ministry of Environment and the Tokyo Metropolitan Government Bureau of Environment at the current stage are summarized in Table 4. With having concluded the Paris Agreement, Tokyo will enact more specific policies with the year 2020 as aim. On the other hand, the Ministry of Environment has indicated that the destruction of the natural environment during the 1964 Olympics, namely the covering of the Shibuya River and the Tachiai River and the construction of the Metropolitan Expressway above the Nihonbashi River, combined with the increase of high-rise buildings due to the relaxation of the 31m height restriction, are factors that generate Tokyo's heat island phenomenon (Ministry of Environment, 2014). The same report indicates that there is an awareness of the differences to the environmental policies of 1964. Concrete policies are the conservation of existing green spaces, the increase of water permeability of roads and parking spaces and application of water-holding coatings, the laying of electric wiring underground, the maintenance of roadside trees and greening of rooftops and walls of surrounding buildings, the reduction of artificial heat disposal by using high-efficiency equipment, the restoration of water surfaces, the improvement of air circulation of surrounding blocks, with the cool air from rivers and green spaces cited as a prime example, and as measures until the hosting of the Olympic Games, planting of trees around event facilities and along streets show a very locally-oriented approach. Such measures of the governmental environment agencies, it can be said that the developmentalism of the period of rapid economic growth has to no small extent changed in introspective Japan, but on the other hand, for example real estate development by railway companies in Japan, cities were formed after the period of rapid economic growth by leadership of private enterprises (Ministry of Land, Infrastructure, Transport and Tourism, 2011: 82). Accordingly, the next part will look at the projects of private sector entities as projects aimed at the 2020 Tokyo Olympics.

PRACTICES TOWARDS SUSTAINABLE CITY IN TOKYO

In Tokyo, private enterprise-led urban redevelopment is gaining momentum at the present due to the estimated growth of the tourism industry in the wake of the hosting of the 2020 Olympics and other events after that, such as the OSAKA Expo, MICE tourism, and integrated resort development (Ikeda, 2017). Among this redevelopment there are many environmentally-friendly projects. This section will provide examples of this eco-friendly redevelopment, such as rooftop and wall greening projects and the renaturation of small and medium-sized rivers by private enterprises, and it will investigate current activities regarding the environment in Tokyo.

Greening Projects

In 2013, the ratio of green space coverage in Tokyo was 19.8% throughout the 23 wards compared to 50.5% in the whole Tokyo Metropolis, showing a decrease by 0.2% since 2008 (Bureau of Environment, 2016). As measure against this decrease of green space coverage 1.01 million roadside trees were planted between 2007

Table 3. Environmental measurements for Olympics from 1990s

Year	Hosting city	Conservations of natural environment	Prevention of environmental pollution	Waste reduction	Re-use of former built environment	Usage of public transit	Energy saving	Measurement of CO2
1994	Lillehammer	✔		✔				
1996	Atlanta					✔		
1998	Nagano	✔	✔	✔	✔			
2000	Sydney	✔	✔	✔	✔	✔	✔	✔
2002	Salt Lake City	✔	✔	✔				✔
2004	Athens	✔	✔			✔		
2006	Turin	✔	✔	✔			✔	✔
2008	Beijing	✔	✔	✔	✔	✔	✔	✔
2010	Vancouver	✔	✔	✔			✔	✔
2012	London	✔	✔	✔	✔	✔	✔	✔
2014	Sochi	✔	✔	✔	✔		✔	✔

Source: (Ministry of the Environment, 2014)

Table 4. Environmental measures towards Tokyo Olympics

Policy subject	Policy objectives	Environmental measures towards Tokyo Olympics	
Ministry of the Environment	2020 Goals/ 2030 Goals/ 2050 Goals	Developing a low-carbon society Mitigation of heat island, heat illness and creation of greenery Realization of clean and comfortable air environment Realization of clean and comfortable aqueous environment Promoting 3R(reduce, reuse, recycle) Developing a sustainable society	
Bureau of Environment	2020 Goals/ 2030 Goals	Climate change and urban energy Sustainable materials and waste management Urban biodiversity and creation of greenery Clean and comfortable air	Sustainable building policy Mitigation of urban heat island Expansion of renewable energy introduction Creation of a hydrogen-based society

Source: (Ministry of the Environment, 2014. & Bureau of Environment, 2017)

and 2015 (Bureau of Environment, 2016). Furthermore, the "New Development of Green Policies" of 2012 specifies the conservation of existing green spaces, the planting of native species, and the raising of public awareness regarding biodiversity as measures. On basis of the 2014 "Tokyo long-term vision" measures for the conservation of biodiversity and the increase of quality of green spaces, and the 2016 "Tokyo environment basic plan" demands the thorough conservation of the city's environment, showing a formulation of measures regarding Tokyo's green spaces in rapid succession during the 2010s.

The reason that lies behind the development of these policies is the heat island effect caused by the loss of green spaces. Especially Tokyo has a remarkable heat island phenomenon, with a temperature increase of 3.2 degrees centigrade over the past 100 years, which is a rise twice as high as in other 15 urban areas of Japan where the increase of the mean temperature amounts to around 1.5 degrees centigrade (Tsunematsu et al., 2016). In addition, in 2013 the time with temperatures above 30 degrees centigrade, which is related to causing heat strokes, has increased in Tokyo's eastern central wards Chuo, Chiyoda, Taito and Bunkyo by 25% and in other wards by 20%, and the number of days with tropical nights (nights with temperatures not below 25 degrees) now lies between 30 and 40 per year (Tsunematsu, 2014). As a countermeasure, the Tokyo government has revised the "Regulations regarding the conservation and recovery of nature in Tokyo" (from here "Regulations for nature conservation") in April 2001, which require newly constructed or reconstructed buildings with a site area of more than 1,000 square meters (public facilities with more than 250 square meters) to green 20% of the roof's area that can be greened with the corresponding size of the rooftop area (Bureau of Environment, 2016). Furthermore, the "Committee on Countermeasures for the Heat Island Effect in Tokyo" was established in 2002 as well as the "Heat Island Countermeasures Policy Initiative" in 2003. In addition, based on a map of the thermal environment of Tokyo's 23 wards that shows the effects of anthropogenic waste heat and ground surface covering on the atmosphere, which are considered to be causes for the heat island effect, "Areas for the Promotion of Heat Island Countermeasures" were dedicated, comprising four areas (center, Shinjuku, Shinagawa train station area, Osaki-Meguro) (Chiyoda, Shinjuku, Minato, Meguro, and parts of Shinagawa Ward). In these areas, apart from greening projects conducted in cooperation by public (state and ward) and private enterprises, the building environment plan system was adopted which promotes the water retention of roads as well as heat insulation of road surfaces and energy-saving measures for private facilities. In other words, whereas greening projects, including the creation of green tracts of land, were until then under the complete responsibility of the government, private-sector initiatives has started to join measuring heat island effects. However, because of the great differences of environmental circumstances between Tokyo's wards and its outskirts, inner-city and suburban area, it is necessary to consider some local practices related to the Olympics as well. Therefore, in the following the greening projects will be divided into two main categories, the greening of tracts of land and the greening of building rooftops and walls, and examples for projects in the wards that are deemed remarkable by the author will be examined in more detail.[4]

Land Greening Projects and Redevelopment in Chiyoda Ward

Compared to rooftop and wall greening projects land greening requires large plots of land, but large-scale greening projects are difficult to realize because many of Tokyo's redevelopment projects are troubled by complicated ground rights. It is Chiyoda Ward that has shown the most activity in land greening projects. Chiyoda, with Tokyo Station as its hub, is lined with offices and commercial buildings and had the most pronounced difference in daytime and nighttime population during the 1990s. Therefore, compared to the outskirts the temperature drop at night is low and the number of tropical nights is high in comparison to other wards, making it a ward with a considerable heat island effect. For this reason, Chiyoda was chosen as main area of Tokyo's "Heat Island Measures Promotion Area" in 2005. Furthermore, in 2009 it was designated as model block in the pilot project "Cool City Center" which is part of the Ministry of Environment's "Low-carbon Society Model Block Formation Promotion Project".[5] In the case of being acknowledged as "cool city model block", half of the costs of related projects are subsidized. In addition, this project is only intended for private enterprises with corporate status.

To begin with, Chiyoda's total area of green space amounts to 244.91ha and has increased by 8.03ha since 2003, leading to a green space ratio of 21.04% (2010). Due to the Imperial Palace and many parks the area of green space is large, but regarding green space ratio Chiyoda stands at number six among all the wards (Shibuya City, 2016). Chiyoda ward is divided into seven administrative areas of which the Fujimi area with the Imperial Palace at its center has the highest green space coverage (41.45%), followed by the Otemachi area (comprising Otemachi, Marunouchi, Yurakucho, Nagatacho) in the south-east of the Imperial Palace (20.70%). In the Otemachi area, approximately 13 kilometers of streets were greened in a roadside greening project. This corresponds to around 20.5% of all the city block streets in this area and represents the highest ratio among all the administrative areas of Chiyoda ward. The above shows clearly that in Chiyoda Ward the improvement of green spaces is remarkable in the Otemachi area. The change of greening along roads shows an increase of 1.8% in all the wards, but Chiyoda exceeds this clearly with an increase of 4.0% (Chiyoda City, 2010). As reason for this can be seen the establishment of the "Council for the Promotion of a Redevelopment Plan in Otemachi, Marunouchi, and Yurakcho" in 1988, centered around local landowners in the Marunouchi area, which lead to a complete redevelopment on a large scale; and in addition, since this development was lead by private enterprises it was adopted by "the cool city model block".

Rooftop and Wall Greening Projects, Redevelopment in Minato and Shibuya Ward

A major role in green space policies play projects for the greening of rooftops and walls of buildings. The greening of rooftops and walls has been acknowledged as having a relaxation effect on the heat island phenomenon and has been promoted and implemented in laws since the year 2000. The change in construction area for the greening of (building) rooftops and walls in Japan shows for both an increase until 2008, then a temporary decrease in 2009, with the former increasing again slowly after 2013, and the latter increasing slowly as well except for the temporary increase in 2011 (Figure 4). The ratio of construction area for each category according to regions shows that Tokyo Prefecture accounts for 38.8% of all rooftop greenings in 2016, followed by the neighboring prefectures Kanagawa with 9.4% and Saitama with 5.2%, making the Tokyo Metropolitan Area account for a total of 53.4% throughout Japan. Apart from that, the prefectures Aichi (8.7%), Ōsaka (7.9%) and Hyogo (6.7%) show nationwide high numbers as well. Regarding wall greening, the same prefectures as before are occupying the first six ranks, with the Tokyo Metropolitan Area accounting for 53.3%. However, Aichi prefecture (12.3%) and the Kinki Metropolitan Area's prefectures Osaka (9.6%) and Hyogo (7.4%) show slightly higher numbers in this category. In other words, most of the rooftop and wall greening projects can be found in and around Japan's three largest cities where the heat island phenomenon is most striking where the city size is the biggest and the number if construction starts is high. In a similar way, the change in greened area of rooftops in the Tokyo Metropolis shows a steady growth in the first half of the 2000s, and a decline in the latter half; from 2010 until 2012 as well as from 2015 on, an increase can be observed (Figure 4). The decline in the latter half of the 2000s can be explained with the economic stagnation caused by the global financial crisis in 2008 which led to a reduction in the number of new building constructions.

In their master plans for parks and open spaces in which each local government establishes its rooftop greening projects, the only wards that have formulated specific goals are Taito (in 2002), Shibuya (2004), Suginami and Kita (both 2010) (Takeuchi, 2012). From this can be inferred that rooftop greening plays an important role in the greening projects of these wards. However, the wards with the largest green areas on rooftops are Minato (12.7ha), Shibuya (11.4ha) and Setagaya (10.2ha) (Nakabayashi, Kishii, Oosawa, 2012). In the following part, examples from the two wards occupying the first two places, Minato and Shibuya, will be examined.

The total area of green space in Minato Ward amounts to 452.26ha, corresponding to a relatively high green coverage ratio of about 21.8%. Among this total area, the ratio of public green spaces is 58.9%, whereas the ratio of public spaces among all greened rooftops is low with 18.7%, and 85.6% of all greened rooftops are privately owned (Minato City, 2017). This data supports that rooftop greening is mainly utilized by private enterprises. Among the five areas in this ward—Shiba, Azabu, Akasaka, Takanawa, Shibaura Konan—green coverage is the highest in Akasaka with 31.7% due to several shrines and the Aoyama Cemetery. In contrast, the highest number of greened rooftops has Azabu with 168.5 places per square kilometer, and wall greening is highest in Shiba with 95.7 places per square kilometer. The Azabu area is home to the world-famous Roppongi Station and in the 1980s land owners and the developer Mori Building Company carried out a large-scale redevelopment of the whole area. Ark Hills, which is located between Minato's Akasaka 1-chome and Roppongi 1-chome, is a high-rise building of 153 meters height completed in 1986 that comprises offices, accommodations, high-class apartments, cultural and media facilities. At this corner, the Ark Garden was created which was a pioneer of rooftop gardens (specifically, hanging gardens) by a private enterprise before the promotion of rooftop greening was started.

On the other hand, with plans for the main sports stadium of the Olympics to be located in Minato's bayside area, vigorous practices related to greening projects have been identified in the ward. For example, the greening plan devised by each ward and the reports based on it are in the case of Minato Ward provided in general every five years. Compared to other wards this is clearly extensive.

The neighboring Shibuya Ward had a total green space area of 311ha in 2003, corresponding to a green space ratio of 20.6% (Shibuya City, 2004). However, since the formulation of the "Master Plan for Green Spaces in Shibuya" in 2004, rooftop greening has been promoted, leading to its propagation and as a result increasing the green space ratio to 21% (Shibuya City, 2016). In 2003, the total area of greened rooftops amounted to 1.35ha, representing less than 1% of the ward's green space coverage (Shibuya City, 2004). The ward is divided into six administrative areas (Honmachi and Sasazuka; Hatsudai, Nishihara and Uehara; Yoyogi; Sendagaya and Jingumae; Omukai and Ebisu; Hikawa and Shinbashi), and with one part of the Shinjuku Imperial Garden, Yoyogi Park and the Meiji Shrine the areas Yoyogi and Sendagaya-Jingumae are the areas with the highest green space ratio among them. Regarding green spaces on rooftops, the largest coverage with 0.88ha has the Omukai-Ebisu area, followed by the Hikawa-Shinbashi area with 0.55ha and the Sendagaya-Jingumae area with 0.52ha. The Omukai-Ebisu area and the Sendagaya-Jingumae area are the center of redevelopment in Shibuya aimed at the Olympics. Both areas are directly located near Shibuya Station, and with the train station at its center the Tokyu Land Corporation, owner of large plots of land, is carrying out

Figure 4. Transition in the area of rooftop and wall greening from 2000 until 2016 in Metropolitan areas
Source: (Ministry of Land, Infrastructure, Transport and Tourism, 2017)

a) Rooftop greening

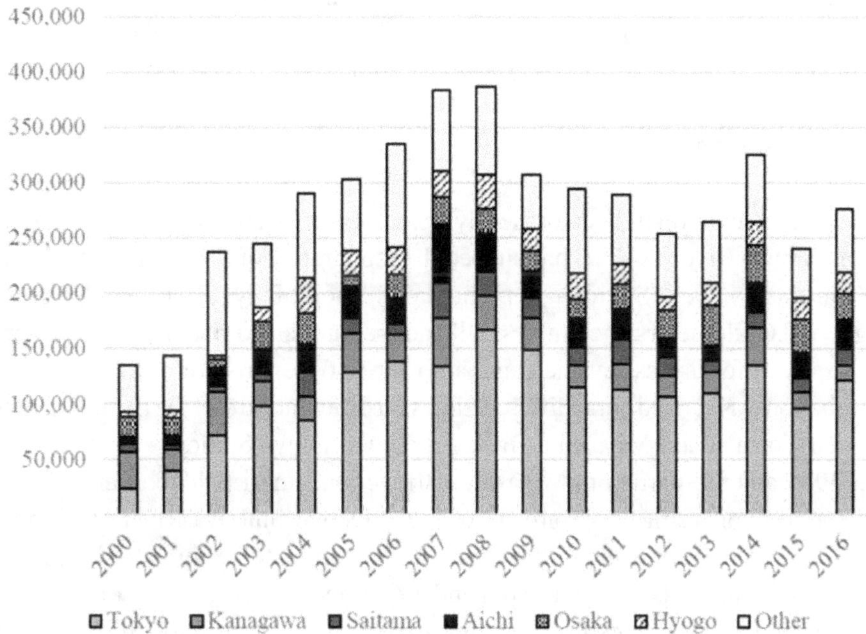

b) Wall greening

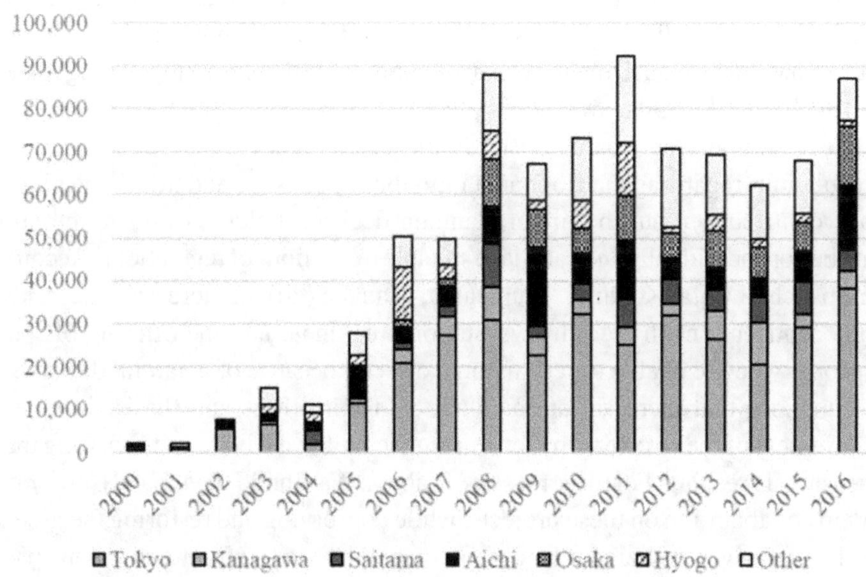

the redevelopment. In addition, around Harajuku Station (Sendagaya- Jingumae area), a well-known sightseeing spot located close to Shibuya Station and visited by many tourists from Japan and abroad, multiple redevelopment projects are carried out. For example, at the Tokyu Plaza which was managed by the same private enterprise responsible for the redevelopment around Shibuya Station—the Tokyu Land Corporation—a green space for recreation named "Omohara Forest" was established for its renewal opening in 2016 (Figure 5).

River Projects

As representative river project for the Olympics the improvement of the hydrophilicity of the Shibuya River—which has shaped the topography of Shibuya (ya of Shibuya 渋谷 means a valley) Station—is pointed out. The Shibuya River is with its total length of 6.6 kilometers one of the small and medium-sized rivers that was diverted into a culvert for use as sewage canal after the "36 Reports" of 1961. The culvert construction was carried out at different times for different parts of the river. The stretch between Yotsuya and Miyashita Bridge (below Shibuya Station) was constructed after 1965 and the part from Miyashita Bridge until Inari Bridge after 1971. The decision to improve the environment of the culverted Shibuya River was made in June 2013, when the hosting of the Olympics was still undetermined (Miki, 2017).[6]

The river projects are not independent projects but are part of the large-scale redevelopment plans for the station area. The planning for this large-scale redevelopment around Shibuya Station began in 2007 and since 2008 urban regeneration special areas (from here on called special areas) were established/specified; 21 areas (districts/zones) in Shibuya (Station) 2-chome in 2008; 21 areas around Shibuya Station, Dogenzaka and 3-chome in 2013; 1 area in Sakuragaokacho in 2014; 15 areas in Udagawacho. [7]

The most important characteristic of the special areas is that for the relaxation of regulations regarding the floor area ratio there is no set standard for the required public contribution, but the urban planning decision holders evaluate the public contribution individually and can make suitable relaxations of regulations. According to Yamauchi, Okata, Koizumi, Murayama, Manabe (2015), there are nine kinds of public contributions in the Shibuya Station area, including the offering of spaces for people who are unable to return home directly because of a natural disaster, the city functions that increase international competitiveness and the environmental improvement of the riverfront by renaturing the Shibuya River and promoting three-dimensional greening. For these reasons, Shibuya Ward and Tokyu Land Corporation work in collaboration on these projects, while conserving and restoring the green of the Shibuya River, utilizing a landscape shaped by a waterfront and green spaces, be the goals of these projects. These projects are ongoing at the present and after

Figure 5. Rooftop garden Omohara-no-mori (Omohara Forest) in Harajuku, Shibuya Ward.
Source: (Photo taken by the author, May 2, 2017)

completion they are planned to provide a green promenade of approximately 600 meters and two public squares (Tokyu Corporation, 2013). Furthermore, not only parts of the Shibuya River's environment but also its upper reach and the green way along the former waterway of the Tama River waterworks are considered for improvements in the future (Shibuya Ward, 2016).

DEVELOPING A SUSTAINABLE ECO-CITY IN PRE-OLYMPIC TOKYO: POTENTIAL OF NEW METHODS AND THEIR LIMITS IN AN URBAN ERA

This chapter reflected about the environmental sustainable city in the urban era by examining the example of the world's most populous city, Tokyo. The city, which will be hosting the Olympic Games for a second time in 2020, has taken environmental problems consciously into consideration since the bidding process for this mega event.

Particularly during the 1964 Olympics, when Japan was in a reconstruction period after the war, the city took a developmentalist approach regarding the improvement of its infrastructure, and especially the rivers and waterways disappeared in this process.

Through introspection about this developmentalism, the administrations of Tokyo and Japan came up with an environmental policy that emphasized a more sustainable urban development. As examples for this, greening projects and the improvement of the waterside environment were given in this chapter. As measures against the heat island effect in Tokyo's inner city, the creation of green spaces was advocated by the metropolitan and the national administration through the enactment of several laws since the 2000s. Propelled by this improved legal situation, land greening projects (Chiyoda Ward) as well as rooftop and wall greening projects (Minato Ward, Shibuya Ward) by private enterprises started to gain momentum throughout Tokyo's 23 wards. In addition, the projects for improving the waterside environment (Shibuya Ward) aim at increasing the amenity value of parts of the city's small and medium-sized rivers that have vanished from the cityscape during Japan's period of rapid economic growth. Private companies are the main actors practicing greening projects, utilizing the administrative legal framework in the pre-Olympic resume, which promotes fundamental urban restructuring as an opportunity.

However, it is uncertain whether these efforts lead to a fundamental solution of the urban heat island issue. Therefore, further research and investigation are eagerly awaited. In addition, as part of the promotion of post-Olympic urban tourism, the promotion of the nighttime economy and night-time activities is currently getting attention (Ikeda, 2017). This could result in the new environmental burden for the city's environment. In this regard, future research is required.

ACKNOWLEDGMENT

This research was supported by the JSPS [grant numbers 17J02079]; the Sasakawa Sports Foundation [grant number 180A2-032].

REFERENCES

Aikawa, M., Hiraki, T., & Eiho, J. (2006). Vertical atmospheric structure estimated by heat island intensity and temporal variations of methane concentrations in ambient air in an urban area in Japan. *Atmospheric Environment*, *40*(23), 4308–4315. doi:10.1016/j.atmosenv.2006.03.044

Arai, T. (1996). Changes in the hydrological environment in Tokyo. *The Journal of Geography, 105*(4), 459–474. doi:10.5026/jgeography.105.4_459

Bornstein, R. D. (1968). Observations of the urban heat island effect in New York City. *Journal of Applied Meteorology and Climatology, 7*(4), 575–582. doi:10.1175/1520-0450(1968)007<0575:OOTUHI>2.0.CO;2

Bureau of Environment. (2016). *Tōkyō-to kankyō hakusho 2016* [White paper on the environment of Tokyo 2016]. Retrieved from http://www.kankyo.metro.tokyo.jp/basic/plan/white_paper/2016.html

Butler, T. (2015). *Overdevelopment, overpopulation, overshoot.* New York, NY: ORO Editions/Goff-Bools.

Chiyoda City. (2010). *Chiyoda-ku midori no jittai chōsa oyobi netsu bunpu chōsa gyōmu hōkokusho* [Report on green space and distribution of heat issues in Chiyoda Ward]. Retrieved from https://www.city.chiyoda.lg.jp/koho/machizukuri/kankyo/ryokuka/heat-island.html

Echizawa, A. (2014). 1964-nen Tōkyō orinpikku to toshi keikaku [1964 Tokyo Olympics and city planning]. *Urban Housing Sciences, 87*, 24 28.

Furukawa, K. (2013). *Shuto kōsoku dōro no saisei* [Renewal of Metropolitan Expressway]. Reference.

Hompo, Y., & Yagasaki, N. (2015). Japan's inbound strategies using 2020 Tokyo Olympic and Paralympic Games based on the experiences of the past games [in Japanese]. *The International Journal of Tourism Science, 8*, 3-11.

Horie, K. (1996). Planning process of expressway in Tokyo [in Japanese]. *Doboku Keikakugaku Kenkyu, Ronbunshu, 13*, 1–22. doi:10.2208/journalip.13.1

Ikeda, M. (2017). A review of nightlife literature and suggestions for future research in Japan. *Geographical Space, 10*(2), 67–84.

International Olympic Committee. (2013). *Report of the 2020 evaluation commission.* Retrieved from https://stillmed.olympic.org/Documents/Host_city_elections/2020_Evaluation_Commission_report.pdf

Ishibashi, N., & Takahashi, T. (2015). *Background of the exclusive use of the elevated Metropolitan Expressway over Nihonbashi River.* The Institute of Civil Engineering of the Tokyo Metropolitan Government Annual Report.

Kondo, H. (2009). City canopy [in Japanese]. *Tenki, 56*(8), 677–678.

Lee, J., Harashima, Y., Lee, D. K., & Morita, T. (1995). A comparative analysis of processes used in developing environmental policies in Japan and Korea. *Environmental Sciences*, *8*(2), 181–192.

Maeda, C., Kondo, A., Akikawa, M., Kaga, A., & Inoue, Y. (2008, March): *Simulation of urban heat island in Hyogo*. Paper presented at the meeting of the Society of Heating, Air-conditioning Sanitary Engineers of Japan, Osaka, Japan.

Maruyama, M., & Nakagawa, Y. (2004). Shutoken seibi keikaku ni okeru ryokuchi seisaku no hensen ni kansuru kisoteki kousatsu [Basic consideration about the transition of green policies in Metropolitan Region development]. In *Proceedings of the Meeting of Japan Society of Civil Engineers* (Vol. 59, pp. 4-334). Tokyo, Japan: Japan Society of Civil Engineers.

Masumoto, K. (2007). Space-time characteristics of air temperature distribution and heat island situation in Osaka City. *Journal of Urban Living and Health Association*, *51*(4), 215–224.

Matsuoka, R. (2017). *Toshi ni okeru ryokuchi ni kansuru seido keikaku no hensen to Tōkyō 23-ku no ryokuchi seisaku no tokuchō* [Transition of institution and planning related to urban green spaces and the features of green policy in Tokyo's 23 wards] (Unpublished graduation thesis). Tokyo, Japan: Waseda University.

Miki, T. (2017). Redevelopment projects in Shibuya Tokyu Corporation [in Japanese]. *Journal of JSEE*, *65*(6), 41–43. doi:10.4307/jsee.65.4_1

Minato City. (2017). *Minato-ku midori no jittai chōsa* [Report of Green Investigation in Minato]. Retrieved from https://www.city.minato.tokyo.jp/ryokukasuishin/kankyo-machi/kankyo/chosa/9midorinojittai.html

Ministry of Land, Infrastructure, Transport and Tourism. (2011). *Shuto-ken seibi hou ni motozuku daitoshi-ken no minaoshi* [Re-examining the Tokyo metropolitan regional policy based on Metropolitan Region Development Law]. Retrieved from http://www.mlit.go.jp/common/000139692.pdf

Ministry of Land, Infrastructure, Transport and Tourism. (2017). Zenkoku okujō hekimen ryokuka sekō jisseki chōsa no kekka hōkoku [Report of investigation of rooftop and wall greening in Japan] [Data file]. Retrieved from http://www.mlit.go.jp/report/press/toshi10_hh_000257.html

Ministry of the Environment. (2014). *2020-nen orinpikku pararinpikku Tōkyō taikai wo keiki toshita kankyō hairyo no suishin ni tsuite* [Report on promoting the environment policy for the 2020 Olympics and Paralympics]. Retrieved from https://www.env.go.jp/press/18532.html

Mizuno, M. (1993). Subjects on environment in 24-hour city: Discussion on urban heat island [in Japanese]. *The Journal of the Acoustical Society of Japan*, *49*(11), 832–838.

Nakabayashi, S., Kishī, T., & Ōsawa, M. (2012). On the rooftop gardening policy in the 23 wards of Tokyo: Focused on the actual condition of the public buildings [in Japanese]. *Journal of the City Planning Institute of Japan*, *47*(3), 475–480.

Nakamura, S., & Oki, T. (2009). The philosophy and chronicle of abolished urban river in 36'report [in Japanese]. *Proceedings of Hydraulic Engineering, JSCE*, *53*, 565–570.

Oke, T. R. (1979). Review of urban climatology 1973-1976. *WMO-Technical Note*, *169*, 1–100.

Ozaki, M. (2002). A Study of the Process of Forming Sport Policy in Japan [in Japanese]. *Hitotsubashi University Research Series Humanities*, *39*, 159–252.

Saito, T. (2014). Orinpikku pararinpikku to kankyō risku no kanri [Olympics, Paralympics and the management of environmental risks]. *Sompo Japan Nipponkoa Risk Management Report*, *126*, 1–11.

Sakakibara, Y. (2001). Comparison between the effect of heating from urban surface and that of mechanical mixing of urban atmosphere to heat island [in Japanese]. *Tenki*, *48*(5), 305–311.

Shibuya City. (2004). *Shizen kankyō chōsa hōkokusho* [Report on the investigation of the environment]. Retrieved from https://www.lib.city.shibuya.tokyo.jp/?action=common_download_main&upload_id=514

Shibuya City. (2016). *Shibuya-ku midori no seibi hōshin* [Green policy in Shibuya Ward]. Retrieved from https://www.city.shibuya.tokyo.jp/kusei/shisaku/ku_keikaku/green_plan.html

Takeuchi, K., & Lee, D. K. (1988). A framework for environmental management planning [in Japanese]. *Journal of the Japanese Institute of Landscape Architects*, *52*(2), 95–104. doi:10.5632/jila1934.52.95

Takeuchi, T. (2012). A study on the potential of Green Structure Plan as comprehensive spatial control plan [in Japanese]. *Journal of the Japanese Institute of Landscape Architecture*, *75*(5), 601–604. doi:10.5632/jila.75.601

Takeuchi, T., & Ishikawa, M. (2008). A study on the green space policies on the fringe of Tokyo wards area in the 1950's and 60's [in Japanese]. *Journal of the City Planning Institute of Japan*, *43*(3), 199–204.

Tokyo Metropolitan Government. (2017). *New Tokyo. New tomorrow. The action plan for 2020*. Retrieved from http://www.metro.tokyo.jp/english/about/plan/index.html

Tokyu Corporation. (2013). *Shibuya-eki minami gaiku purojekkuto ni kansuru toshi keikaku no kettei ni tsuite* [Notification of the decision about the south district project at Shibuya Station]. Retrieved from https://www.tokyu.co.jp/file/130123-3.pdf

Tsuboi, S. (2003). A study on change of water area and water activity projects in Tokyo ward area [in Japanese]. *Comprehensive Urban Studies*, *82*, 19–34.

Tsunematsu, N. (2014). *Tōkyō ni okeru hiito airando no jittai to shonetsu taisaku ni tsuite* [Heat island in Tokyo and countermeasures against heat]. Retrieved from https://www.tokyokankyo.jp/kankyoken/wp-content/uploads/sites/3/2014/12/4cc271adc7839d797e861f8940f5e06c.pdf

Tsunematsu, N., Yokoyama, H., Honjo, T., Ichihashi, A., Ando, H., Yamagata, Y., Murakami, D., & Shigyo, N. (2016). Relationship between land use variations and spatiotemporal changes in amounts of upward infrared radiative energy in downtown Tokyo at midday on hot summer days. *Annual report of the Tokyo Metropolitan Research Institute for Environmental Protection*, 76-82.

UN Habitat. (2016). *World cities report 2016: Urbanization and development: emerging futures*. Retrieved from http://wcr.unhabitat.org/main-report

United Nations. (2016). *The World's Cities in 2016 Data Booklet*. Retrieved from http://www.un.org/en/development/desa/population/publications/pdf/urbanization/the_worlds_cities_in_2016_data_booklet.pdf

Uno, I., Wakamatsu, S., & Ueda, H. (1988). Behaviour of nocturnal urban boundary layer and air pollutants [in Japanese]. *Journal of Japan Society of Air Pollution*, *23*(2), 103–114.

Yamauchi, T., Okata, J., Koizumi, H., Murayama, A., & Manabe, R. (2015). A study of the process to determine public contribution elements in spatial districts for urban generation – Case studies in Ochanomizu, Shibuya, Ginza districts [in Japanese]. *Journal of the City Planning Institute of Japan*, *50*(3), 904–911.

ENDNOTES

1 The general explanation for this effect is that bulks of hot air rise from the surface and expand, and as a result the difference to the ambient air temperature disappears leading to a stagnation at a constant altitude, forming the dome of urban air (Mizuno, 1993).

2 The Basic Law for Environmental Pollution Control was changed into the Basic Environment Law in 1993 and has specified the principal policy of Japan's environmental administration since then.

3 These projects were carried out by private railroad companies during the development of the railway lines around Tokyo and consisted mainly of housing land development for company managers and senior bureaucrats.

4 In addition, actions regarding greening projects in the wards are not easy to compare merely by their numbers. Matsuoka (2016) compares the public and private practices for each greening category and shows that projects for the greening of rooftops are also easy to handle for private enterprises and that in the wards Chiyoda, Koto and Edogawa public greening projects and in the wards Minato and Itabashi private projects are most active.

5 The criteria for the selection as model block are a significant heat island phenomenon, the easy attraction of general attention with a high effectiveness to promote further spreading, and the high probability to enforce concentrated measures.

6 However, according to developer Tokyu Land Corporation there was the formation of a consensus saying that "The city functions around Shibuya that were shaped during the last Olympics in 1964 have not much changed since then and the people concerned had a common understanding that a reconstruction was necessary", which suggests an awareness that the city's structure and functions were determined by the Olympics (Miki, 2017).

7 These areas are, based on the Urban Renaissance Special Measure Law enacted in 2002, in the areas for city regeneration urgent improvement based on the existing zoning, excluded from regulations regarding usage and floor area ratio, making a high flexibility in planning possible.

This research was previously published in Bioeconomical Solutions and Investments in Sustainable City Development edited by Jose G. Vargas-Hernandez and Justyna Anna Zdunek-Wielgolaska, pages 196-223, copyright year 2019 by Engineering Science Reference (an imprint of IGI Global).

Chapter 11
The Role of Urban Green Spaces in the Transformation of Community Ecosystem in Developing Countries

José G. Vargas-Hernández
University of Guadalajara, Mexico

Justyna Zdunek-Wielgołaska
Warsaw University of Technology, Poland

Karina Pallaggst
Technische Universität Kaiserslautern, Germany

ABSTRACT

This chapter aims to analyze the review the existing literature on some important features of urban green spaces such as the components, functions, services, community involvement, initiatives, and actions from an ecosystem perspective. The analysis begins from the assumption that urban green spaces are ecosystems of vital importance in enhancing the quality of life in an urban environment and supplying ecosystem services such as biodiversity and climate regulation. Meeting the needs of users is related with the functions and services that urban green spaces provide to communities. The methods employed in this analysis are the empirical literature and documents review, analysis of existing data on uses and users, interviews with authorities, and more detailed examination of a case specific data. Also, as some concluding remarks, some wider environmental, economic, and social initiatives for local authorities and communities are suggested that can justify initiatives by all represented and involved stakeholders. Finally, the study advances some opportunities, challenges, and further research.

DOI: 10.4018/978-1-7998-0441-3.ch011

Copyright © 2020, IGI Global. Copying or distributing in print or electronic forms without written permission of IGI Global is prohibited.

INTRODUCTION

The history of life on earth is one of the living things surrounded by a natural environment which supplies water, fresh air, minerals, plants, vegetation, animals and all the fruits of nature, and so on to enjoy everyday life. However, these natural and environmental resources are not lasting forever and ever, must of them are either polluted and corrupted or extinguished by consumption. Urban populations are facing ecologically vital threats from over urbanization, such as they are water and air pollution, agricultural and forest lands are urbanized with vegetation removal and ground water overdraft. The ecological dimension of urban green spaces considers the objective and subjective components of a place providing a supportive habitat of biological diversity.

The renewal and increasing of urban green spaces considering the fast demographic growth and agglomeration should be accompanied by inhabitants' participation on environmental, social, cultural and economic actions and objectives to promote bio economy in urban biodiversity and sustainable development. Inhabitants should be informed and motivated to participate in environmental, cultural and educational activities and become active in designing and planning of urban green spaces. Public support and political involvement of citizens for urban green space development needs different actions. Public urban green spaces are accessible to and used by all the citizens. Arrangements of public activities and action on planned urban green spaces raise awareness amongst inhabitants of the city. For example, a public event can have the purpose to make users aware and educate them through experiencing and enjoying different activities organized by the urban green space.

In 2014, 54 percent of the world's population were living in urban settings and it is projected to reach 70 percent by 2050 (United Nations Department of Economic and Social Affairs – Population Division, 2014). By the year 2020, around 62 percent of world's population will live in urban areas covering 2% of world land space and consuming 75 percent of nature resources. By the year 2030, two thirds of urban area that will exist has to be built in sustainable urban environments. In 1953, the seminal report on Park Life recognized urban green spaces as a vital component of urban environment and their role in social renewal (Comedia & Demos, 1995).

This chapter aims to contribute to the main topic of the book Optimizing Regional Development through Transformative Urbanization by presenting and analysis of the role of urban green spaces in the transformation of community ecosystems with emphasis in developing countries. The chapter begins by describing the components of urban green spaces to focus in the ecosystems functions and services. The methodological approaches are reviewed to understand the different functions that users give to urban green spaces and some of the most important factors contributing to a successful community involvement.

The analysis of the role of urban green spaces is centered on a discussion on the challenges and opportunities for public community action deriving from there some public initiatives and actions as concluding remarks. Finally, the analysis projects some future research directions.

COMPONENTS OF URBAN GREEN SPACES

An urban green space system is an important component in any community development, in housing, business, leisure areas, etc. (Baycan-Levent, 2002). Components of urban green areas are vegetation, water, accessibility, services of shelters, toilets, seating, playgrounds and sport areas, events and activities, environmental quality conditions and resources such as lighting, safety, litter bins, friendly staff, artistic features and artifacts such as sculptures, etc. The quality assessment of green spaces is measured by some factors such as infrastructure, vegetation, accessibility, security, equipment. Conditions that favor the use of urban green spaces are the distance walking time (Herzele & Wiedeman, 2003), location and distribution, easy access, proximity. Environmental enhancement makes urban green spaces of more quality and attractive by promoting inward investments, increasing the land value and economic stimulation of the community.

Urban green spaces are urban areas which were natural or semi natural ecosystems that were converted on urban spaces by human influence (Bilgili, 2012). Urban green spaces are public and private open spaces in urban areas primarily covered by vegetation, which are active or passive recreation or indirectly positive influence on the urban environment available for the users (Tuzin, Leeuwen, Rodenburg, & Peter, 2002). Urban green spaces provide recreational and sports activities. Urban green spaces provide sustainable diverse places where, according to the classic report Park Life *"people will find a sense of continuity, of relief from the pressure of urban living, places to be in touch with the natural cycle of the seasons and of wildlife and also places to meet and celebrate with others"* (Comedia & Demos, 1995).

The broader notion of green space connotes turf grass-related residential, commercial and institutional surfaces and public facilities such as parks and playing fields. Turf grass is associated with the notion of green space that connotes turf-related surfaces as residential, commercial and institutional lawns and turf surfaces. Urban green spaces connect the urban and the nature while caring for the environment, social and economic elements. Public forests and green roofs in public and community buildings, and vacant and derelict land also provide ecosystem services. Productive land use ensures long-term regeneration initiative to use properly if green spaces for economic revenue by implementing sustainable urban initiatives such as drainage schemes. Green spaces include more wild, woodland-type and untamed elements, with more area for child development.

An urban green space can be considered as a continuum without fences, hierarchies and horizontally maintained at the same level community oriented service and use-oriented approach. Green spaces are the spirit of the community. Urban green spaces are a focal point for communities (Greenspace, 2007). Urban green spaces contribute to build a sense of community among residents being more likely to enjoy strong social ties. Green spaces promote interaction between people; develop social ties and community cohesion. Greenness in neighborhood is one of the most important predictor of neighborhood satisfaction (Van Herzele & de Vries, 2011).

Green Spaces has as a role stimulating creative/innovative activities in urban area and in making cities innovative engines generating socio-economic growth.

ECOSYSTEMS FUNCTIONS AND SERVICES OF URBAN GREEN SPACES

Urban green spaces are ecosystems of vital importance in enhancing the quality of life in an urban environment. Urban green spaces supply ecosystem services such as biodiversity, climate regulation. Urban green spaces are essential for the quality of life, health and well-being of citizens. Urban green spaces are critical for protecting wildlife, watersheds, meads vegetation; provide air quality for a dense urban environment and recreational activities. Cool islands in dense urban areas can be provided between spaced green spaces. Dense green space is more effective in preventing nitrogen run-off, untreated human and industrial waste, toxic materials, and debris. Urban green spaces provide ecosystem services that can improve conditions of environment, pollution and congestion of the large metropolitan areas.

The reasons of visiting urban green spaces are mostly for enjoying a wide range of environmental elements like flowers, trees, nature, fresh air, wildlife, watching cascades, educational opportunities, social activities, taking children to play, social interaction, to meet friends, picnics, meeting people, getting away from it all, passive walking and activities, shelter and sitting, etc. Shared parks and gardens may be setting up to facilitate social links, collective participatory projects and collective cultural interventions, well-being recreational areas and walkways connecting attractions and facilities.

Urban green spaces have a beneficial physical, psychological and health effects through physical activities, green exercises. Environmental determinants affect the use of green spaces, physical activities and leisure. The amount of green spaces available to users in the living environment correlates with socioeconomic, demographic, and self-perceived health. Higher levels of greenness have been positively associated with lower stroke mortality. Perceived neighborhood greenness is associated with physical and mental health. Socioeconomic and cross-cultural variations may result of unequal distribution of green spaces.

An important reason to visit urban green spaces is also the use of facilities such as cafes, restaurants environmental centers, libraries, museums, sports like football, tennis, etc., biking, skateboarding, cycling and other forms of active enjoyment. Events as other important motive to visit urban green spaces like group music performance, concerts, Christmas carol concerts, orchestral performance, craft fairs, fun fairs, opera, circus, firework displays, bands playing, dance up, etc.

Making inhabitants aware of the urban green spaces existence and use values contributing to urban inhabitant lives in a more balanced quality of life, lifestyle, encouraging physical and mental fitness, reduces tensions and conflicts, relieving the harshness of the urban environment, providing places for social and cultural interaction in informal contacts and more formal participation in social events, social inclusion, recreation, aesthetic pleasure and wildlife and fostering community development.

Urban green spaces are natural meeting points for local inhabitants facilitating social inclusion and integration, community cohesion, social capital, civic society, supported by an increasing sense of identity and belonging (Konijnendijk et al., 2013; Abraham et al., 2010). By providing a meeting place for social interaction and integration between community users, green spaces influence social capital. Inhabitants living near urban green spaces reduce health inequalities and have lower circulatory diseases (Mitchell & Popham, 2008). Inequitable distribution of green spaces is correlated with distribution of disadvantaged inhabitants. Surveys have shown that people are less stressful, communicate well and make sensible decisions by the earth surrounded by green spaces.

Urban green spaces can be linked as wildlife corridors to facilitate the movement of fauna preventing fragmentation and isolation of wildlife (Rouquette et al., 2013; Hale et al., 2012). Urban green spaces are the home of many species including those that are rare and threatened and the habitat for pollinators. More urban green spaces sustain more wildlife and biodiversity providing a more favorable habitat, therefore requiring more protection from human interference (Cornelis & Hermy, 2004; Fuller et al., 2009; Schwartz et al., 2002; Baldock et al., 2015). Creation, protection and development of urban green spaces is a relevant element of sustainable urban development.

Urban green spaces have an impact on human thermal comfort and air quality in human health (Cohen et al., 2012; Weber, Haase & Franck, 2014; Nowak et al.,

2014). Green spaces reduce urban heat islands and air pollution improving air quality in urban settings (Bowler et al., 2010). Community-level air quality depends of tree population (Morani, Nowak, Hirabayashi & Calfapietra, 2011).

The cooling capacity of green spaces is affected by multiple variables such as density, size and shape associated to increase of air quality. Urban green spaces reduce heat, ozone and ultraviolet (UV) radiation and improve air quality. Absorbing pollutants improve air quality. Research based on modeling has weak evidence that capturing pollutants and particles by urban green spaces improve air quality. Wong et al. (2013) reviewed the evidence of the relationship between green spaces, heat and air quality considering variables such as green space type, climate, method, etc. Building orientation and heights affect cooling and air quality from green spaces.

Air temperatures in warm humid climates are significantly cooler within the urban green spaces (Oliveira et al., 2011) when compared to non-green areas (Armson et al., 2012). Humidity tends to be higher in urban green areas than in inhabited zones. Tight inhabited areas without green spaces usually have inadequate climate. Green spaces mitigate the effect of climate warming by providing shade. Replacing paved yards with urban green spaces reduce the heat-island effect during the summer by moderating temperatures expected with climate change.

The role and behavior of urban green spaces and gardens on improving climate and reducing air pollution. Pollution in urban areas distributes on the type of architecture and proximity to green spaces. An avenue with a green space is less polluted because dispersion is better, while narrow streets tend to be more polluted. Walkways with large green spaces are more protected from pollution (IngegaÈrd, 2000). Planting more trees on street canyons may not be a good prescription where may increase concentration of pollutants

Trees in urban green infrastructure capture and sequester carbon mitigating the negative effects of emissions. Carbon sequestration is the removal of the greenhouse gas carbon dioxide and its incorporation into plants. Any green spaces balance carbon taking more than return to the atmosphere (Nowak et al., 2013; Nowak & Crane, 2002). A forest in a green space maximizes carbon sequestration (Strohbach et al., 2012).

METHODOLOGICAL CONSIDERATIONS

An analysis of the green space deals with the physical and quantitative, functional, ecological, environmental, economic and quality aspects. Economic aspects are the expenses of development, costs of maintenance, financing and budget sources. The quality of urban green space experience requires to be studied from an interdisciplinary perspective drawing from both natural and social sciences. Some of the physical quantitative indicators are the supply and distribution of natural and landscapes resources of public green as percentage if the city area, the m2 per capita, structural and morphological characteristics. Quantitative evaluation of the relationship between urban population and urban green spaces takes into account functionality, green space ratio, green space coverage and green space area per capita (Xion-Jun, 2009). The quality aspects of urban green spaces are the suitability and quality of site structure, design and provision, quality conditions.

Finding meaningful information on uses and users of urban green spaces is hampered by inconsistencies of information from local authorities. The use of model surveys to collect information from users of green spaces regarding satisfaction of needs and aspirations need to be developed through pilot studies and consultation by researchers and local authorities. Consultation and involvement on environmental issues identify the community needs. Also, results of research finds evidence on differentiation of needs of having green space close to the living place as opposed to where they work (Greenspace, 2007).

User's perceptions on urban green areas matter for the community´s image and deciding to make use. Perceptions on image of urban green spaces affect uses and user aspirations and value creation of the community in designing, meeting the needs and sustainable managing. The more related issues to designing are the variety, activities, spaces, sensory stimulation, vegetation, water, birds and animals, etc.

Levels of satisfaction on uses of urban green spaces are related with provisions for safety, cleanliness and tidiness, well kept, peace and quiet, not noisy, fresh air, flowers, lakes, well signposted, wildlife, catering and bar facilities, toilets, historic setting, pageantry, trees and greenery, playgrounds, visitor center, route marking and signposting, reduction of traffic, drinking fountains, etc.

A sound basis on collection and analysis of data is the mean to find out the priorities. A pool of data should be collected and analyzed to find out priorities in terms if type, quantity and quality, location, accessibility. Quantity of green space is reduced by the trend towards more compact urban environment (Burton, 2003). The observed current trends on urban green spaces suggest an increasing degradation and without support it is not likely to reverse the process. An already available expert study may help to compare and checking the planning context and legislation.

USERS OF URBAN GREEN SPACES

Meeting the needs of users is related with issues of awareness of needs, the nature of facilities and its conditions, opportunities for activities, events and playing, provisions of comforts like toilets, shelters, seating, refreshments. Users of urban green areas develop some patters of use of informal and passive activities, with peaks in the afternoons, weekends and holidays on a daily basis. Involvement in urban green spaces leads to create facilities to meet the needs of users with quality use and experience. Facilities of urban green spaces must meet environmental, socioeconomic and psychological of user needs and attitudes (Balram & Dragicevic, 2005). To meet the users' needs at local environmental, social and economic levels require the development of local standards, such as provision of urban green space per head.

In one research, users of urban green spaces manifested psychological reasons (Dunnett et al., 2002). It was found significant relationship between the use of green spaces and levels of stress (Grahn & Stigsdotter, 2003). Green spaces in the living environments also positively affect stress and quality of life. Urban green spaces reduce stresses for users and provide them a pleasant positive distraction (Ulrich, Quan, & Zimring, 2010). Use if green spaces are associated with less stress. Viewing nature and urban green spaces ameliorates stress (Ulrich, 2002).

Natural green environments have restorative effects and pleasing stimuli promoting soft fascination (Forest Research, 2010). Also users are happier and have higher well-being when they live in an urban area with large green spaces (White et al., 2013). The evidence between green space and physical activity is strong although it has been reported beneficial links between urban green spaces and emotional, psychological and mental health and well-being the evidences are weak. Large urban green spaces contribute to physical and mental health and well-being of users.

To meet the user´s needs of urban green areas are relevant factors the nature of activities conditions of facilities, opportunities for playing and equipment designed to develop creativity, social skills and confidence, provision of activities, suitable seating, shelters, drinking fountains, café, etc.

Accessibility to urban green spaces are more related to ease of access by proximity and no physical barriers, transportation, open fences an early hour, accessibility to disable people, information on cues and way-finding features, maps, information on entrance, path junctions, slopes and cambers, inadequate parking, gravel car park surfaces, heavy gates, attendance for those with disabilities, visual impairment. Urban green areas are safer gathering places for children and young people, at least more than they are while being on the street.

Unequal distribution of green spaces and less access to green environments is related to health inequalities, increasing pollution and intense heat. Deprivation levels are linked to access to green spaces. Distance from the green spaces is related to physical activity, thus users living nearby report higher physical activities although there is no correlation between accessibility to green spaces. Proximity to green spaces is associated with self-reported health.

Access to urban green spaces for elderly, disabled, children, women and minority ethnic groups' concerns issues such as ease entrance, proximity, social inclusion, provision for the visually impaired, public transport, parking, moving safely and surfaces design. Awareness and understanding for social inclusion in urban green areas is recognition of the particular social and cultural needs and aspirations of users. Psychological issues related to the use of urban green areas prevent users of not going alone for the feeling of vulnerability, fears, safety concerns, laziness, loneliness, lack of confidence, inertia, etc.

Some negative economic impact of urban green spaces may have fall in profits and potential problems derived of gentrification of the area as a result of increasing the added value of land and increasing of housing prices. Negative impacts identified with green space are the increased green density that increase street canyon air pollution.

Some personal issues that deter from using urban green spaces are factors such as not having enough time, working unsocial hours, poor health and mobility, preferences for visiting other places, issues related to the location of urban green spaces, accessibility, user experience, environmental quality. Other personal issues can deter users of going to urban green spaces such as preferences of visit other places, having own park, health problems, changing circumstances, family and parental restrictions. To increase parental responsibility, training sports sessions of children and young, encourage active participation of parents. Users of urban green spaces are deterred by lack or deficient facilities, low environmental quality, the influence of undesirable people, safety issues and psychological concerns, dog mess, litter, graffiti, vandalism.

Deterrent effects of "other people" users are related to conflicts between children and young people, teenagers, with adults, drug users, undesirable characters, users drinking alcohol, verbal abuse, gay men, bikes and skateboards, gamblers, noisy people, crowded, etc. The study of urban environment combines the sound levels, biodiversity and green spaces. The results of this study confirms that planning and designing of urban green spaces is enhanced by the ecological quality in issues such as noise levels of livable and sustainable communities (Williams et al., 2000; Girardet, 2004). The soundscapes of green urban spaces have been less well-studied.

The declining quality of urban green spaces contributing to a decline of urban quality of life has been studied by Irvinea, Devine-Wrightb, Payneb, Fullerc, Paintera, and Gastonca (2009). Dog mess is a critical concern in urban green space and required special attention like dog free areas and the dog areas, good positions of dog bins in suitable locations, dogs on the lead and controlled, dog toilets, proper use of fines, etc.

Some barriers that prevent users from using urban green spaces are lack of or deficient facilities, lack of events and activities, poor physical and psychological safety, lack of information boards or center, poor maintenance, the presence of undesirable users, dogs and other poor environmental conditions such as vandalism, lick of litter bins, unease accessibility, lack of toilets, lack of other services like café, lack of play areas for children, elderly and disable users, etc. Elderly and disable people have concerns to ease of access and moving around urban green spaces safely.

FACTORS OF SUCCESSFUL COMMUNITY INVOLVEMENT IN URBAN GREEN SPACES

Some relevant factors to improve the use of urban green areas are less dog mess, improved having more things to do, dogs being on a lead, dog free areas, more staff, provision of more seats, no motor vehicles, no cycling, no roller skating and roller blading, play areas, lower planting near paths. Accessibility to an urban green space and being easier to get in close to home, with parking facilities o good public transportation. An information center and information boards, displays boards, braille signs, maps signing posts with directions, etc.

The increasing application of digitals technologies plays an important role not only to enhance these factors but also as part of the benefits users receive from a public space. In the near future and with increasing urbanization, digital technologies may provide interactive information about the activities but also for entertainment and educative matters. It also may facilitate matters of governance.

Urban green spaces provide opportunities for all people to meet, regarding their cultural, religious, ethnic origin, political ideology, courting couples, businessmen, etc. Urban green spaces are sites for community spirit, although different friends and user's groups have different levels of involvement, engagement and development which are from frankly adversarial onto the ones that have formed already partnerships. Some community activities on children and young people could spring out around the urban green spaces providing a free open gathering place, usually more on weekends, summer time and holidays. Members of the friends group can clean-up the urban green space every weekend.

As a result of a consultative process, local authorities committed to partnerships must consider demonstrations of willingness of community groups and residents to get involved in some initiatives and to take responsibility for some tasks at the urban green spaces. In fact, some participants are willing to volunteer and engage in some green space proposals given the provision for unique experience.

Partnerships offer an opportunity for coordination of efforts for environmental regeneration programs at potential low financial cost. For this purpose, a priority proposal is to establish a user community group to include representative volunteer local members to work as a charity in partnership. Provision of other facilities that encourage the use of urban green areas are café, information centers, boards, toilets, sports areas, sporting events, dog litter bins and seating, parking, staffing including attendants, park wardens, keepers, first aid facilities, boating, water features, water plants.

Different models of partnerships between urban green spaces and communities require cultural change to move the emphasis on community involvement and sense of ownership which results in caring, resourcing, involvement, creativity and innovation. Some factors contributing to a successful involvement are the institutional culture of local authorities, community groups and users, resources and capabilities, sense of funding, investing and ownership, voluntary commitment and communication between stakeholders. The responsibility and ownership of urban green spaces should not be fragmented between different authorities and different structures to achieve more innovation, efficiency and community involvement. Local authorities develop approaches to engage and involve users through discussion groups, consultations, artistic events, sport activities, ethnic minority background activities, leisure programs, environmental and horticultural activities, community gardens, organic food growing projects, etc.

Community involvement and engagement in urban green spaces leads to enhance quality of experiences and uses meeting the needs of users and long term sustainability, give access to additional funding and expertise. User groups usually set up priorities for urban green spaces where funding is available and consultation from local communities is required, tied as they are into yields tangible results. Groups should be active to complement capabilities of local authorities. One way to motivate is to provide grants for urban green spaces projects available to all groups. Urban green spaces are a catalyst for community projects because they resolve around the most relevant community issues and they potential for environmental, social and economic change derive from adequate funding and managing of resources. Therefore, the prominence of urban green spaces in communities is in the promotion for funding in community regeneration initiatives.

Creative and innovative approaches to funding and resourcing of urban green spaces require if designing the appropriate arrangement to make the best with the available resources. An innovative process is not exempt of conflicts. Conflicts arise between users and community organizations and groups who set up the trust. More innovative and creative local authorities are able to achieve more and better resources with less financial investment and spending. There are different methods of allocating, administering and using the funding to be spent according to creative approaches aimed to enhance quality of life.

Local authorities should ensure that the backgrounds, culture and environmental resources, new expertise, skills and interests brought are in harmony to develop the potential is self-fulfilled without leaving aside the commitment and voluntary efforts of traditional users. Urban green spaces are critical for community volunteer groups to achieve change providing facilities and activities to local users and involving and engaging other users. Activities developed by community groups in urban green space are most essentially voluntary in actions such as conservation and maintenance tasks, although volunteer maintenance is coordinated by rangers and the feeling of ownership upkeep is the responsibility of local authority.

Volunteers and trainees can be in charge of maintenance. Usually voluntary community groups get involved on some routine operations and maintenance such as planting, grass cutting, cleaning, etc. Volunteering activities are more common in business groups conducted through staff initiatives. More active volunteers in the community need to be more motivated and negativity managed to achieve more active involvement, engagement and collaboration in a task-orientation approach with local authorities. A green space watch scheme run by volunteers can be set in partnership with the police.

Urban green space service delivery from environmental authorities may have a more holistic approach of policy and budget implementation. Available resources to local authorities and their efficient use make better provision of quality service delivery. One of the main problems facing the urban green spaces is the capital and financial resources and budget decline in real terms by the spending per head of population for funding urban green space projects spending per head not necessarily takes into account the area of green space. Comparison can be made on the spending per head and per hectare of green space, despite that there is not a consistent methodology. Urban green space officers must have expertise in community involvement and engagement, environmental training.

Community engagement and involvement occurs with a change of institutional culture of local government and changes in users' culture. Sensory stimulation with things to touch, see and experience such as gardens to smell and touch, statues to touch, warning sounds and colored items such as seats, litter bins, lights shining on the water at night stimulation through running water, scented planting, quacking ducks, calm flowery area, etc. Planting programs with volunteering users from a community not only inject color but the sense of identity.

Determining the economic value of urban green spaces is considering their natural resources. Some economic factors of urban green spaces provide production of wood, supply of fruits, economic value of the area, jobs creation, tourism attraction, etc. Urban green spaces are ecological based value (Bilgili, 2012) (that has become a necessity together with aesthetic and recreational values. Evidence on the value of green ecological networks on wildlife is limited, although have become an element of urban planning (Tzoulas et al., 2007). Ecological and environmental aspects are the biodiversity and ecological values, urban climate and natural corridors.

This vision must be agreed and shared with all the users and stakeholders and local authorities. A vision can develop and protect the quality standards of using urban green spaces, in healthy and pleasant environment and improve new kinds of use and ensuring sustainability with high ecological and environmental value for healthy living, offering well designed and maintained green space meeting the demands of users, ensuring participative action and accessibility, stimulating socioeconomic development and quality of lifestyle in the community, contributing to the spatial identity.

The spatial concept of urban green space incorporates of green in the urban structure is related to the concept of a green system, network of corridor. A spatial concept for urban green space development describes and incorporates green issues, interconnects the existing urban spaces and the future desired network and their relationships with the hinterland of the city. Green spaces are in relationship and connected with green networks and green corridors defining preservation, improvement and development areas, neighboring countryside, regional green network, and pedestrian and cycling paths, etc.

Urban green areas can provide countryside activities and educational activities to children outside school hours and to adults through training programs, workshop and cultural events on urban regeneration initiatives ranging from horticulture, maintenance, school education visits on nature, art activities, lectures and training on environmental education, vocational qualifications in horticulture, animal husbandry a four-week summer play scheme. Also school children supported by their schools in some activities related with environment, ecology, tree planting, etc., as for example, providing an eLearning module to increase awareness and knowledge. Educational institutions can get advantages by making use of urban green spaces for educational, sporting programs and community-based education activities for children young and adult people. Urban green spaces offer children the development of social environment to improve cognitive and motor skills, higher levels of creative play, socialization, more collaboration, and emotional resilience (Forest Research, 2010).

A partnership structure that enables a crosscutting integration of community groups initiatives with officers of local authorities and the urban green space in a network to coordinate responsibilities developing action plans and activities to developing biodiversity, improve the environment. The action plan describes the specific tasks for implementing and achieving each type and each issue, actions, timescale, potential funding sources, partners. Local authorities of urban green spaces working closely with an attitude of acting as the eyes and ears with the friends and user's groups are positive to work on partnership shaped by a community orientation. Unintended consequences of urban green spaces are avoided with community-based decision-making. Partnerships raise the quality of urban green space.

Partnerships with business, agencies and communities with local authorities bring available base-line funding to achieve higher and better added value far more than can achieve a local authority alone. Effective partnerships between local government, business, agencies, neighborhood organizations and community groups can add financial and quality values to the green spaces. The identification of spatial, organizational and financial problems on the planning and managing of urban green spaces, such as distribution, changing use, green corridors and networks. Among the organizational problems are the communication and cooperation problems. Financial problems are related with funding. Other important arrangements to increase and make more efficient financial resources are among others, partnerships with gran making foundations, private financial initiatives, community and business groups, targeted grant funding and creative initiatives to increase revenue spending.

External funding and resources from external funded capital programs are required to complement the budget required for maintenance, although they are essentially crucial for capital works. Other forms of external funding are the so call landfill tax credit scheme and private and business sponsorship that enable creation and operation of facilities and a wide range of financial private initiatives as a means of injecting private capitals. Partnering to achieve external funding and expertise from community and business involvement is a form to lift quality standards. An active sports program of events can attract funding to be able to be financially self-supporting.

Creative and innovative approaches for external funding from community and business groups are usually selective in their applications such as tackling deprivation. Local governments have to change make use of the best opportunities available for external funding through partnerships. Other relevant factors important for the success are the political support and networking support. Some factors contributing to external funding are the political will of the local authorities to match funding to urban green spaces by embracing an entrepreneurial culture and creativity of external funding officers to investigate sources and resources through partnership opportunities. Voluntary activities enable volunteers with creative, innovative and entrepreneurial capabilities and skills to contribute to urban renewal by pursuing personal development. Bringing the necessary external resources to the urban green spaces by managing change through the involvement and engagement of local residents requires professional input expertise to discuss and accept the evolving structure.

Private sponsorship should make more significant contributions in budgeting and enabling more facilities. The transference of the bulk of urban green space is transferred by contract to a private contractor but retaining accountability and quality monitoring roles essentially through consultation mechanisms and to ensure public accountability and quality of service delivery.

Financial values result in increasing the land prices, attracting more inward investments, economic growth and development, community economic spin-offs, etc. Urban green space is one of the main drives to attracting investments and multinational corporations that usually choose to build facilities taking into consideration the urban environment and landscape (Baycan-Levent & Nijkamp, 2009; Wuqiang et al., 2012).

Urban green spaces are the catalyst for social economic spin-offs in the community, such as sport activities, community centers, training programs, job creation, etc. Urban green space-based groups counting on right individuals involved, have potential to have spin-off effects in the community. Quality values are more intangible and may result in community strengthening and environmental quality. These programs and projects can be in partnerships with local business, industry, companies, and financial organizations in a continuing involvement with local schools, universities, research centers, museums, heritage organizations, local authorities, local community, neighborhoods and people, green and environmental societies and organizations, etc.

Partnerships between local authorities, funding agencies and institutions, community groups and business can contribute time and resources to adding value and quality. Partnerships can be of voluntary sector support, voluntary sector led-managed, environmental-regeneration projects and finally partnerships around a hub. Ground work trusts are partnerships locally-based committed to national organizations as an area-wide player as a network with local operators, although sometimes have difficulties to secure the long term commitment and leave the community with aspirations to continue the project.

Trusts are an alternative for recreation and amenity facilities, environmental and wildlife, potential new business and urban opportunities. Urban green spaces have the capacity to be attractive to local, national and international leisure visitors while playing a beneficial role for the brand of the city. Thus, indirectly plays a role in location business decisions. Trusts and private finance initiatives are a kind of partnerships with communities in different situations, with different role of partners, with appropriate safeguards. Trust partnerships provide assistance at the level of friends groups.

Value-added benefits that essentially are coming from community involvement and engagement are contradicted by the costs and problems derived from involving groups not participative because requires capacity building and development. Costs in urban green spaces for local authorities include all kinds of resources such as human capital, financial and investments, material, knowledge, etc. Other costs are conflicting demands. Communities face the costs of responsibilities, skills balance in services, commitment of volunteers, etc.

Some costs associated to involvement of local authorities in community development are lack of long term vision, the increase on workload without a complimentary resource, major demand on resources, greater expectations, motivation and maintaining moment in capacity and supporting groups, over-reliance on volunteers and jealousy, identification of good leaders and representativeness in the community, lack of appropriate capabilities and skills, a hard learning process, volunteers commitment and responsibility, community development and maintenance, managing demands that conflicts and contradicts constructive engagement, extending and delaying the process, job security, successful community development may be seen as a substitution of budgeting and investment. Volunteers receive training and are hired when there is funding, thus building capacities and promoting employment and ensuring commitment to the project.

CHALLENGES AND OPPORTUNITIES FOR PUBLIC AND COMMUNITY ACTION

Some of the important challenges faced by urban green spaces are the structural and organizational changes in services delivery, reengineering the staffing, redesigning and refurbishment, involving communities, agency to promote the cause, providing guidance and sharing experiences in good practices, avoiding duplication of responsibilities and stimulating active cooperation. Good practices are successful initiatives that are transferred and utilized in similar situations.

One of the most important challenges that mankind faces is sustainable urban green spaces where more than half of the world population live and need to improve the lifestyle. Urban green spaces provide environmental, economic, social, political, cultural and psychological services for the wellbeing and have an environmental sustainability impact on human activities, a role that cannot be ignored by policy makers. Uses of urban green spaces improve wellbeing, reducing anxiety and depression. These impacts raise awareness to the natural rational use of nature resources. An environmental sustainability for any urban agenda must consider for designing, planning, managing and maintaining the distribution and qualitative improvements of urban green spaces as an integrated approach to provide quality service delivering to users.

Designing of urban green spaces may result in good quality and variety of activities in open spacious areas, trees and spaces, exit points, quiet areas, good network of paths, meadows, water, formal areas, meeting places, monuments, hills, mazes, etc. Design play areas with the community engagement provide opportunities for skill developments and major satisfaction of final users. External experts in design and mediation are very welcome to the task.

Designing and planning of urban green spaces must move from traditional park railings and webs interconnected to planning land uses for multiple purposes, that is, recreational and conservation uses with other uses such as wildlife corridors beside streams and roads, public gardens on top of buildings, reservoirs and waterparks; flood prevention with canoe courses, hides and ornithological habitats in conjunction with sewage farms. Conservation planning of urban green spaces should ensure natural flora, fauna, landforms, water, air, soil, etc. and protect them from other land uses. Urban green spaces comprise habitats supporting a wide range of species some of them with a conservation concern (Park & Lee, 2000; Mo¨rtberg & Wallentinus, 2000).

A variety of social and physical opportunities are available at urban green spaces, a particular recreational activity and getting away from daily life demands and relieving stress. Some social factors of urban green spaces to be included are biodiversity of land uses, health and active lifestyles, inclusiveness and social justice, opportunities, etc. (Scottish Executive, 2001). Urban green spaces offer opportunities for aesthetic experiences with positive psychosomatic effects that reduce attention deficit and other cognitive disorders, stress and mental fatigue, lowering blood pressure (Tzoulas et al., 2007).

Urban green spaces are the opportunities for community dwellers to be surrounded by nature and biodiversity enjoying more from spending time (Dallimer, 2012). Users are willing to pay access to green spaces if they receive in return the opportunity to see the richness of nature in species (Dallimer et al., 2014). Urban green spaces provide opportunities for physical exercise and better air quality. Opportunities and potential for green urban spaces are the spatial, structural and morphology, functional and ecological. The ecological aspects of urban green spaces are the maintenance of a healthy environment with water, air and soil, diversity of wildlife and resources, reduction of the impact of human activities, preserving natural and cultural heritage. Promote private investments for preserving cultural heritage.

The spatial system of the urban green space requires conservation, restoration, maintenance, improvements and protection of existing and developments of new spatial forms (Kong et al., 2010) considering the impedance in the habitats posed by land use and landscape. Green infrastructure in a green network provides and overall value by offering opportunities for wildlife to various species of animals, birds and insects. Urban green space infrastructure provides natural drainage, water interception, infiltration, storage, pollutant removal, surface flow and rainwater runoff reduction and water quality. Storage of water in green areas is superior in quality to the run-off from streets, roads and roofs (Hou, 2006; cited in Zhang et al.,2012).

Spaces for musical group's performance provide opportunities for more cultural awareness in the community, such as active event programs of musical performance, theatre, fairs, bandstand, etc. An inventory of resources, maintenance and building facilities is required for which the urban green space is responsible. Finding new ways of reducing costs of maintenance without a loss of green assets is relevant, such as using woodland spaces for recreation and protection from climatic conditions. Also, involving the extensive agricultural and agro ecology use of urban green spaces and afforests derelict land green spaces.

Evaluation of the current state conditions and economic perspectives are important to achieve results in relation to costs. Valuations of urban green spaces combine both monetary and non-monetary to assess the value. Green spaces are linked to residential and commercial property values. Cost-benefit analysis of urban green spaces is a critical factor policy action (McPherson et al., 2011). The cost-benefit analysis of green spaces is a trade-off. A cost-benefit analysis of green spaces should take into consideration all the significant benefits such as energy consumption, reduction of pollutant materials, heat stress island effects, greenhouse gas emissions reduction, storm water runoff, mortality rates, etc., (Mackey, Lee & Smith, 2012).

AS A CONCLUSION: PUBLIC INITIATIVES AND ACTIONS

After the role of the urban green spaces is analyzed, the components, functions, users, and the factors of success such as the community involvement, some important public initiatives and actions are worth to be concluded.

Urban green spaces are wider initiatives of local authorities and communities with environmental, social and economic objectives that can justify any funding for all represented and involved stakeholders. The institutional structure framework of urban green spaces is a design concern of local authorities in response to provide services for satisfaction of user's needs. Urban green spaces have to be large enough to satisfy the urban users' needs and aspirations and distributed throughout the total urban area in such a way can sustain better relationships with the environment

Therefore, usually it is a political issue of high priority and commitment for local authorities promoting cooperation relationships through networks between urban green spaces and groups of friends. Some driving forces of urban green space initiatives behind the community development are to improve the poor state of development, to generate employment and return to the greens as it was in the past.

Public initiatives and actions supported by local authorities addressed to inhabitants to urban green spaces, parks and gardens in public spaces should demonstrate their attachment to sustainable development and the environment. Grass roots initiatives usually form community groups to work towards achieving better provision of services. Local and community initiatives in a green space develop as a result of inadequate provision for users' needs and aspirations from the local authorities, or had not been developed because of lack of resources.

There is also beyond this, an economic stimulation with the regeneration of the community. Urban green space stimulates social and economic regeneration of communities in a multi-agency area, beyond landscaping that can be considered only cosmetic change. There is a link between environmental regeneration and economic stimulation, providing substantial infrastructure in housing, job creation, etc. Usually policymakers underestimate the role of urban green spaces beyond the greenery or landscaping in urban regeneration to reinvigorate communities and neighborhoods, by increasing the lifestyles, making them more pleasant and attractive, increasing the land value, strengthening the community spirit and social networks, economic stimulation, etc. Urban green spaces play a relevant environmental, economic, social and cultural role.

Designing, planning and sustainable managing should address the resources for improving urban green areas focusing on meeting the user's needs in location, access and environmental quality, lighting, security and safety, play grounds for children, elderly, disables and youngers, dog-free areas, no vehicles, cycling or roller skating. Urban green spaces should be accessible, uniformly distributed, in optimal in quality and quantity and large enough to accommodate the population needs (Haq, 2011), sustainable and livable. These features should be considered at the stages of designing, managing, maintaining and protecting.

Clean outdoor green spaces should provide facilities for pedestrian and cycle routes to promote well-being and health and encourages physical activities such as walking, jogging, trim trails, running, cycling. Walking in green spaces self-guided or lead by guides for health walks on prescriptions is a service that can be provided other offers are to promote healthy living by providing safe routes to school or business, facilitating journeys between home, the school and the community. A green space surrounding schools lowers the levels if traffic-related pollution (Dadvand et al., 2015).

To investigate creativity and innovation in creating alternative models and developing funding partnerships may increase budgeting, the use and spending of resources. On a plot scheme, green spaces can be adopted by groups or individuals, residents and tenant's associations in agreements with local authorities. Any plot can be of any size across to a large area of green space, taking over bits of a street by the participant of the program adopt a plot, for example, and who have responsibilities for clearing the site and maintenance. Despite that resident's upkeep the green area, however, local authorities have the responsibility of the bulk of planting and landscaping and the community may use it for public events

Urban green space renewal action plans require political attention and have a wide range of environmental, economic and social regeneration, an increase in resources and investment and educational benefits, contributes to improve the urban landscape and its use for recreation and enjoyment. Planned activities in urban green spaces for development operations may include: Plant trails, biological corridors, botanical gardens, shared gardens, play areas for recreational uses, etc.

Urban spaces that could potentially receive biodiversity should be inventoried, including open spaces, abandoned, wastelands, etc. Regarding water as an element of urban green spaces should include fountains and waterfalls with sound, water for children's play, ponds with wildlife, rivers, streams, boating lakes, etc. Also, the provision of a skate park, wheelchair activities, a graffiti wall, outdoor chess and draughts and a community fishing space are important. Sport facilities in urban green spaces requires changing facilities in good conditions, dress correctly, probation of football pitches, access to bowling greens, tennis, basketball courts, facilities for organized clubs and for casual teams and the option to join in activities without being a member. Programs involving the elderly, children, women, etc., like playing bowl contributes to more social cohesion, as it is football for Youngers.

FUTURE RESEARCH DIRECTIONS

There is a research gap in the empirical literature including lack of data in green spaces on the optimal size, characteristics, distribution and the influence on health effects (Bowler et al. 2010). Research on inequalities in urban greener environments is necessary to improve health equity. However, there is little research on the effects of green areas on carbon capture, although research on green spaces and pollution looks at the link carbon capture in green spaces capacity for pollution and absorption of particles. Urban green infrastructure contributes to carbon capture by building up soil carbon reserves (Forest Research, 2010).

Green barriers are useful in protecting from traffic emissions but require further research to clarify the effects green street canyon geometries, wind speeds, air pollutants velocity, types of vegetation, etc.

A gap in the research on urban green space is the reduction of habitat to one independent variable such as levels of vegetation (Kuo et al., 1998a, 1998b), overlooking the structural complexity of biodiversity patterns interact with social and psychological benefits and bypassing the intangible benefits associated with socioeconomic factors (Pickett et al., 2001; Hope et al., 2003; Martin et al., 2004; Kinzig et al., 2005).

REFERENCES

Abraham, A., Sommerhalder, K., & Abel, T. (2010). Landscape and wellbeing: A scoping study on the healthpromoting impact of outdoor environments. *International Journal of Public Health, 55*(1), 59–69. doi:10.100700038-009-0069-z PMID:19768384

Alberti, M., & Marzluff, J. M. (2004). Ecological resilience in urban ecosystems: Linking urban patterns to human and ecological functions. *Urban Ecosystems*, *7*(3), 241–265. doi:10.1023/B:UECO.0000044038.90173.c6

Armson, D., Stringer, P., & Ennos, A. R. (2012). The effect of tree shade and grass on surface and globe temperatures in an urban area. *Urban Forestry & Urban Greening*, *11*(3), 245–255. doi:10.1016/j.ufug.2012.05.002

Baldock, K. C. R. (2015). Article. *Philosophical Transactions of the Royal Society of London. Series B, Biological Sciences*, *282*, 20142849.

Balram, S., & Dragicevic, S. (2005). Attitude towards Urban Green Spaces; Integrated Questionnaire Survey and Col-laborative GIS Techniques to Improve Attitude Measurement, Elsevier. *Landscape and Urban Planning*, *71*(2-4), 147–162. doi:10.1016/S0169-2046(04)00052-0

Baycan-Levent, T., & Nijkamp, P. (2009). Planning and management of urban green spaces in Europe: Comparative analysis. *Journal of Urban Planning and Development*, *135*(1), 1–12. doi:10.1061/(ASCE)0733-9488(2009)135:1(1)

Bilgili, B. C., & Gökyer, E. (2012). *Urban Green Space System Planning, Landscape Planning*. InTech. Available from: http://www.intechopen.com/books/landscape-planning/urban-green-space-system-planning

Bowler, D., Buyung-Ali, L., Knight, T., & Pullin, A. (2010). Urban greening to cool towns and cities: A systematic review of the empirical evidence. *Landscape and Urban Planning*, *97*(3), 147–155. doi:10.1016/j.landurbplan.2010.05.006

Burton, E. (2003). Housing for an urban renaissance: Implications for social equity. *Housing Studies*, *18*(4), 537–562. doi:10.1080/02673030304249

Cohen, P., Potchter, O., & Matzarakis, A. (2012). Daily and seasonal climatic conditions of green urban open spaces in the Mediterranean climate and their impact on human comfort. *Building and Environment*, *51*, 285–295. doi:10.1016/j.buildenv.2011.11.020

Comedia and Demos. (1995). *Park Life: Urban Parks and Social Renewal*. Stroud, UK: Comedia.

Cornelis, J., & Hermy, M. (2004). Biodiversity relationships in urban and suburban parks in Flanders. *Landscape and Urban Planning*, *69*(4), 385–401. doi:10.1016/j.landurbplan.2003.10.038

Dole, J. (1994). Greenscape 5: Green Cities. Architects' Journal, 61-69.

Dunnett, N., Swanwick, C., & Woolley, H. (2002). *Improving urban parks, play areas and green spaces*. Department for Transport, Local Government and the Regions. Available at: http://www.ocs.polito.it/biblioteca/verde/improving_full.pdf

Dadvand, P., Rivas, I., Basagaña, X., Alvarez-Pedrerol, M., Su, J., De Castro Pascual, M., ... Nieuwenhuijsen, M. J. (2015). The association between greenness and traffic-related air pollution at schools. *The Science of the Total Environment*, *523*, 59–63. doi:10.1016/j.scitotenv.2015.03.103 PMID:25862991

Dadvand, P., De Nazelle, A., Triguero-Mas, M., Schembari, A., Cirach, M., Amoly, E., ... Nieuwenhuijsen, M. (2012). Surrounding greenness and exposure to air pollution during pregnancy; an analysis of personal monitoring data. *Epidemiology (Cambridge, Mass.)*, 1286–1290. PMID:22647671

Dallimer, M., Tinch, D., Hanley, N., Irvine, K. N., Rouquette, J. R., Warren, P. H., ... Armsworth, P. R. (2014). Quantifying Preferences for the Natural World Using Monetary and Nonmonetary Assessments of Value. *Conservation Biology*, *28*(2), 404–413. doi:10.1111/cobi.12215 PMID:24372643

Dallimer, M., Irvine, K. N., Skinner, A. M. J., Davies, Z. G., Rouquette, J. R., Maltby, L. L., ... Gaston, K. J. (2012). Biodiversity and the Feel-Good Factor: Understanding Associations between Self-Reported Human Well-being and Species Richness. *Bioscience*, *62*(1), 47–55. doi:10.1525/bio.2012.62.1.9

Forest Research. (2010). *Benefits of green infrastructure. Report to Defra and CLG.* Available at: http://www.forestry.gov.uk/pdf/urgp_benefits_of_green_infrastructure_main_report.pdf/$file/urgp_benefits_of_green_infrastructure_main_report.pdf

Fuller, R. A., Tratalos, J., & Gaston, K. J. (2009). How many birds are there in a city of half a million people? *Diversity & Distributions*, *15*(2), 328–337. doi:10.1111/j.1472-4642.2008.00537.x

Girardet, H. (2004). *Cities people planet: liveable cities for a sustainable world.* London: John Wiley & Sons.

Gregory, R. D., & Baillie, S. R. (1998). Large-scale habitat use of some declining British birds. *Journal of Applied Ecology*, *35*(5), 785–799.

Girardet, H. (1996). *The Gaia Atlas of Cities: new directions for sustainable urban living*. UN-HABITAT.

Grahn, P., & Stigsdotter, U. A. (2003). Landscape planning and stress. *Urban Forestry & Urban Greening*, *2*(1), 1–18. doi:10.1078/1618-8667-00019

Gregg, J. W., Jones, C. G., & Dawson, T. E. (2003). Urbanization effects on tree growth in the vicinity of New York City. *Nature*, *424*(6945), 183–187. doi:10.1038/nature01728 PMID:12853954

GreenSpace. (2007). The park life report. The first ever public satisfaction survey. *Green Spaces: The Benefits for London Bibliography of Britain's parks and green spaces.* Available at: http://www.greenspace.org.uk/downloads/

Grimm, N. B., Faeth, S. H., Golubiewski, N. E., Redman, C. L., Wu, J., Bai, X., & Briggs, J. M. (2008). Global change and the ecology of cities. *Science*, *319*(5864), 756–760. doi:10.1126cience.1150195 PMID:18258902

Haq, S. M. A. (2011). Urban green spaces and an integrative approach to sustainable environment. *Journal of Environmental Protection*, *2*(5), 601–608. doi:10.4236/jep.2011.25069

Hale, J. D., Fairbrass, A. J., Matthews, T. J., & Sadler, J. P. (2012). Habitat Composition and Connectivity Predicts Bat Presence and Activity at Foraging Sites in a Large UK Conurbation. *PLoS One*, *7*(3), e33300. doi:10.1371/journal.pone.0033300 PMID:22428015

Hamada, S., Tanaka, T., & Ohta, T. (2013). Impacts of land use and topography on the cooling effect of green areas on surrounding urban areas. *Urban Forestry & Urban Greening*, *12*(4), 426–434. doi:10.1016/j.ufug.2013.06.008

Herzele, V., & Wiedeman, T. (2003). A Monitoring Tool for the Provision for Accessible and Attractive Green Spaces, *Elsevier Sciences. Landscape and Urban Planning*, *63*(2), 109–126. doi:10.1016/S0169-2046(02)00192-5

Hough, M. (2004). *Cities and natural process*. Routledge.

Hope, D., Gries, C., Zhu, W., Fagan, W. F., Redman, C. L., Grimm, N. B., ... Kinzig, A. (2003). Socioeconomics drive urban plant diversity. *Proceedings of the National Academy of Sciences of the United States of America*, *100*(15), 8788–8792. doi:10.1073/pnas.1537557100 PMID:12847293

Eliasson, I. (2000). The use of climate knowledge in urban planning. *Landscape and Urban Planning*, *48*(1-2), 31–44. doi:10.1016/S0169-2046(00)00034-7

Irvinea, K. N., Devine-Wrightb, P., Payneb, S. R., Fullerc, R. A., Paintera, B., & Gastonca, K. J. (2009, February). Green space, soundscape and urban sustainability: An interdisciplinary, empirical study. *Local Environment*, *14*(2), 155–172. doi:10.1080/13549830802522061

Kinzig, A., Warren, P., Martin, C., Hope, D., & Katti, M. (2005). The effects of human socioeconomic status and cultural characteristics on urban patterns of biodiversity. *Ecology and Society*, *10*(1), 23. doi:10.5751/ES-01264-100123

Kong, F., Yin, H., Nakagoshi, N., & Zong, Y. (2010). Urban green space network development for biodiversity conservation: Identification based on graph theory and gravity modeling. *Landscape and Urban Planning*, *95*(1-2), 16–27. doi:10.1016/j.landurbplan.2009.11.001

Konijnendijk, C. C., Annerstedt, M., Nielsen, A. B., & Maruthaveeran, S. (2013). Benefits of urban parks: A systematic review. *IFPRA World*, *2012*(6), 10–12.

Kuo, F. E., Bocacia, M., & Sullivan, W. C. (1998a). Transforming inner-city neighbourhoods: Trees, sense of safety, and preference. *Environment and Behavior*, *30*(1), 28–59. doi:10.1177/0013916598301002

Kuo, F. E., Sullivan, W. C., Coley, R. L., & Brunson, L. (1998b). Fertile ground for community: Inner-city neighbourhood common spaces. *American Journal of Community Psychology*, *26*(6), 823–851. doi:10.1023/A:1022294028903

Mackey, C. W., Lee, X., & Smith, R. B. (2012). Remotely sensing the cooling effects of city scale efforts to reduce urban heat island. *Building and Environment*, *49*, 348–358. doi:10.1016/j.buildenv.2011.08.004

Martin, C. A., Warren, P. S., & Kinzig, A. P. (2004). Neighborhood socioeconomic status is a useful predictor of perennial landscape vegetation in residential neighborhoods and embedded small parks of Phoenix, AZ. *Landscape and Urban Planning*, *69*(4), 355–368. doi:10.1016/j.landurbplan.2003.10.034

McPherson, E. G., Simpson, J. R., Xiao, Q., & Wu, C. (2011). Million trees Los Angeles canopy cover and benefit assessment. *Landscape and Urban Planning*, *99*(1), 40–50. doi:10.1016/j.landurbplan.2010.08.011

Mitchell, R., & Popham, F. (2008). Effect of exposure to natural environment on health inequalities: An observational population study. *Lancet*, *372*(9650), 1655–1660. doi:10.1016/S0140-6736(08)61689-X PMID:18994663

Moore, M., Gould, P., & Keary, B. S. (2003). Global urbanization and impact on health. *International Journal of Hygiene and Environmental Health*, *206*(4), 269–278. doi:10.1078/1438-4639-00223 PMID:12971682

Morani, A., Nowak, D. J., Hirabayashi, S., & Calfapietra, C. (2011). How to select the best tree planting locations to enhance air pollution removal in the MillionTreesNYC initiative. *Environmental Pollution*, *159*(5), 1040–1047. doi:10.1016/j.envpol.2010.11.022 PMID:21168939

Mo¨rtberg, U., & Wallentinus, H.-G. (2000). Red-listed forest bird species in an urban environment – assessment of green space corridors. *Landscape and Urban Planning*, *50*(4), 215–226. doi:10.1016/S0169-2046(00)00090-6

Newman, P., & Jennings, I. (2008). *Cities as sustainable ecosystems: principles and practices*. Island Press.

Nowak, D. J., Greenfield, E. J., Hoehn, R. E., & Lapoint, E. (2013). Carbon storage and sequestration by trees in urban and community areas of the United States. *Environmental Pollution*, *178*, 229–236. doi:10.1016/j.envpol.2013.03.019 PMID:23583943

Nowak, D. J., & Crane, D. E. (2002). Carbon storage and sequestration by urban trees in the USA. *Environmental Pollution*, *116*(3), 381–389. doi:10.1016/S0269-7491(01)00214-7 PMID:11822716

Oliveira, S., Andrade, H., & Vaz, T. (2011). The cooling effect of green spaces as a contribution to the mitigation of urban heat: A case study in Lisbon. *Building and Environment*, *46*(11), 2186–2194. doi:10.1016/j.buildenv.2011.04.034

Park, C. R., & Lee, W. S. (2000). Relationship between species composition and area in breeding birds of urban woods in Seoul, Korea. *Landscape and Urban Planning*, *51*(1), 29–36. doi:10.1016/S0169-2046(00)00094-3

Pickett, S. T. A., Cadenasso, M. L., Grove, J. M., Nilon, C. H., Pouyat, R. V., Zipperer, W. C., & Costanza, R. (2001). Urban ecological systems: Linking terrestrial, ecological, physical, and socioeconomic components of metropolitan areas. *Annual Review of Ecology and Systematics*, *32*(November), 127–157. doi:10.1146/annurev.ecolsys.32.081501.114012

Rouquette, J. R., Dallimer, M., Armsworth, P. R., Gaston, K. J., Maltby, L., & Warren, P. H. (2013). Species turnover and geographic distance in an urban river network. *Diversity & Distributions*, *19*(11), 1429–1439. doi:10.1111/ddi.12120

Schwartz, M. W., Jurjavcic, N. L., & O'Brien, J. M. (2002). Conservation's Disenfranchised Urban Poor. *Bioscience*, *52*(7), 601–606. doi:10.1641/0006-3568(2002)052[0601:CSDUP]2.0.CO;2

Scottish Executive. (2001). Rethinking Open Space. The Stationery Office, Kit Campbell Associates.

Stadt Leipzig. (2003). *Umweltqualitatsziele und -standards fur die Stadt Leipzig*. Stadt Leipzig, Der Oberburgermeister, Amt fur Umweltschutz, Leipzig. Retrieved from http://www.leipzig.de/imperia/md/content/36_amt_fuer_umweltschutz/umweltziele.pdf

Strohbach, M. W., Arnold, E., & Haase, D. (2012). The carbon footprint of urban green space—A life cycle approach. *Landscape and Urban Planning*, *104*(2), 220–229. doi:10.1016/j.landurbplan.2011.10.013

Tuzin, B., Leeuwen, E., Rodenburg, C., & Peter, N. (2002). *Paper*. Paper presented at the 38th International Planning Congress on "The Pulsar Effect" Planning with Peaks, Glifada, Athens, Greece.

Tzoulas, K., Korpela, K., Venn, S., Yli-Pelkonen, V., Kazmierczak, A., Niemela, J., & James, P. (2007). Promoting ecosystem and human health in urban areas using green infrastructure: A literature review. *Landscape and Urban Planning*, *81*(3), 167–178. doi:10.1016/j.landurbplan.2007.02.001

Ulrich, R. S. (2002). *Health Benefits of Gardens in Hospitals*. International Exhibition Floriade. Retrieved from: http://www.planterra.com/research/research_3.php

Ulrich, Q. X., & Zimring, C. (2010). *The Role of the Physical Environment in the Hospital of the 21st Century: A Once-in-a-Lifetime Opportunity*. Report prepared for TriPoint Hospital Center. Retrieved from: tinyurl.com/healthdesignstudy

United Nations, Department of Economic and Social Affairs – Population Division. (2014). *World Urbanization Prospects: The 2014 Revision, Highlights* (ST/ESA/SER.A/352). Author.

Van Herzele, A., & de Vries, S. (2012). Linking green space to health: A comparative study of two urban neighbourhoods in Ghent, Belgium. *Population and Environment*, *34*(2), 171–193. doi:10.100711111-011-0153-1

Weber, N., Haase, D., & Franck, U. (2014). Zooming into temperature conditions in the city of Leipzig: How do urban built and green structures influence earth surface temperatures in the city? *The Science of the Total Environment*, *496*, 289–298. doi:10.1016/j.scitotenv.2014.06.144 PMID:25087062

White, M., Alcock, I., Wheeler, B., & Depledge, M. (2013). *Would you be happier living in a greener urban area? A fixed effects analysis of panel data*. European Centre for Environment and Human Health. Available at: http://www.ecehh.org/publication/would-you-be-happier-living-greenerurban-area

Williams, K., Burton, E., & Jenks, M. (2000). *Achieving sustainable urban form*. London: E&FN Spon.

Wong, G., Greenhalgh, T., Westhorp, G., Buckingham, J., & Pawson, R. (2013). RAMESES publication standards: Meta-narrative reviews. *BMC Medicine*, *11*(1), 20. doi:10.1186/1741-7015-11-20 PMID:23360661

Woolley, H., & Rose, S. (n.d.). *The value of public space. How high quality parks and public spaces create economic, social and environmental value*. CABEspace. Available at: http://webarchive.nationalarchives.gov.uk/20110118095356/http:/www.cabe.org.uk/files/the-value-of-publicspace.pdf

Wuqiang, L., Song, S., & Wei, L. (2012). *Urban spatial patterns based on the urban green space system: A strategic plan for Wuhan City*. Retrieved from www.intechopen.com

Xion-Jun, W. (2009). Analysis of Problems in Urban Green Space System Planning in China. *Journal of Forestry Research*, *20*(1), 79–82. doi:10.100711676-009-0014-2

Zhang, B., Xie, G., Zhang, C., & Zhang, J. (2012). The economic benefits of rainwater runoff reduction by urban green spaces: A case study in Beijing, China. *Journal of Environmental Management*, *100*, 65–71. doi:10.1016/j.jenvman.2012.01.015 PMID:22366359

KEY TERMS AND DEFINITIONS

Community Ecosystem: An ecological system used by members that reside in a specific locality.

Community Involvement: The active involvement of people living in one locality in the process of improving the quality of an urban environment.

Sustainable Urban Green Spaces Development: A process of synergy of different aspects of city development that results in forming balanced and eco-friendly urban green spaces fulfilling the needs of current and future generations.

System of Green Spaces: Vegetated areas found in an urban environment that form a system of functional, esthetical, or other interrelations on a larger scale.

Urban Ecosystem: An ecological system located within a city or densely populated area.

Urban Green Space Renewal: The rehabilitation of substandard city green areas or the conversion of wasteland into green land open to the public and for the public good.

Urban Green Spaces: An area within an urban environment of grass, trees, shrubs or other vegetation.

This research was previously published in Optimizing Regional Development Through Transformative Urbanization edited by Umar G. Benna, pages 204-224, copyright year 2019 by Information Science Reference (an imprint of IGI Global).

Chapter 12
Assessing Urban Ecosystem Services:
Different Methodological Approaches Applied in Brazil, Germany, and Portugal

Marise Barreiros Horta
Federal University of Minas Gerais, Brazil

Maria Inês Cabral
Martin Luther University, Germany & iDiv, Germany

Iva Pires
Nova University of Lisbon, Portugal

Laura Salles Bachi
Federal University of Minas Gerais, Brazil

Ana Luz
Nova University of Lisbon, Portugal

Geraldo Wilson Fernandes
Federal University of Minas Gerais, Brazil

Maria Auxiliadora Drumond
Federal University of Minas Gerais, Brazil

Sónia Carvalho-Ribeiro
Federal University of Minas Gerais, Brazil

ABSTRACT

By integrating social, ecological, and economic perspectives, the assessment of ecosystem services (ES) provides valuable information for better targeting landscape planning and governance. This chapter summarizes different participatory approaches for assessing ES in urban areas of three countries. In Belo Horizonte (Brazil), a conceptual framework for the vacant lots ES assessment is presented as an attempt to integrate landscape, social, and political dimensions. In Leipzig (Germany), a

DOI: 10.4018/978-1-7998-0441-3.ch012

Copyright © 2020, IGI Global. Copying or distributing in print or electronic forms without written permission of IGI Global is prohibited.

INTRODUCTION

There is a growing recognition that the assessment of the Ecosystem Services (ES) can provide valuable information for better targeting landscape planning and governance, especially for shaping innovative adaptation strategies in the context of global environmental change (Brendan, Costanza, Turner, & Morling, 2007; de Groot, Alkemade, Braat, Hein, & Willemen, 2010; Gómez-Baggethun et al., 2013). Many authors agree that the concept of ES (designated as the benefits people obtain from nature) is currently widespread and accepted, providing a useful framework that aggregates social, ecological, and economic perspectives (Burkhard, Petrosillo, & Costanza, 2010; Gómez-Baggethun et al., 2013; Koschke, Fürst, Frank, & Makeschin, 2012; Millenium Ecosystem Assessment [MEA], 2005; Primmer et al., 2015).

Although ES offer valuable insights concerning the human and nature connections and dependencies, the concept has however remained away from the political applications (Burkhard et al., 2010). In this sense, empirical applications and tools are required for the development and improvement of the ES concept and its insertion in the decision-making process (Burkhard & Müller, 2015).

Over the past decades the assessments of ES have been favoring the biophysical and economic aspects, leaving aside the social dimension formed by the services beneficiaries and the institutions (Martín-López, Gómez-Baggethun, Garcia-Llorente, & Montes, 2014; Primmer et al., 2015). The biophysical approach involves the ecosystem services supply and demand, and it is influenced by Land Use and Land Cover (LULC) structures and processes, which change in space and time (Burkhard & Müller, 2015). The economic view comprises the monetary valuation of services varying from local to regional and global scales (de Groot et al., 2012).

More recently, the accumulated knowledge in the field has raised the need for integration, which enables the overcoming of the limitations through the incorporation of applied methods and tools to bring ES assessments closer to the political and decision-making agendas. One avenue for moving forward in the decision-making process is the insertion of the socio-cultural dimension through the utilization of participatory approaches (Bixler, Dell´Angelo, Mfune, & Rob, 2015). Participatory approaches are based on interactivity and include social interaction, mutual learning, and communication (de Montis, 2007).

In this context, attempts have been made focused on the use of participatory approaches for broadening the traditional biophysical ecosystem services perspective into a set of social and political processes (Haines-Young & Potschin, 2014; Martín-López et al., 2014; Turnpenny, Russel, & Jordan, 2014). Among the tools and methods utilized, it is possible to highlight the social surveys and interviews organized for collecting data on ecosystem services, concerning the mode in which multiple users or beneficiaries acknowledge ecosystem´s capacity to deliver services and the economic value attached to it (Martín-López et al., 2012; Martín-López et al., 2014; Haines-Young & Potschin, 2014).

For urban areas, there is a set of studies evaluating ES that have used participatory approaches. Participatory ES assessment often focuses on social processes and governance, working with city actors that interacts with urban ecosystems (Ernston, Barthel, Anderson, & Borgström, 2010). In this context, vacant lots in the cities, allotments, and community gardens are types of landscape components that provide important ecosystem services to urban communities, such as local climate and water regulation, as well as habitat provision for biodiversity. However, the ways urban dwellers perceive ES need further attention.

In this chapter, it is explored how different participatory processes have been developed in the distinct socio-cultural contexts of three countries (Brazil, Germany, and Portugal), through the ES assessment in urban areas. The goal of the study is therefore to characterize different participatory methods that were used to appraise urban dwellers perception on spaces composed of vacant lots, allotments and community gardens.

The overall chapter structure includes a literature review on participatory approaches and on how those participative methods have been developed for assessing ES in urban areas and, a presentation of case studies comprising participatory approaches for assessing ES in urban sites. The first case study addresses the public preferences for land cover (a feature associated with ES) in vacant lots in the city of Belo Horizonte, Brazil. It is followed by the assessment of ES in two different types of urban gardens, namely allotments and community gardens, in the city of Leipzig, in Germany. The third case illustrates the issues involved in the management of community gardens in Lisbon, Portugal. The discussion and conclusion presented at the end, summarize and offer reflections on the use of participatory approaches for assessing ES in urban areas.

PARTICIPATORY METHODS FOR URBAN ECOSYSTEM SERVICES ASSESSMENT

Although traditionally associated with rural development, participatory approaches have become popular as a tool for governance in urban areas (Mitlin & Thompson, 1995). The utilization of participatory methods in urban environments has been viewed as a suited approach for designing neighborhoods, promoting more than a formal consultation, as much as it fosters dialogues and interactions between actors (users, experts, decision makers) throughout the whole planning and governance processes (Montreal Urban Ecology Centre [MUEC], 2015). Participatory tools and methodologies are flexible and focused on the processes aiming at enabling local people to get dominance over the development process and governance (Mitlin & Thompson, 1995).

The participative policy processes normally involve a three-stage cycle composed of planning, implementation, and evaluation. Participatory approaches can be utilized in some or all of those phases (Elliot, Heesterbeek, Lukensmeyer, & Slocum, 2005). It is particularly important to consider that participation shall not be imposed but rather developed through dialogues and joint analysis among the various stakeholders (Mitlin & Thompson, 1994).

Participatory approaches integrate local people in exchange and discussions on resource distribution and control. This has the purpose of increasing awareness about key actors and groups involved; improving information concerning local conditions; enabling local people to recognize limitations, establish priorities and take initiative; and empowering local organizations and developing mechanisms to solve local conflicts (Mitlin & Thompson, 1995).

Elliot et al. (2005) synthesized the participatory methods in four groups according to the motivations and outputs desired: democratization, advising, mapping out diversity and reaching consensus. In the case of the democratization, the objective of using participatory methods is to capacitate participants to use their knowledge to generate their options, so that they can exert influence in the decision-making process. The advising perspective has the goal to expose stakeholder´s values, impressions and opinions pertinent or as a support to the decision-making process. Mapping out diversity has the objective to generate a range of information and choices allowing a group to reveal one sort of knowledge explicitly. The reaching consensus allows a group to get to a single agreement on a subject.

According to Drumond, Giovanetti, & Guimarães (2009), different methods and tools can be used to search information, to get insights, to scan problems, to plan jointly, to evaluate processes and to mobilize communities. The authors indicated a set of participatory methods and tools available for assessing, analyzing, facilitating, and coordinating the participative process comprising: brainstorming, semi-structured interviews, participative mapping, Venn diagram, crosswalks, seasonal calendar, daily routine or clock of activities, historic diagram, matrix, flux diagrams or problem tree, line of ideas, and SWOT analysis. Those methods are summarized in a comparative frame for participatory assessments presented next (Table 1).

Some other participatory approaches combine methods and tools such as interviews, mapping, and scenarios development (Villamor, Palomo, Santiago, Oteros-Rosa, & Hill, 2014). Mapping and scenarios building are promising tools for the ecosystem assessment with the public participation. Ecosystem services mapping facilitates incorporation into policies and decision-making as far as maps promote interpretation through visualization and synthesize complex information (Maes et al., 2012; Troy & Wilson, 2006). Scenario construction is an integrative and interdisciplinary framework that provides an examination of the ecosystem services spatial and temporal dimensions (Syrbe, Rosenberg, & Vowinckel, 2015). Using this tool, one can indicate desirable future changes and landscapes transformations for the improvement of the ecosystem services provision (Rosenberg, Syrbe, Vowinckel, & Walz, 2014).

CONCEPTUAL FRAMEWORK FOR BELO HORIZONTE´S VACANT LOTS ECOSYSTEM SERVICES ASSESSMENT

The participatory methodology under development for the vacant lots in Brazil has the purpose to outline a set of future scenarios and subsidize the design of politics, programs and strategies for a better urban sustainability (Carvalho-Ribeiro, Lovett, & O'Riordan, 2010; Carvalho-Ribeiro & Lovett, 2011; Carvalho-Ribeiro, Migliozzi, Incerti, & Pinto Correia, 2013a; Carvalho-Ribeiro et al., 2013b; van Berkel, Carvalho-Ribeiro, Verburg, & Lovett, 2011). The choice on participatory scenarios development relates to the fact that it can foster a social appreciation of the results, which helps in implementations and contributes to educational effects for the participants (Syrbe et al., 2015).

The vacant lots ecosystem services assessment in the city of Belo Horizonte has encompassed a conceptual framework guided by the utilization of different tools and methodological approaches. To this end, it seeks to characterize the landscape dimension and ecological attributes of the vacant lots and integrate them with the social, political and economic contexts. The assessments include spatial and socioecological aspects, potential ecosystem services mapping based on the Geographic Information Systems (GIS) utilization and the adoption of a participatory approach for the generation of ecosystem services scenarios in the city.

The framework delineated is, therefore, an attempt to integrate the landscape dimension - through the biophysical attributes and vacant lots - with the social and political ones (Figure 1). A similar approach was presented by de Groot, Alkemade, Braat, Hein, & Willemen (2010) aiming at integrating ecosystem services assessment, planning, and management. In that approach, it was indicated as an initial step the landscape and ecosystems characterization with the definition of the functions, including goods and services. That coincides with the present vacant lot´s ecosystem services assessment framework that begins with the study of the biological and physical characteristics of the generation of the vacant lots potential ecosystem services. The biological component is represented mainly by the vegetation land cover which is one the greatest determinants of the type and amount of ecosystem services (Burkhard et al., 2012). The physical, LULC factors or biophysical attributes

Table 1. Comparative frame of participatory methods and tools (Drummond et al., 2009).

Methods/Tools	Objectives	What for?
Brainstorming	To stimulate participants to express their point of view on a topic.	For assessing opinions on topics such as the effectiveness of the participatory process; planning, monitoring and process evaluation.
Semi-structured interviews	To discuss openly and informally through conversations with individuals or small groups specific issues and important facts for the project.	For the diagnosis phase, when information can be obtained from the various stakeholders; useful for building trust.
Participatory mapping	To visualize features and issues spatially, such as natural resources, infrastructure, LULC; to obtain clearly the participant perception of the environment.	For discussing environmental aspects and social issues in a geographically explicit manner; to take advantage of the use of maps, that synthesize complex information through visualization and interpretation.
Venn diagram	To get to know the different formal or informal stakeholders and groups and their participation on the evaluated topic.	For monitoring and evaluation phases; enables to examine the possibilities of participation and acting of each one of the actors and institutions.
Crosswalks	To explore the spatial characteristics of the study area; the work team and participants walk through the area to get to know the soils, production systems, vegetation, infrastructure, water resources, and other relevant aspects.	For the investigation, monitoring and evaluation phases; allows the generation of detailed environmental profiles; important for the discussion and promotion of reflexive activities on the conservation and use of the natural and sociocultural assets.
Seasonal calendar	To broaden the knowledge on the variation of environmental phenomena during the year (rains, floods), on the utilization of natural resources (hunting, fishing, vegetable extractivism); on the plant cultivation (planting and harvesting); on the cultural ceremonies (popular parties, religious ceremonies) and other local life´s aspects.	For the investigation and monitoring phases; allows generating information on the seasonal variation of the constraints and opportunities.
Clock of activities	To clarify the performance of the day to day tasks; the differences among men and women's routine; the differences among different social classes routine; the time availability for performing other tasks.	For the investigation and monitoring phases; used to understand the activities routine relevant in processes of searching economic alternatives as a means for improving life quality and for reducing pressure on natural resources.

continues on following page

Table 1. Continued

Methods/Tools	Objectives	What for?
Historic diagram	To understand the temporal shifts concerning the land use, population, pollution levels, vegetation cover, natural resources availability, land tenure, etc.	Indicated for the investigation and monitoring phases; it helps in the understanding of the present situation through the knowledge of the community history and of the causes that might have led to the situation in the present.
Matrix	To stimulate the reasoning on the importance of the topics and themes according to a sequence.	For the investigation, planning, monitoring, and evaluation phases; allows organizing priorities and criteria according to its importance.
Flux diagrams	To assess the interrelations among the various elements for the identification of the problems´ effects and causes, as well as of the potential solutions and actions.	For the investigation, systematization, planning and monitoring phases; allows the construction of cause and effect problem-oriented diagram constituting a simple and easy-to-use tool.
Line of ideas	To present and discuss the results of a project in a systematic way.	For the systematization and problems organization; for the discussion of the positive and negative aspects of an issue.
SWOT analysis	To systematize people´s opinion on the strengthens and opportunities, as well as on the weakness and threats of a topic.	For the investigation, planning, monitoring and evaluation phases; it is a powerful instrument for monitoring and evaluating implemented actions, for the identification of new projects to be implemented and for the work performance evaluation.

are then coupled with the utilization of a spatially explicit multicriteria analysis, through the features overlapping.

The cooperation of the stakeholders starts with the generation of the vacant lots potential ecosystem services map. Before that, it will be verified all the groups and actors to participate. Then, plenary sessions will occur, aiming at the building of the scenarios with vacant lots ecosystem services. Maps discussion and (re) construction will take place during the plenary meetings and workshops. Scenario workshops are indicated to interdisciplinary and social issues, involving choices on different situations and exchanges among professionals and locals. The urban

Figure 1. Conceptual framework for the Belo Horizonte´s vacant lots ecosystem services assessment

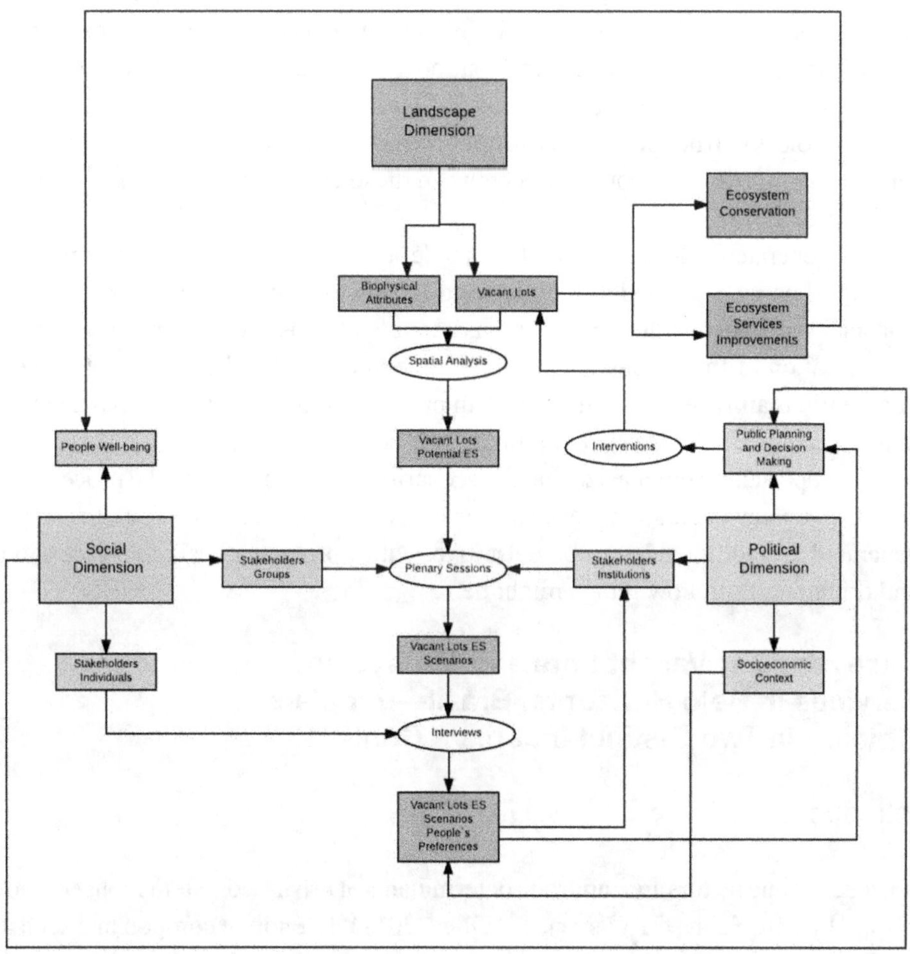

sustainability is, therefore, a perfect subject for using the scenarios approach, given its transdisciplinary nature and various stakeholders involved (Street, 1997).

The vacant lots ecosystem services scenarios generated will be incorporated next into questionnaires to be utilized in the surveys and interviews to assess the individual's preferences over the city ecosystem services scenarios. This phase comprises, besides the semi-structured questionnaires building and ethics license acquisition, information gathering over the city populations characteristics, statistics for participants number definition, as well as segmentation of the population into categories regarding gender, age, education, income, and origin.

At the later stage, after conducting the interviews, it will be used ecological, economic valuation methods, but not for the monetary valuation of the ecosystem services itself, but rather for understanding what is there beyond the people´s preferences and choices. For that, the choice modeling approach, from the ecological economics revealed preferences methods, will be utilized. This approach agrees with Martín-López et al. (2012) that highlighted the need of addressing ecosystem services from the perspective of human outlooks and values to understand the implicit motivations for the social preferences on ecosystem services.

The choices will be confronted with the socioeconomic contexts of the respondents for checking if there is a correlation between those characteristics and the people´s preferences.

The generated information will be made available in the political dimension domain to be incorporated in the planning process, normalization and placed into financial mechanisms, to work as a landscape and ecosystem instrument, for the incorporation of those considered needed interventions for the vacant lots. Those interventions aim at ecosystem services improvements and ecosystem conservation, that can, in turn, be converted in human well-being.

It is important to emphasize that the scenarios developed during the process will have a spatial representation, be open to changes and modifications. Thus, the results generated are not intended to be definitive, but to provide a basis for discussions and orientations on how things might be.

Case Study 1: Vacant Lots and Ecosystem Services in Belo Horizonte, Brazil—People´s Opinion in Two Distinct Income´s Context

Introduction

Socioeconomic factors are important determinants of urban ecosystems conservation (Swan, Pickett, Szlavecz, Warren, & Willey, 2011). Economic components such as family income can influence issues related to that, such as plant diversity patterns in urban landscapes, that may respond positively to income´s increase (Hope et al., 2003).

The income factor is usually used in the economic literature as a representation of human well-being, considering that it increases the ability to consume goods and services (Vemuri, Grove, Wilson, & Burch Jr., 2011). Beyond that, human well-being is composed by elements such as security, basic material for a good life, health, good social relations, freedom, and choice, that expresses not only what a person values, but also a collective experience (McMichael et al., 2005).

The human well-being components rely on a great extent on ES and the links among them bring about concerns regarding the inequity of their distribution in space, which calls for the adoption of approaches that integrate the ecological, social, and economic aspects. The incorporation of the ES interests and benefits of the groups affected can offer insights into the options in the present and those that will be accessible to future generations (McMichael et al., 2005).

In the present study case, it was verified the variations of people´s views on vacant lots, their characteristics regarding land cover, ES associated and multiple uses, in two different economic contexts.

Methodology

The two neighborhoods evaluated - Palmeiras and São Bento - are located in Belo Horizonte county (Figure 2) where it is evidenced, as in other Brazilian cities, socio-spatial inequalities with segregated spaces for the poor and rich (Costa & Mendonça, 2010).

The São Bento is an upper-middle-class neighborhood comprised of 3947 inhabitants, and the average family income is around 1000 euros, according to the last Brazilian official census, in the year 2010 (Instituto Brasileiro de Geografia e Estatística, 2010). Palmeiras is a popular middle-class neighborhood comprising around 5339 people and family income about 220 euros; the census year is 2010. Both are composed predominantly of houses, instead of buildings with apartments, and were founded during the 1960 – 1970´s decades.

The methodological approach used for assessing peoples' opinion on vacant lots and ES comprised face to face interviews applied to local dwellers, using a semi-structured questionnaire. A total of 80 interviews equally distributed among the two neighborhoods were undertaken.

The questionnaire was structured into three topics concerning the following themes: a) vacant lots definitions and associations; b) land cover types preferences for a hypothetical vacant lot neighbor; c) the reasons for preferences on land cover types of neighbor's vacant lot; d) multiple uses and ES options for the vacant lots.

Figure 2. Study area location

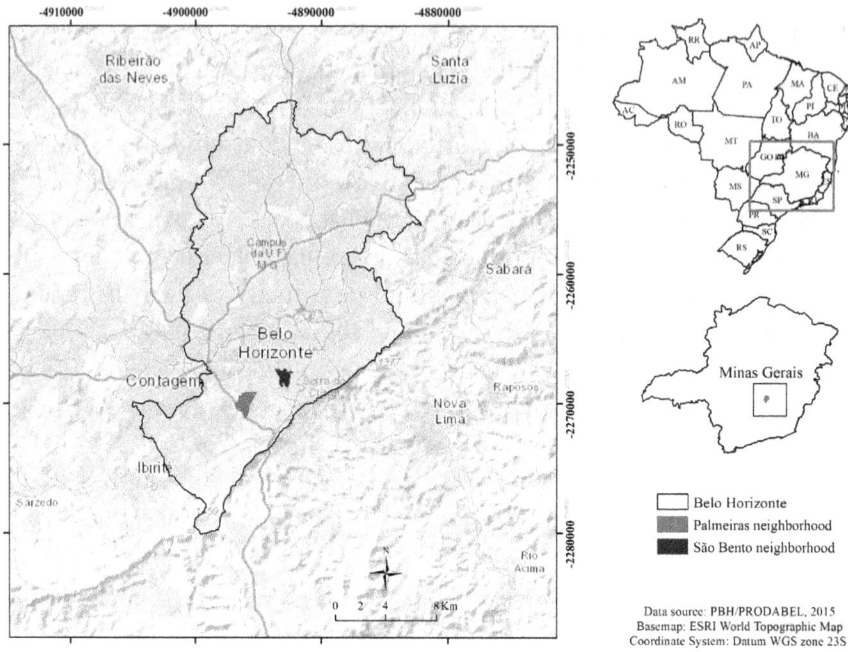

In the case of the land cover types preferences, images of the vacant lots situation in the neighborhoods were used, comprising the three more common land covers types in both areas: trees/woods; grasses/low vegetation and bare soil (Figures 3 and 4).

The answers from the questionnaires were organized and grouped into frequencies classes. The statistical Z-test was performed for verifying the significance of the differences in the answers frequencies for the land cover types preferences.

To contextualize the vacant lots in terms of distribution and size, maps made available by the municipal administration (PBH/PRODABEL) were processed in the GIS environment.

Results and Discussion

In general, both neighborhoods have similar characteristics about the quantity and size of the vacant lots (Table 2). The São Bento neighborhood congregated 148 vacant lots and the Palmeiras area 88 plots. São Bento presented the largest vacant lot (11250m^2) and a higher average and medium size. The bigger plot of land found in the Palmeiras area totalized 5863m^2.

Figure 3. Sights used to illustrate for people´s selection of the three vacant lots land cover types in the São Bento's neighborhood: left - trees/woods; center - grasses/ low vegetation and right - bare soil (Source: Google Earth Image – 2016)

Figure 4. Sights used to illustrate for people´s selection of the three vacant lots land cover types in the Palmeira's neighborhood: left - trees/woods; center - grasses/low vegetation and right - bare soil (Source: Google Earth Image – 2016)

Table 2. Vacant lots general characteristics in the São Bento and Palmeiras neighborhoods.

São Bento Neighborhood		Palmeiras Neighborhood	
Number of vacant lots	148	Number of vacant lots	88
Mean Area (m^2)	571	Mean Area (m^2)	686
Median Area (m^2)	457	Median Area (m^2)	512
Standard Deviation (m^2)	896	Standard Deviation (m^2)	780
Minimum Area (m^2)	322	Minimum Area (m^2)	312
Maximum Area (m^2)	11250	Maximum Area (m^2)	5863

The São Bento neighborhood is one of the most recent populated in the city. The settlement began from 1960 to 1970, when the rural aspects of the regions changed by the creation of big avenues. The recent foundation may explain the occurrence of large vacant lots in that neighborhood. The Palmeiras area was also occupied from the 1970s when it still had rural characteristics. The pressure from the surroundings has determined a fast growing of the neighborhood and the reduction of the vacant lots (Arreguy & Ribeiro, 2008).

The São Bento territory comprises a total area of 0,97 km^2. Most vacant lots (113) are distributed in the size class ranging from 407m^2 to 545m^2. The distribution of the plots among the other classes consists: 322-407 (11); 545-900 (21); 900-2040 (2); 2040-11250 (1) (Figure 5).

The Palmeiras region sums up around 1,4 km^2 presenting vacant lots majority (43) distributed in the size class ranging from 442m^2 to 609m^2. The other size distribution classes are: 312-442 (19); 609-1461 (23); 1461-3156 (1); 3156-5863 (2) (Figure 6).

In both neighborhoods, the respondents defined the vacant lots first as a place without construction (Figure 7). The abandonment was also highlighted as an important feature, expressed through the carelessness, compounding those plots commonly a wasteland. Related to that, dirtiness was pointed out as a usual component of those plots. Some people reported their frequent conversion in places for waste disposal. The lack of use was also referred as a means of definition of those landscape elements.

Figure 5. Vacant lots distribution according to size, in the São Bento area (Source: PBH/PRODABEL, 2015)

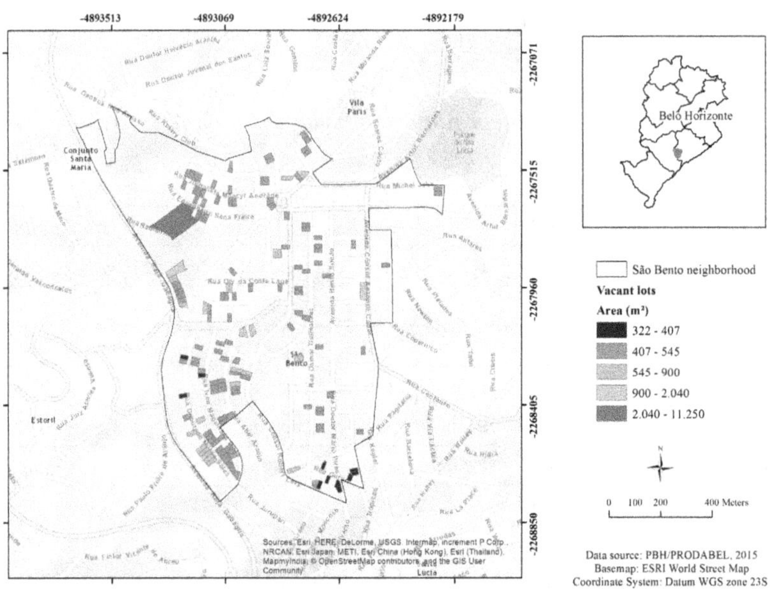

Figure 6. Vacant lots distribution according to size, in the Palmeiras´ area (Source: PBH/PRODABEL, 2015)

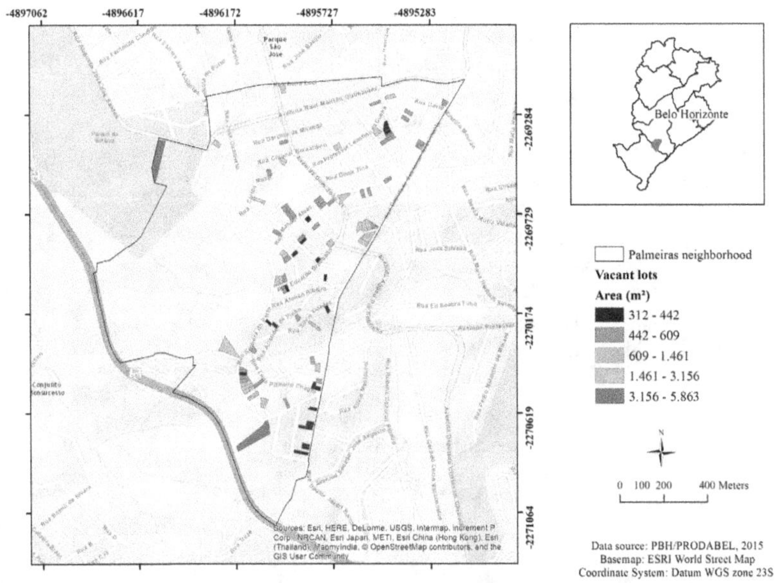

Insecurity generated by the presence of outsiders (marginals, drug addicts, and thieves) in the vacant lots was mentioned more often by the respondents from the São Bento area. The presence of bush and animals (insects, rats, etc.) was cited a little more by the Palmeiras dwellers. The referred bush vegetation is mainly comprised of an invasive tall grass species (*Panicum maximum*).

The association of the vacant lots with diseases relates mainly to the dengue and chikungunya fevers recent epidemics, functioning the plots as diseases outbreaks.

Some people from São Bento associated the vacant lots with investment (business opportunity or sells) and inheritance process (which keeps the plot idle while the process is in progress). Divestment was stated by a few people of Palmeiras region. For them, in some cases the plots´ owners refuse to sell, to construct or to give it a proper use, considering their wealth and their condition of living outside the neighborhood.

Some people associated the vacant lots with the presence of a physical delimitation (walls, fences), a most common condition currently, whereas others with its absence, a situation prevalent in the past.

The respondent's preferences on the land cover types of a hypothetical vacant lot neighbor to their house were significant different for the trees/woods cover (60% of the people preferred this cover type in São Bento neighborhood and 40% in the

Figure 7. Vacant lots definition according to the people´s opinion in the two neighborhoods

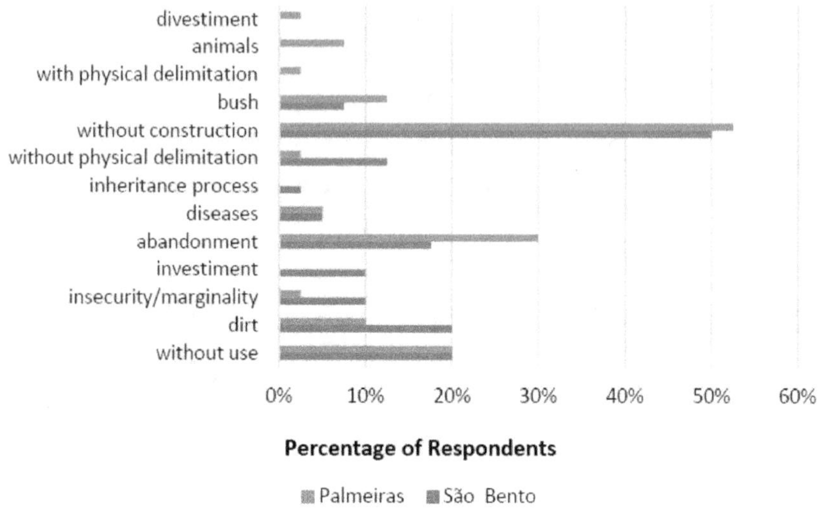

Palmeiras area) and for the bare soil cover (17,5% of the people preferred this cover type in the São Bento area and 35,0% in Palmeiras neighborhood). Grasses/low vegetation cover had similar preferences frequencies for the two neighborhoods (Table 3).

The reasons pointed out for the preferences on the land cover type trees/woods revealed the people´s knowledge, concerns, and appreciation of a cluster of ES. A great emphasis was placed, in both neighborhoods, on the regulation service of air purification (Figures 8 and 9).

For the São Bento area heat alleviation, another regulation ecosystem service, and aesthetic beauty, an immaterial cultural benefit from the ecosystems, were indicated as major reasons for having woods nearby (Figure 8). Another regulation

*Table 3. People´s preferences on land cover types of neighbor's vacant lot, in the São Bento and Palmeiras areas (*Z >1,282 significant at p = 0,10; Z > 1,960 significant at p = 0,05)*

Land cover types	São Bento	Palmeiras	Z test
Trees/woods	60,0	40,0	**1,83***
Grasses/low vegetation	22,5	25,0	0,26
Bare soil	17,5	35,0	**1,81***
Total	100,0	100,0	-

Figure 8. People´s reasons for preferences on land cover types of neighbor's vacant lot, in the São Bento area

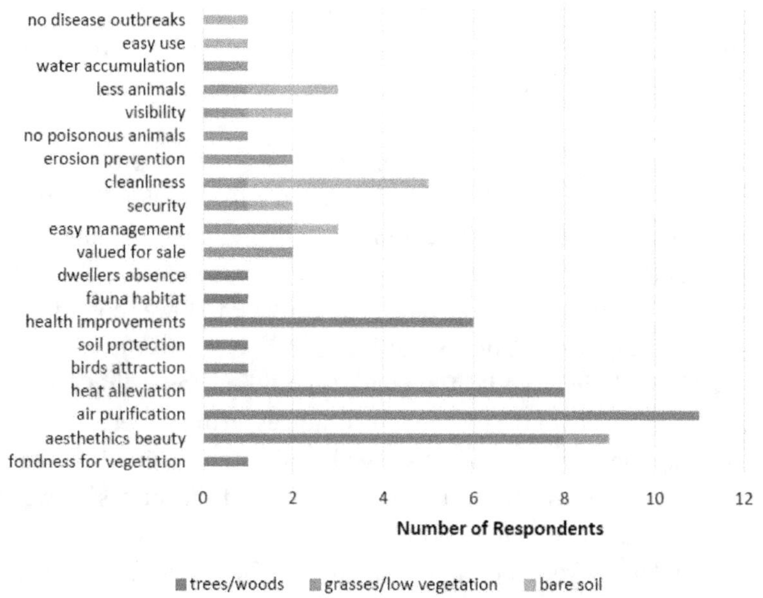

Figure 9. People´s reasons for preferences on land cover types of neighbor's vacant lot, in the Palmeiras area

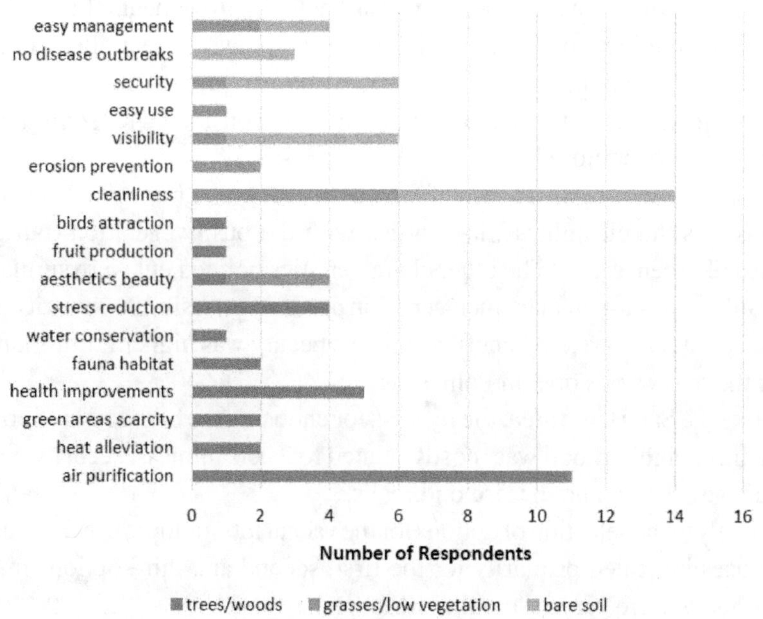

service associated with the trees/woods land cover for the São Bento´s was the soil protection provided by the vegetation. Ecosystems services of support and habitat were also mentioned such as fauna habitat and bird's attraction.

Other positive features and appreciation associated with the trees/woods cover for the São Bento´s were: health improvements, dwellers absence, and fondness for vegetation.

The land cover type grasses/low vegetation was given preference by some respondents in São Bento mainly due to its easy management and all the facilities to care and to maintain. Erosion prevention was a regulation ES associated with this cover type. The plots with grasses/low vegetation were also considered most valuable to sell since they are probably more attractive for having lower costs for removing the plot´s vegetation, for example. Security was also a positive feature pointed out for this cover considering that it offers fewer conditions for hiding criminals. The visibility, another reason for the respondent's choices on this cover favors, according to them, not only the security against fringe but also from animals. Other positive characteristics pointed out by the respondents comprised: cleanliness, fewer animals, no poisonous animals, less water accumulation, aesthetic beauty.

The bare soil land cover type was considered as a better choice for a neighbor for fewer informants in São Bento´s mostly for its cleanliness, providing more security, visibility, fewer animals, easy use, easy management and no disease outbreaks.

For the Palmeiras neighborhood besides air purification, the land cover type trees/woods was selected as a choice for neighbor given the possibility of bringing stress reduction from the city pace as well as health improvements (Figure 9). Other ES related by the informants to this land cover type were: heat alleviation, fauna habitat, bird's attraction, water conservation, and fruit production. The increasing scarcity of green areas in the neighborhood was also considered for some as a reason for having this cover close by.

The main reason for the choice of the land cover type grasses/low vegetation in Palmeiras was the cleanliness, in contrast with the bush vegetation (tall grasses) that is usually seen as dirt. The other characteristics pointed out were similar to São Bento´s including easy management, erosion prevention, visibility, security and easy use. The cultural ecosystem benefit aesthetics beauty was higher for this land cover than for the tree/woods one, in Palmeiras area.

Just like the São Bento area, the reasons for choosing the bare soil land cover type in Palmeiras neighborhood was mostly related to its cleanliness, security, visibility, easy management and no disease outbreaks.

Relatively to the selection of options for the vacant lots multiple uses and functions, the four uses indicated primarily for the first, second and third options in the two neighborhoods were very resembling (Figure 10).

Figure 10. People´s options for the vacant lots multiple uses and ecosystem services in the two neighborhoods. Abbreviations: CA- construct apartment building; CH - construct house; BC - build something commercial; GA - preserve a green area; HE - heat alleviation; FS - favor shading; BO - build a backyard orchard; VG - build a vegetable garden; RR - rainwater retention; BP - build a parking; BPG - build a backyard with pool and grill; PR - air pollution reduction; SE - prevent soil erosion; PG - plant a garden; DN - donation

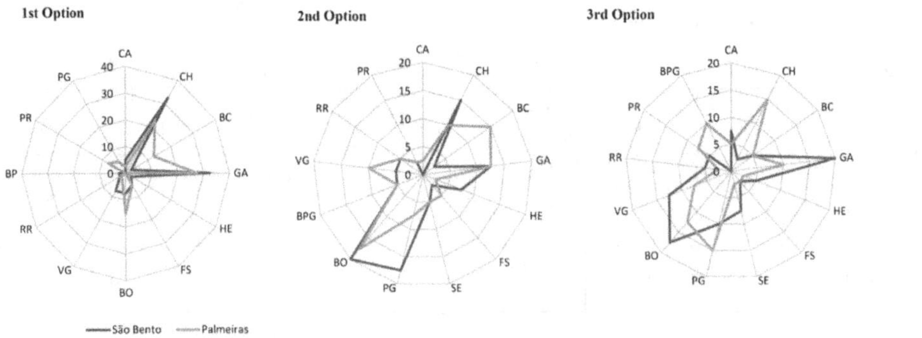

The most mentioned preferences for the vacant lots use in the two neighborhoods were to preserve a green area, construct a house, build a backyard orchard, build a vegetable garden, plant a garden and build something commercial. The choices to preserve a green area and to build a backyard orchard appeared among the most selected in both areas for the first, second and third options (Tables 4 and 5).

In the São Bento area, the first options comprised mainly: preserve a green area (GA), a preference for 32,5% of the people; construct house (CH), a better choice for 32,5% of the respondents; build a backyard orchard (BO) with 7,5% of the preferences and build a vegetable garden (VG), also with 7,5% of the answers (Table 4).

For the Palmeiras neighborhood the highlights from the first options were: preserve a green area (GA), congregating 27,5% of the respondents; construct house (CH), totaling 22,5% of the answers; build a backyard orchard (BO), with 15,0% of the preferences and build something commercial (BC), with 12,5% of the selection (Table 5).

About the definition of the vacant lots it is important to mention that there is not only one and different plots characteristics may include: open or closed spaces, undisturbed sites, abandoned land, plots contaminated, with bare soil, undeveloped land, abandoned sites with construction and infrastructure traces, land with physical limitation for development, areas utilized for parking, and others (Kremer, Hamstead, & MacPhearson, 2013; McPhearson, Kremer, & Hamstead, 2013; Pagano & Bowman,

Table 4. People´s preferences percentages for the multiple uses and ecosystem services in the São Bento neighborhood.

First options		Second options		Third options	
Uses	**Pref (%)**	**Uses**	**Pref (%)**	**Uses**	**Pref (%)**
preserve a green area	32,5	build a backyard orchard	20,0	preserve a green area	20,0
construct house	32,5	plant a garden	17,5	build a backyard orchard	17,5
build a backyard orchard	7,5	construct house	15,0	build a vegetable garden	12,5
build a vegetable garden	7,5	preserve a green area	12,5	plant a garden	10,0
construct apartment building	5,0	heat alleviation	7,5	construct apartment building	7,5
favor shading	5,0	prevent soil erosion	5,0	prevent soil erosion	7,5
build a parking	2,5	build a backyard with pool and grill	5,0	build something commercial	5,0
build something commercial	2,5	build a vegetable garden	5,0	heat alleviation	5,0
heat alleviation	2,5	rainwater retention	5,0	rainwater retention	5,0
rainwater retention	2,5	build something commercial	2,5	air pollution reduction	5,0
-	-	favor shading	2,5	construct house	2,5
-	-	air pollution reduction	2,5	favor shading	2,5
Total	100,0	-	100,0	-	100,0

2011). The plots are usually seen as a social, economic, environmental or aesthetical problem being symbols of decadence, degradation, and negligence (Kim, 2016)

The answers provided by the respondents from the São Bento and Palmeiras neighborhoods broaden the definitions above adding some peculiarities to the already known characteristics of the vacant lots, although most of them associated with unfavorable images of those landscape components.

In general, there were more similarities then contradictions among the answers and opinions of the respondents of the two neighborhoods, but some distinctions are feasible to present. For the assessment undertaken there was a greater preference in

Table 5. People´s preferences percentages for the multiple uses and ecosystem services in the Palmeiras neighborhood.

First options		Second options		Third options	
Uses	Pref (%)	Uses	Pref (%)	Uses	Pref (%)
preserve a green area	27,5	build a backyard orchard	17,5	construct house	15,0
construct house	22,5	build something commercial	15,0	plant a garden	15,0
build a backyard orchard	15,0	preserve a green area	12,5	build a backyard orchard	12,5
build something commercial	12,5	construct house	10,0	preserve a green area	10,0
air pollution reduction	7,5	build a vegetable garden	10,0	build a backyard with pool and grill	10,0
plant a garden	5,0	plant a garden	7,5	build a vegetable garden	7,5
construct apartment building	2,5	favor shading	5,0	air pollution reduction	7,5
favor shading	2,5	prevent soil erosion	5,0	construct apartment building	5,0
build a vegetable garden	2,5	build a backyard with pool and grill	5,0	build something commercial	5,0
rainwater retention	2,5	rainwater retention	5,0	heat alleviation	2,5
-	-	construct apartment building	2,5	favor shading	2,5
-	-	heat alleviation	2,5	prevent soil erosion	2,5
-	-	air pollution reduction	2,5	rainwater retention	2,5
-	-	-	-	donation	2,5
Total	100,0	-	100,0	-	100,0

São Bento for the land cover type trees/woods that provides a greater spectrum of ES. The land cover bare soil was most preferred in Palmeiras area.

In the Palmeiras neighborhood, there seems to be an interest in the immediate use of the vacant lots which may explain the preference on bare soil as a land cover type, that is ready to be built without expenses on taking off the vegetation.

The fact that the choice “build something commercial” appeared among the first preferred ones for the first and second uses options coupled with “construct house” that came out in all the options can corroborate to that. In São Bento, more ecological uses replaced those choices in the first positions, such as “build a vegetable garden” or “plant a garden.” Considering that Palmeiras is a neighborhood with the lower income it is reasonable to think that there is an interest in producing and improving this condition through investing in something commercial.

The option for constructing apartments building was not selected among the primary ones in both areas though in São Bento neighborhood it has shown closer to the first choices. As mentioned previously both neighborhoods are dominated by houses what seems to please the dwellers, although in Palmeiras the buildings from the neighbor Buritis area are increasingly close.

Some authors argued that despite income, environmental quality issues are concerns for both wealthy, popular or poor urban residents (Pickett at al., 2008). In the present study case, the fact that many respondents from both areas could point out some ES provided by the trees/woods cover reveals the dweller´s knowledge and appreciation of ecological features. The most recognized ecosystem services in the studied urban areas comprised: air purification, aesthetic beauty, heat alleviation, health improvements, stress reduction, fauna habitat, birds attraction, water conservation, soil protection and fruit production.

ASSESSING ECOSYSTEM SERVICES AND GOVERNANCE IN URBAN GARDENS

Allotment gardens are recognized for their contribution to provisioning and regulating ecosystem services (ES) (Borysiak, Mizgajski, & Speak, 2016; Speak, Mizgajski, & Borysiak, 2015). Community gardens, on the other hand, are often recognized for providing several cultural services (Rall & Haase, 2011). This opposition is rarely compared in studies (Cabral, Costa, Bonn, & Weiland, 2017; Cabral & Weiland, 2016). A comparative study of this nature can support governance strategies, by setting out how each garden complements the other one functions. In fact, allotment gardens are primarily top-down initiatives, and community gardens are bottom-up ones, promoting therefore different levels of governance (Cabral & Weiland, 2016). The assessment of Leipzig`s urban gardens presented next (Case Study 2) intends to appraise the allotment and community gardens ES contribution as well as people´s ecological knowledge on them, in a city that has one of the highest ratio of gardening area per capita in Europe, as well as a very high number of urban voids, turned into community gardens. As for Lisbon (Case Study 3), the assessment is primarily based on a sociological approach, although the city has promoted urban gardens to increase ES in a compact city, according to the Green Plan Strategy of 2010.

Case Study 2: Assessment of Urban Gardens in Leipzig, Germany

Introduction

Leipzig has been an important trade city in central Europe for several centuries. For 40 years it was also part of German Democratic Republic (GDR). The city is located in a floodplain and features an extensive forest in the heart of the city (Figure 11) which has been a source of wood and hunting species for more than 500 years (Strohbach, Haase, & Kabisch, 2009)

Currently, the city has around 570.000 inhabitants. From 1930 to 2010, Leipzig lost almost one-third of its population due to several economic crises and political changes (Strohbach et al., 2009). After 2010, the economic recovery has resulted in demographic regrowth, while Leipzig still has to cope with large urban voids.

The city is the birthplace of the *Schrebergärten* movement since the first allotment gardens or *Schrebergärten* were created during the Industrial Revolution as a means to mitigate poverty, malnutrition and health problems among kids and factory workers. These *Shrebergärten* estates (now called *Kleingärten*- KG) are

Figure 11. Location of study area (credit: Christian Langer)

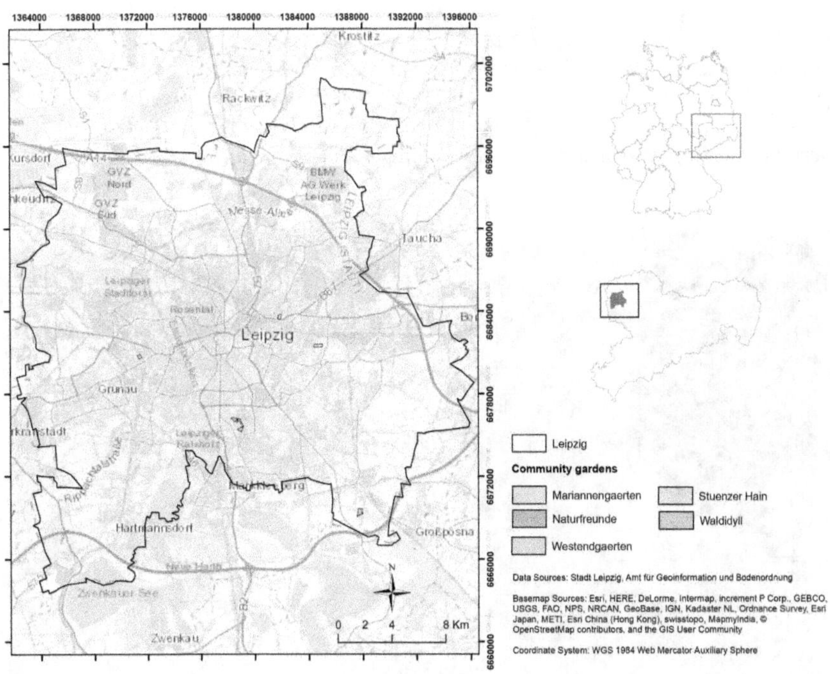

usually larger than one hectare and include more than 100 plots or parcels (Figure 12a). The estate usually provides space for a children playground (Figure 12b) and a clubhouse. Each *Kleingärten* (allotment garden) club or association enforces the regional/national code or open laws (*Kleingärtengesetzes*), which establishes that each plot is divided into three parts allocated to three types of land use: 1/3 ornamental species, 1/3 edible species, 1/3 lawn or build-up area (Figure 12 c).

Those code or open laws promote homogeneity and aesthetics along the aisles (Figure 13), enforce weeding in common areas and recommend weeding in private areas, as well as forbid large mature trees within the plots because these tend to block the sun in the adjacent plots. Historically, the allotment gardens in Leipzig have provided food and recreation. Three of the oldest ones, dated to the 19th century, are classified as heritage: the KG *Johannistal* estate since 1832, the KG *Südvorstadt* estate since 1874, and the KG Dr. Schreber estate since 1864. The city is proud of its pioneering role and hosts the only *Kleingärten* museum in Europe (*Deutsches Kleingärtnermuseum*) which is located within Dr. *Schreber* estate.

There is a large number of allotments in Leipzig partially because the city has only recently started facing real-estate pressure. In total there are 276 allotment gardens covering 1229 ha (Bürgerservice und Verwaltung, 2015) resulting in 23 m^2 of gardening area per inhabitant, which is one of the highest ratios among European cities (Cabral et al., 2017).

Figure 12. Allotment gardens, Leipzig
a) Suburban allotment garden with an area of 6 ha and 120 plots (adapted from Google maps).
b) Mapping of mature trees (above 5m) showed a total of 181 trees. Evergreen trees are shown as full-circles, deciduous trees as circumference, common playground as a circle.
c) An aerial view of a plot showing vegetation in infra-red light (credit: Roland Krämer).

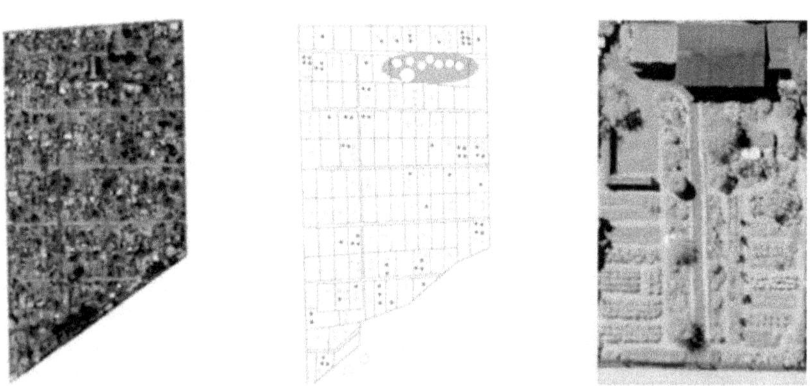

A B C

Figure 13. Allotment garden aisles bordered by fences and by hedges (Credit: Inês Cabral)

More recently, due to urban shrinkage, the municipality decided to initiate a program that allowed the interim-use of its urban voids by local initiatives of community gardeners (Rall & Haase, 2011). These spaces prime goal is recreational since they promote social meetings and sometimes pedagogic actions for school children (Figure 14). In 2016 the city accounted for 11 community gardens, scattered all over the city (see map on https://garten-findet-stadt.de/en/about-us/allotment-and-community-gardens-in-leipzig/).

Allotments in Leipzig have traditionally provided food during the two world wars as well as during the communist times. Nowadays allotment gardens in Leipzig still contribute for food provisioning but are also valued for services such as local climate mitigation and water regulating. In fact, allotment estates host up to 22% of all shrubs and small trees of the built-up area and 11% of the city´s grassland (Banzhaf & Kollai, 2015). This way, allotments contribute for air cooling and air filtration. Water infiltration, on the other hand, is provided by the plots permeable area, as the code limits the buildup area to a maximum of 1/3 for each plot.

The comparative case study undertaken in Leipzig addressed the contribution of urban gardens for biodiversity and ES within 12 gardens: 6 (six) traditional allotment gardens (figure 15) and 6 (six) community gardens. It was used an urban-suburban gradient to understand the effect of proximity to the city center on ES and biodiversity. It was also employed a stratified sampling design, including allotment plots of different management intensity and community gardens along a gradient of urbanity (Figure 16).

Methodology

The methodological approach included field surveys, visual image interpretation of high resolution, multi-spectral images (aerial photos with visible light and near

Figure 14. Stadtgarten Connewitz (run by NGO Ökolöwe) has several types of insect hotels on display and an adobe oven for pedagogic activities (Credit: Inês Cabral)

Figure 15. Surveyed allotment gardens selected across an urban-suburban gradient (2 urban; 2 semi-urban and two suburban allotment estates)

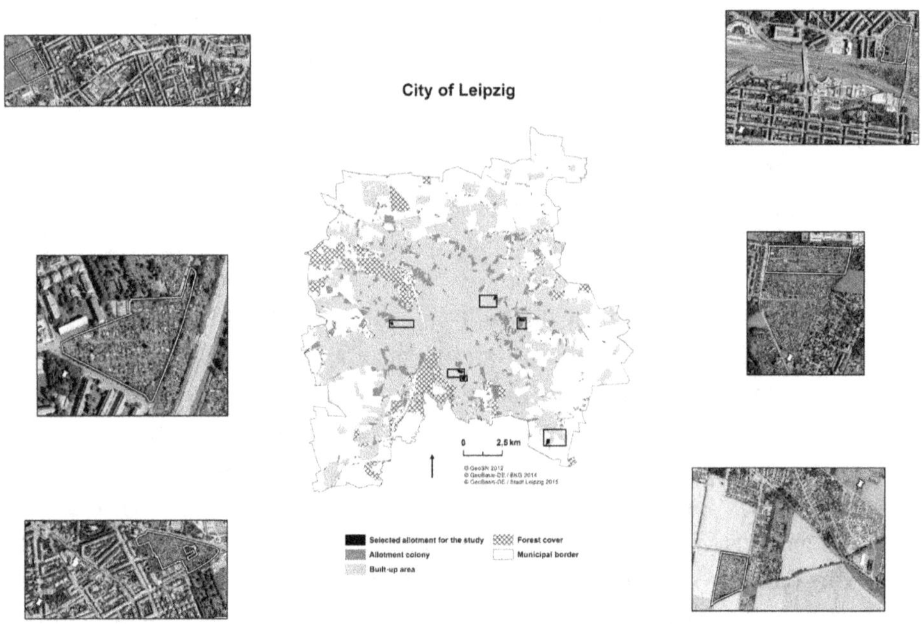

infrared, 3 cm resolution for LULC) collected by drone flights (senseFly Ltd`eBee drone) at an altitude of 400 m during July 2015) and interviews for assessing ecological knowledge among groups. The interviews encompassed a total of 30, with 24 of them applied in six allotments and 6 (six) undertaken in six community gardens. The indicators employed to assess each ES category and type are presented in Table 6.

Figure 16. Diagram showing management intensity across plots in Mariannengarten and plots selection according to distance to vacant plot. (V- Vacant plots; M1 - Medium managed plots; H1- High managed plots; High-intensity plots are shown in dark grey; Medium-intensity plots are shown in light grey)

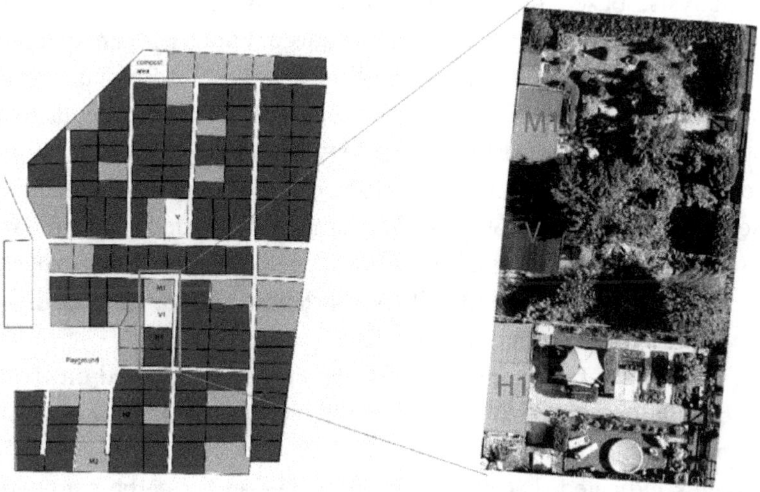

Table 6. Indicators employed to assess the various ecosystem services (Adapted from Cabral et al., 2017)

Ecosystem services categories	Ecosystem services	Indicators
Supporting services	Habitat provision	Land cover and manmade structures, such as ponds, rainwater deposits, stumps, beehives, insect hotels and bird houses.
	Biodiversity	Species richness and distribution patterns. Willingness to provide space for spontaneous species/weeds.
Provisioning services	Food provision	Species composition and proportional land cover of edible plant species patches.
Regulating services	Climate regulation	Above-ground carbon storage in trees.
	Water regulation	Permeable surface cover as a proxy for water infiltration potential.
Cultural services		Interviews on gardening motifs and ecological awareness.

Results and Discussion

Supporting services such as biodiversity, based on on-site botanical surveys and interviews, showed species richness distribution patterns as a bimodal relationship of overall vascular plant species richness varying with management intensity, with highest species richness in medium intensity managed plots. This result is related to the fact that the plots managed in medium intensity provide space for both cultivated, edible and ornamental species as well as native and spontaneous species (Cabral et al., 2017).

The analysis, based on remote sensing, provided information on habitat provisioning represented by the number of pools or ponds per garden, among other features. It demonstrated that suburban allotments have more man-made habitats than urban allotments and that community gardens host components not found in the allotments such as livestock shelters, beehives and stone walls.

As for food provisioning, the detailed botanical survey revealed that the number of edible species grown in allotments and community gardens was similar. Despite that, there was different species composition, i.e., allotments tended to have a higher diversity in fruit tree cultivation, while the variety of culinary and medicinal herbs tended to be higher in community gardens.

In the case of the regulating services, the comparison based on field surveys denoted that community gardens provide a higher ratio of permeable soil surface. Furthermore, calculations of above ground storage in trees based on fieldwork measurements indicated that allotments contribution for carbon storage is more significant if more mature trees are present in their common playground. Furthermore, this benefit is higher on suburban estates than on urban estates (Figure 17).

Figure 17. Abundance of mature trees within six allotments according to urban gradient (U-Urban; SM- Semiurban; SU- Suburban; deciduous trees are shown in black; evergreen trees are shown in grey)

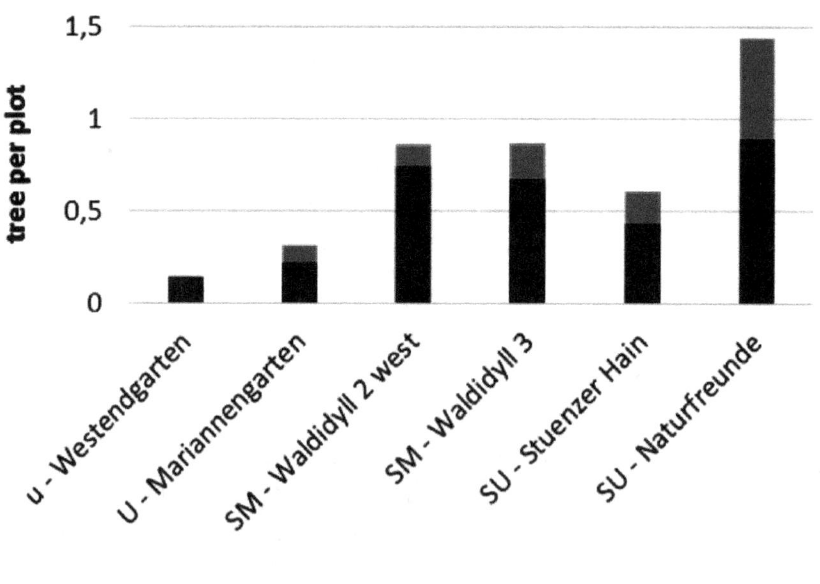

The assessment of ecological knowledge among the groups resulted in a greater knowledge among community gardeners on the meaning of ES and biodiversity (Figure 18), which is correlated to its larger amount of wild/unmanaged areas (Figure 19).

Given to the fact that the sampling method included allotment and community gardens that are adjacent to each other, it was possible to assess if knowledge exchange occurred between the respective gardeners. Regular interaction within allotment gardens was confirmed mainly due to the regional gardens association network. Interaction among the surveyed community gardeners also occurred. In few cases, there was also the interaction between allotment and community gardeners. One particular interaction was due to the educational goals of a community garden (Stadtgarten Connewitz) managed by a local NGO (Ökolöwe), which provided experience and know-how to the adjacent allotment garden.

The ES assessment in Leipzig clarified the need for knowledge exchange between community and allotment gardens to improve their contribution for ES, while at the same time promote social cohesion by increasing cooperation development.

Case Study 3: Farming to Horticultural Parks— Benefits and Farmer´s View on the Regularization Process in Lisbon, Portugal

Introduction

The benefits of urban gardening for the city and its inhabitants are well reported, when it comes to its contribution for social, cultural, economic, and institutional

Figure 18. Are gardeners familiar with the term "Ecosystem services"?

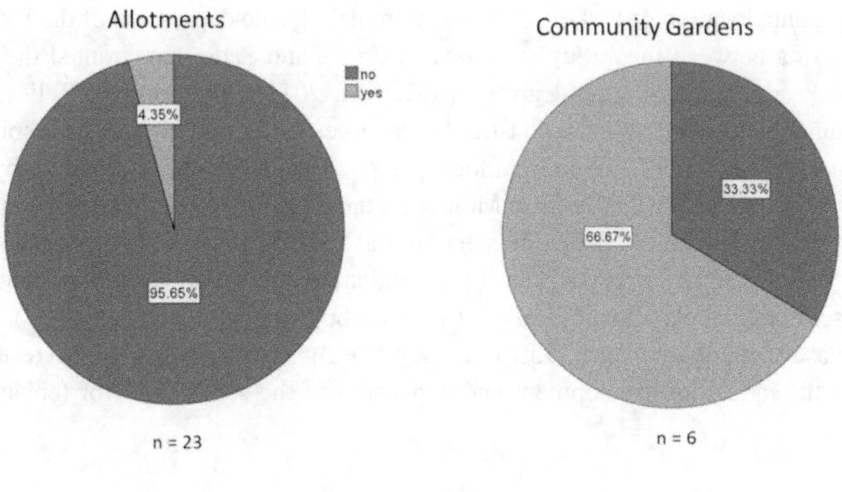

Figure 19. Do gardeners keep wild/unmanaged areas in their gardens?

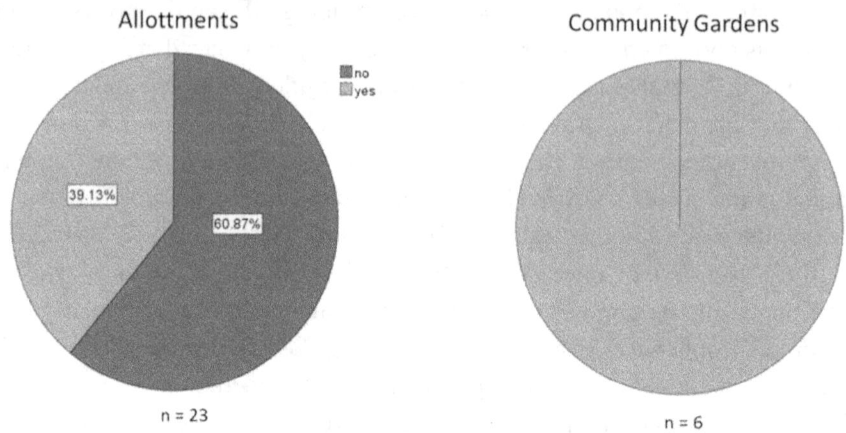

sustainability of the communities, thus creating social capital (Barthel, Folke, & Colding, 2010; Barthel & Isendahl, 2013; Bendt, Barthel, & Colding, 2013; Cabannes & Raposo, 2013; Firth, Mayde, & Pearson, 2011; Kingsley & Townsend, 2006; Milbourne, 2012; Ramos, 2011).

At the same time, urban gardens reportedly act as reservoirs of "socioecological memory", fighting "social amnesia" regarding our dependence on nature; contribute to biophysical ES; promote social justice and public participation; and contribute to food security and generate income for farmers (Andersson, Barthel, & Ahrné, 2007; Barthel et al., 2010; Barthel & Isendahl, 2013; Milbourne, 2012; Bendt et al., 2013; Martins, 2012, Matos & Batista, 2013; Ramos, 2011; Roth, Frixen, Tobisch, & Scholle, 2015). All in all, it might be affirmed that urban gardening contributes to socioecological urban resilience.

Agriculture has always been present in the structure of the city of Lisbon. The geographic location of Lisbon, set in an extremely fertile depression of the Iberian Peninsula between the valleys of the river Tagus and Sado, has granted the city close productive agricultural grounds (Martins, 2012). According to Telles, the conjunction of both estuaries constitutes the largest extension of flat fertile soils in Portugal (Telles, 1997). In a morphologic perspective, Lisbon can be delimited by the territory confined by the Serra de Monsanto, the Odivelas' stream and the Trancão river (Telles, 1997). In 2003, almost half of the land of Lisbon's Metropolitan Area was used for agricultural purposes (48%) and around 39% was woodlands, natural or 'semi-natural' landscapes, only 13% was a built-up area (Martins, 2012). The increase of population in the 1950's and the related urbanization process that resulted from the industrial development and expansion of the services sector (enhanced later by the return of 600 000 people from the ex-colonies) led to the loss of most of the productive farming spaces (Martins, 2012). Some small pockets remain now dispersed in-between the urban fabric as part of leisure areas, and some of them in the form of urban parks (Bernardo, 2013; Henriques, 2009; Lima, 2012; Madaleno & Armijo, 2004; Martins, 2012; Matos & Batista, 2013). According to the Statistics National Institute, there are currently 547 773 people living in the city of Lisbon (data from 2011).

Urban voids that resulted from this process of urban expansion were occupied for decades for the practice of subsistence agriculture, especially by low-income immigrant communities, most coming from former Portuguese colonies (Cabannes & Raposo, 2013; Martins, 2012; Matos & Batista, 2013). The majority of these spontaneous gardens were located on public lands owned by the Lisbon Municipality (Câmara Municipal de Lisboa (CML)) with its tacit acceptance but without any regulation or legal status (Bernardo, 2013; Cabannes & Raposo, 2013; Madaleno, 2003; Henriques, 2009).

In 2007 Lisbon´s Green Plan, designed by the landscape architect Ribeiro Telles, became the ecologic structure of the municipality of Lisbon. The plan proposed the creation of "green corridors" that would cross the city connecting various land uses, such as urban parks, gardens, bicycle lanes and footpaths and also urban gardening. Within this framework, agriculture was assumed as an important component of the ecological structure of the city, and a Strategy for Urban Agriculture was defined (Henriques, 2009).

This Strategy for Urban Agriculture aimed at responding to several challenges among which:

- The scarcity of available green spaces in the city - green space availability: 27.8 m^2 per capita (Santos, 2015);
- The objectives of the Green Plan to link most of the green spaces through ecological corridors;
- The impact mitigation of channeling waterlines by creating water basins in strategic valleys (Telles, 1997);
- The need to reorganize allotments that were growing fast and unorganized (partly due to the economic crisis and rising unemployment);
- The increasing demand for allotments.

After the Strategy was approved, the municipality started the process of regularization of former illegal gardens on void spaces and initiated the construction of several new allotments in poorly maintained municipal green areas (http://www.cm-lisboa.pt/fileadmin/VIVER/Ambiente/Hortas_Urbanas/Parques_Horticolas_2016.pdf) (Câmara Municipal de Lisboa [CML], 2016).

In the previously informal allotments, the municipality displaced the people working there to reorganize and unify the plots, to provide infrastructures, water access, and small sheds for tools storage. The municipality also defined the rules of use to ensure that all farmers complied the organic mode of production (after technical training offered by the municipality), composting, prohibition of using chemical pesticides, herbicides, chemical fertilizers, Genetic Modified Organisms (GMO) and infesting species. It guaranteed the supervision as well.

A typology of regulated urban gardens was created including three types, according to their main objective, either social, pedagogical or recreational. Social gardens are institutional gardens that emerge as a social support project, intended to minimize social injustice and urban poverty, contributing to food security. The annual fee established by the municipality for the use of these spaces is lower than the one practiced in recreational farms in which the emphasis is given to the promotion of public health, both physical and psychological. Another distinction is the size of the plots, i.e., 50 m^2 within recreational gardens and 100 m^2 or more in the case of social gardens, where the sale of surpluses is allowed, underlining the subsistence character associated to them.

After the reorganization process, plots were redistributed to people who were already occupying the place, and the remaining was reserved for public tenders. The first public tender occurred in 2013, and the eligibility conditions were mainly based on the distance from the place of residence to the plots.

Methodology

In order to understand the motivations behind the practice of agriculture in the city of Lisbon (Figure 20) and the course of the process of integration of informal uses into the city planning, interviews were undertaken with farmers in two of the settlements where this transition occurred, namely Quinta da Granja and Vale de Chelas - Figure 21 (Krylová & Luz, 2017; Luz & Pires, 2015).

This empirical study is based on 26 qualitative interviews. Around 26 interviews were performed with farmers in both parks (eight in Quinta da Granja and 18 in Vale de Chelas) in three different periods: between March 5th and April 15th and between October 1st and 3rd, 2014, in Vale de Chelas; and between 20th and 25th May, 2015, in Quinta da Granja. From the 26 respondents 20 were men and 6 (six) were women, 5 (five) were unemployed, 12 were retired, and 7 (seven) were employed. For the remaining 2 (two) no information was obtained.

Figure 20. Location of study area

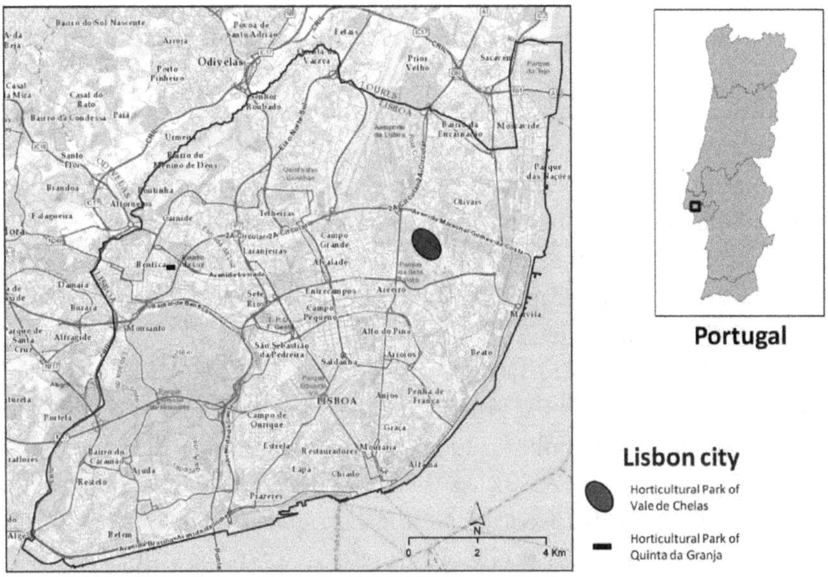

Sources: Esri, HERE, DELorme, Intermap, increment P Corp., GEBCO, USGS, FAO, NPS, NRCAN, GeoBase, IGN, Kadaster NL, Ordnance Survey, Esri Japan, METI, Esri China(Hong Kong), swisstopo, MapmyIndia, © OpenStreetMap contributors, and the GIS User Community; Sources: Esri, DeLorme, USGS, NPS

Figure 21. Quinta da Granja, a recreational horticultural park (left) and Vale de Chelas, a social horticultural park (right)

Results and Discussion

In Vale de Chelas, the gardeners were originally from 4 (four) countries (Angola, Cape Verde, India, and Portugal) while in Quinta da Granja the people interviewed were all born in Portugal. As for the practice duration, it varied from 1 (one) year (a newcomer in Quinta da Granja that occupied a vacant plot) up to 41 years (the "oldest farmer" in Quinta da Granja). As for Vale de Chelas, the "oldest farmers" are working since around 30 years and the new farmers are arriving since the end of 2013.

The Horticultural Park in Vale de Chelas is the city's largest park. The first CML project was conceived for 15 hectares of land, of which 6,5 hectares were meant for agricultural use. However, due to changes in the project budget, the area had to be reduced, and the number of plots decreased from 400 to slightly more than 200 each with 150 square meters. Part of the plots was attributed to the nearly 100 people who were already occupying the place, and the remaining reserved for a public tender that occurred in 2013.

In Vale de Chelas, before farmers started to build their plots, through very challenging tasks (e.g., clear the ground, cut the grass and reeds and work the land to create a suitable soil for cultures to grow), the place was a "space leftover after planning" (Krylová & Luz, 2017). As the name indicates, they are not part of the urban planning but rather consequences of it. Later, with the people's initiative, the place gained a new meaning, with obvious social and cultural benefits. The possibility to grow their own food and products from their homeland (as part of the farmers are from tropical African countries), allowed them to continue their gastronomic traditions and practices but mostly avoid starvation.

Furthermore, they could create social bonds, through the sharing of an activity, space, and through knowledge exchange. Like some of the oldest farmers said, informally, in one interview: "F: (…) you know, at this age, instead of hanging around, I come here…; P: I don't even know how to play cards…; F: me neither…; P: nor domino…; F: here we take the chance to have a few conversations and all…". A healing effect was also mentioned, to fight stress or ease preoccupations. This is clear in this farmers' discourse: "I am unemployed, for a long time already. Then, it coincided with my mother having an AVC (…) I had done some volunteer work also, and when I became aware of the gardens I thought 'this is something I like' (…)" (personal communication, March 5, 2014).

When asked about their view on the main changes introduced by the reorganization process, interviewees reported that water supply and accessibility improvement to be the most significant changes. A farmer said, "For instance, if somebody gets hurt, first aid can easily reach us now, before it didn't happen" (personal communication, March 5, 2014). Nevertheless, the water supply was the most consensual factor. As a farmer in Vale de Chelas said "It is a good price I pay for the plot because of the water, I have as much water as I want, that is the greatest advantage, no doubt about that (…)" (personal communication, March 5, 2014). Previously, gardens in Vale de Chelas produced mainly winter crops. Water, when necessary, was collected in sewers (creating potential health problems related to the consumption of the products), in very small springs with insufficient pressure, or through the accumulation of rainwater. Presently, each cluster of four plots has access to the public city water.

Quinta da Granja was the first park being built. Part of the current area of the park was owned by a family, who allowed farmers to use the land, although without any legal compromise. Some decades ago the property was sold to the municipality, and in 2011 a project was developed for the renewal of the area. Until then people were allowed - by the CML - to continue using the land. Here farmers also had to cope with the initial conditions by themselves. Water was conducted through a system they implemented, and which is still operational today. With the implementation of the farming park, not only new farmers arrived to occupy the plots made available by the Municipality but also there was a clear change in the local uses. With the implementation of the farming park, there was a clear change in the local uses. Today, besides urban farming, there are cycling ways, walking paths, coffee shops, and kids' playgrounds, all together.

Two "types" of farmers were found in both parks: i) the ones that already farmed there informally, having a kind of a tacit agreement with the owner, and ii) the newcomers that arrived through CML's tenders. Their perspectives toward the project are different once the old farmers have prior experiences in these places. The construction of the farming parks brought new conditions, especially regarding access to water, but farmers were also submitted to limitations in their use of the land. At the same time, a fee of 80 euros per year is now required to farm in those lots.

Nevertheless, most of, many of the interviewees considered it a fair value, taking into account that the intervention greatly improved their working conditions: "Let's see, the municipality also invested, also spent money, the infrastructures, the shelters..." (personal communication, March 5, 2014). However, two of them complained about that obligation, and one said he would stop farming in the regulated areas "I do not spend 80 euros in vegetables per year, right?" (personal communication, March 5th, 2014).

Even though it seems that there is a change of approach from the municipality toward a more bottom-up intervention – seen in the priority given to the distribution of the plots between "old farmers", giving them also the opportunity to choose the plots' location –, the municipality's will to keep control over the farms' uses emerges. Aesthetics aspects were defined by the municipality and imposed to the farmers. In that sense, the recreational and aesthetic role set by the municipality for the farming parks does not allow the simultaneous presence of seemingly incompatible activities.

The accumulation of stuff, plastics or other materials as fences are not compatible with the existence of a beautiful, ordered and attractive garden; keeping dogs and organizing gatherings collide with the possibility of contemplative silence (Krylová & Luz, 2017). As an "old" farmer in Quinta da Granja said: "Before [when the use was informal] I had a huge table, more than 10 people could sit there, we made barbecues, gathered friends and family… now, I tried to create some shadow by hanging one canvas near the storage house, but they said we can't" (personal communication, May 25th, 2015). Moreover, another farmer says, after admitting he does not follow all rules: "Who sets the rules in my land? Them? I am the boss in my land; I am paying to be here" (personal communication, May 25th, 2015).

Finally, although farmers regret the changes imposed by the new rules in the control of the uses (e.g., limitations in the size of the plot area, in social uses, in the planting of trees, etc.), farmers clearly see the benefits of this initiative. As an "old" farmer in Quinta da Granja, resumes: "well, it is a good project, after all, they could have just thrown us out, and it is more beautiful now, more organized."

DISCUSSION

On the first case presented, the vacant lots in Brazil have played a secondary role in the creation of strategies for sustainability. However, these underutilized plots of land can offer perspectives for the urban land use transformation especially in economic crisis periods. In this context, positive vacant lots shifts can contribute to the community development and ES provision, establishing a socioecological infrastructure in cities (Anderson & Minor, 2017; Kim, 2016; Kremer et al., 2013 McPhearson et al., 2013).

The studies on vacant lots in Belo Horizonte are still in a tailoring and diagnosis phase. The conceptual approach encompasses the incorporation of the social and political aspects, especially considering the importance of taking into consideration the people´s view and concerns about ecological features, nature and ecosystem services. It is expected that the research on vacant lots can bring about some insights into the problem and contribute to the policies designs and normalization, favoring ES provision.

The choice of scenarios and mapping can help on capturing the situation and on clarifying the chart of options available, providing a basis for discussions and orientations. The studies are being designed therefore to contribute to the decision-making process and to highlight the importance of the incorporation of people´s preferences on the vacant lots ES provision in the city.

Belo Horizonte is a city comprising around 2,6 million people, with 22047 vacant lots registered for the year 2015. The county has faced a center to periphery population growth in the last years, presenting a larger number of vacant lots in the administrative unity named Pampulha (5235 plots), that showed the highest population growth in the last census (Costa & Mendonça, 2010; Instituto Brasileiro de Geografia e Estatística [IBGE], 2010), This means that the presence and amount of vacant lots in the city are not related to shrinkage but rather with the city expansion.

In Germany, on the other hand, the city of Leipzig was faced with the process of shrinkage, and the municipality decided to use its voids for more public green spaces, including leasing vacant lots for community gardens (Rall & Haase, 2011). The city hosts 276 allotment estates covering 1229 ha (Bürgerservice und Verwaltung, 2015). Overall, the county has over 23 m^2 of gardening area per inhabitant, which is one of the highest ratios among European cities. The city also hosts eight community gardens.

The city of Leipzig has experienced two types of governance regarding urban gardens: bottom-up and top-down initiatives. The assessments undertaken in Leipzig addressed a combination of ES with a spatial dimension, thus producing planning guidelines. Allotment gardens and community gardens are used by different social groups whose interaction is still very limited, requiring the use of participatory methods to develop even further the garden's contribution to ES, by exchanging knowledge.

As for Portugal, the capital city Lisbon is facing a population migration to the suburbs as tourism and financial business have taken the center. As a compact city, its green infrastructure does not provide enough ES, such as water and local climate-regulating services. As the municipality developed its green strategy, the implementation of urban gardens became a multifunctional solution potentially contributing to mitigate the economic crisis and rising unemployment rates. As such, horticultural parks were developed in either old farm estates or low-income urban parks.

In Lisbon, the urban gardens are adapting to a new normalization and governance. In fact, allotment gardens were created for ecological and social benefits. The green infrastructure of Lisbon has been growing under the Green Strategic Plan that it is solely a top-down initiative. With some exceptions, the groups of gardeners require further education, and so far, participatory methods have not been used yet.

FUTURE RESEARCH DIRECTIONS

The study, development, and use of appropriate tools and methods for assessing ecosystem services can contribute to improvements in urban sustainability, as far as it provides the generation of technical-scientific knowledge applied to the cities policies and governance. Beyond that, there is an urgent need for the democratization of the produced knowledge so that citizens and communities can make own of issues related to the planning and management of their city spaces, which are part of their lives.

This means that the researchers and city planners play an important role in the provision of guidance concerning the suitable means to reach measures and goals agreed collectively, but should avoid centralizing, contributing to the dissemination, exchange and discussion of the research results (Souza, 2010).

For the study presented here, the expectations at this point are to contribute with tools and approaches that offer insights into the importance, benefits and peoples' opinions on ecosystem services provided by nature in the cities context. It is a concern, however, that those assessments are just a starting point placed in the earlier planning phases.

For the peoples' appropriation of the ecosystem services and their benefits for themselves, cities sustainability and life quality improvements there is a need for the studies based on participatory approaches to follow up continuously through time, incorporating other methods (i.e. action participant) that facilitate people's inclusion, and the movement from a theoretical basis into action.

CONCLUSION

The chapter presented assessments of ES in three urban regions of Brazil, Germany, and Portugal that included different methodological approaches, incorporating the socio-cultural dimension, ecological features and a conceptual framework based on participatory methods.

In general, the studies demonstrated the potential benefits of utilizing ES assessments approaches on urban landscapes, especially for better understanding the interactions between people and nature, in urban gardens and vacant lots. At the same time, it provided a set of tools to improve approaches to the urban ES assessment, useful for a wider community of academics and practitioners.

For Belo Horizonte, Brazil, the conceptual framework presented aims to be used to characterize the landscape dimension and ecological attributes of the vacant lots and integrate them with the social, political and economic contexts. The study case on people's opinion on ES revealed a set of definitions associated with the plots and the inhabitants' knowledge, recognition and appreciation of ecological features. The regulating service of air purification was the most recognized by them. Other conspicuous ES reported comprised: air purification, aesthetic beauty, heat alleviation, health improvements, stress reduction, fauna habitat, bird´s attraction, water conservation, soil protection and fruit production.

In Germany, the assessment of urban gardens in Leipzig was based on a myriad of methods aiming to provide insights on governance levels and new forms to enhance it. The initiatives promoted different levels of governance, ranging from top-down on allotment gardens to bottom-up in the community gardens. Provisioning, regulating, and cultural ES are the most conspicuous provided by the gardens. It was found a greater knowledge among community gardeners from Leipzig on the meaning of

ES and biodiversity, correlated to its larger amount of wild and unmanaged areas. Governance on the studied area could be improved by creating synergies between allotment gardeners and community gardens.

The study in Lisbon, Portugal, demonstrated that although regulating ES are an important motif for the city to implement new and re-organized allotments, the interviews performed in the largest and poorest horticultural parks, resulted in provisioning services (food production) being the most significant service for the users. It also revealed the existence of a social conflict between the gardeners and the existing policies, which needs to be addressed in the future, probably through a participatory process.

ACKNOWLEDGMENT

Marise Barreiros Horta received a scholarship from CAPES (Brazilian Federal Agency for Support and Evaluation of Graduate Education). The authors would like to thank Jessica Keim. Rolf Engelmann, Roland Kraemer, Aletta Bonn for their contribution to the Leipzig case study. The authors would also like to thank the PBH (Belo Horizonte Municipal Administration), PRODABEL (Belo Horizonte Informatics and Information Municipal Company) and Felipe Antônio Carneiro Rodrigues for the provision of the Belo Horizonte county spatial dataset.

REFERENCES

Anderson, E. C., & Minor, E. S. (2017). Vacant lots: An underexplored resource for ecological and social benefits in cities. *Urban Forestry & Urban Greening*, *21*, 146–152. doi:10.1016/j.ufug.2016.11.015

Andersson, E., Barthel, S., & Ahrné, K. (2007). Measuring social-ecological dynamics behind the generation of ecosystem services. *Ecological Applications*, *17*(5), 1267–1278. doi:10.1890/06-1116.1 PMID:17708207

Arreguy, C. A. C., & Ribeiro, R. R. (Eds.). (2008). Histórias de bairros [de] Belo Horizonte: Regional Barreiro / Regional Barreiro / Regional Centro-Sul / Regional Leste / Regional Nordeste / Regional Noroeste / Regional Oeste / Regional Pampulha / Regional Venda Nova. Belo Horizonte, Brazil: Arquivo Público da Cidade de Belo Horizonte.

Banzhaf, E., & Kollai, H. (2015). *Monitoring the Urban Tree Cover for Urban Ecosystem Services - The Case of Leipzig, Germany*. Paper presented at the meeting of the ISPRS -International Archives of the Photogrammetry, Remote Sensing and Spatial Information Sciences, Berlin, Germany.

Barthel, S., Folke, C., & Colding, J. (2010). Social-ecological memory in urban gardens—Retaining the capacity for management of ecosystem services. *Global Environmental Change*, *20*(2), 255–265. doi:10.1016/j.gloenvcha.2010.01.001

Barthel, S., & Isendahl, C. (2013). Urban gardens, agriculture, and water management: Sources of resilience for long-term food security in cities. *Ecological Economics*, *86*, 224–234. doi:10.1016/j.ecolecon.2012.06.018

Bendt, P., Barthel, S., & Colding, J. (2013). Civic greening and environmental learning in public-access community gardens in Berlin. *Landscape and Urban Planning*, *109*(1), 18–30. doi:10.1016/j.landurbplan.2012.10.003

Bernardo, J. M. (2013). O campo na cidade: as hortas e os hortelões de Lisboa. In I. L. Cardoso (Ed.), *Paisagem e Património - Aproximações Pluridisciplinares* (pp. 195-218). Oporto: Dafne Editora I CHAIA. Retrieved from https://dspace.uevora.pt/rdpc/bitstream/10174/10444/1/Bernardo2013 Campo na cidade Hortas e horteloes de Lisboa.pdf

Bixler, R. P., Dell'Angelo, J., Mfune, O., & Rob, H. (2015). The political ecology of participatory conservation: Institutions and discourse. *Journal of Political Ecology*, *22*, 164–182. Retrieved from https://journals.uair.arizona.edu/index.php/JPE/article/view/21083/20671

Borysiak, J., Mizgajski, A., & Speak, A. (2016). Floral biodiversity of allotment gardens and its contribution to urban green infrastructure. *Urban Ecosystems*, *20*(2), 1–13. doi:10.100711252-016-0595-4

Brendan, F., Costanza, R., Turner, R. K., & Morling, P. (2009). Defining and classifying ecosystem services for decision making. *CSERGE Working Paper EDM, 7*(4). Retrieved from https://projects.eionet.europa.eu/leac/library/international_classifica/background_documents/costanza-brfisher-edm/download/en/1/Costanza-BrFisher-edm_2007_04.pdf

Bürgerservice und Verwaltung. (2015, November 27). *Statistisches Jahrbuch 2015 veröffentlicht*. Retrieved from http://www.leipzig.de/news/news/druckversion-des-statistischen-jahrbuches-2015-liegt-vor/

Burkhard, B., & Müller, F. (2015). Ascertainment and Assessment of ES - Indicators and Quantification Approaches. In K. Grunewald & O. Bastian (Eds.), *Ecosystem Services – Concept, Methods and Case Studies* (pp. 75–143). Berlin, Germany: Springer; doi:10.1007/978-3-662-44143-5_4

Burkhard, B., Petrosillo, I., & Costanza, R. (2010). Ecosystem services – Bridging ecology, economy and social sciences. *Ecological Complexity*, *7*(3), 257–259. doi:10.1016/j.ecocom.2010.07.001

Cabannes, Y., & Raposo, I. (2013). PeriUrban Agriculture, social inclusion of migrant population and Right to the City - Practices in Lisbon and London. *City: analysis of urban trends, culture, theory, policy, action, 17*(2), 235-250. doi:10.1080/13604813.2013.765652

Cabral, I., Costa, S., Bonn, A., & Weiland, U. (2017). Urban gardens as multi-functional nature-based solutions for societal goals in a changing climate. In N. Kabisch, H. Korn, J. Stadler, & A. Bonn (Eds.), *Nature-Based Solutions to Climate Change Adaptation in Urban Areas—Linkages Between Science, Policy and Practice* (pp. 237–253). Cham, Switzerland: Springer; doi:10.1007/978-3-319-56091-5_14

Cabral, I., Keim, J., Engelmann, R., Krämer, R., Siebert, J., & Bonn, A. (2017). Ecosystem services of allotment and community gardens: A Leipzig, Germany case study. *Urban Forestry & Urban Greening*, *23*, 44–53. doi:10.1016/j.ufug.2017.02.008

Cabral, I., & Weiland, U. (2016). Urban gardening in Leipzig and Lisbon: A comparative study on governance. In S. Tappert (Ed.), *Growing in Cities: Interdisciplinary Perspectives on Urban Gardening. Conference Proceedings* (pp. 66–79). Basel, Switzerland: University of Applied Sciences. Retrieved from http://www.academia.edu/28449844/Urban_gardening_in_Leipzig_and_Lisbon_A_comparative_study_on_governance

Carvalho-Ribeiro, S. M., & Lovett, A. (2011). Is an attractive forest well managed? Correlating public preferences for forests across rural/urban gradients. *Forest Policy and Economics*, *13*(1), 46–54. doi:10.1016/j.forpol.2010.09.003

Carvalho-Ribeiro, S., Ramos, I. L., Madeira, L., Barroso, F., Menezes, H., & Pinto Correia, T. (2013b). Is land cover an important asset for addressing the subjective landscape dimensions? *Land Use Policy*, *35*, 50–60. doi:10.1016/j.landusepol.2013.04.015

Carvalho-Ribeiro, S. M., Lovett, A., & O'Riordan, T. (2010). Multifunctional forest management in Northern Portugal: Moving from scenarios to governance for sustainable development. *Land Use Policy*, *27*(4), 1111–1122. doi:10.1016/j.landusepol.2010.02.008

Carvalho-Ribeiro, S. M., Migliozzi, A., Incerti, G., & Pinto Correia, T. (2013a). Placing land cover pattern preferences on the map: Bridging methodological approaches of landscape preference surveys and spatial pattern analysis. *Landscape and Urban Planning*, *114*, 53–68. doi:10.1016/j.landurbplan.2013.02.011

Costa, H. S. M., & Mendonça, J. G. (2010). *Urbanização recente e disputa pelo espaço na dinâmica imobiliária metropolitana em Belo Horizonte*. Paper presented at the meeting of the XVII Encontro Nacional de Estudos Populacionais, ABEP, Caxambu, MG, Brasil.

De Groot, R., Brander, L., Ploeg, S., Costanza, R., Bernard, F., Braat, L., ... Van Beukering, G. P. (2012). Global estimates of the value of ecosystems and their services in monetary units. *Ecosystem Services*, *1*(1), 50–61. doi:10.1016/j.ecoser.2012.07.005

De Groot, R. S., Alkemade, R., Braat, L., Hein, L., & Willemen, L. (2010). Challenges in Integrating the Concept of Ecosystem Services and Values in Landscape Planning, Management and Decision Making. *Ecological Complexity*, *7*(3), 260–272. doi:10.1016/j.ecocom.2009.10.006

De Montis, A. (2007). Participative and Interactive Evaluation: A Review of the Methodologies. In M. Deakin, G. Mitchell, P. Nijkamp & R. Vreeker (Eds.), Sustainable Urban Development Volume 2: The Environmental Assessment Methods (pp. 473-491). London: Routledge.

Drumond, M. A., Giovanetti, L., & Guimarães, A. (2009). Técnicas e Ferramentas Participativas para a Gestão de Unidades de Conservação. Brasília, Brazil: Ministério do Meio Ambiente (MMA). Retrieved from

http://www.icmbio.gov.br/educacaoambiental/images/stories/biblioteca/gestao_participativa/Caderno_4_-_ARPA.pdf

Ernston, H., Barthel, S., Anderson, E., & Borgström, S. T. (2010). Scale-crossing brokers and network governance of urban ecosystem services: The case of Stockholm. *Ecology and Society*, *15*(4), 28. doi:10.5751/ES-03692-150428

Firth, C., Mayde, D., & Pearson, D. (2011). Developing community in community gardens. *Local Environment*, *16*(6), 555–568. doi:10.1080/13549839.2011.586025

Gómez-Baggethun, E., Gren, A., Barton, D. N., Langemeyer, J., McPhearson, T., O'Farrel, P., ... Kremer, P. (2013). Urban Ecosystem Services. In Urbanization, Biodiversity and Ecosystem Services: Challenges and Opportunities (pp. 175-251). New York, NY: Springer. doi:10.1007/978-94-007-7088-1

Haines-Young, R., & Potschin, M. (2014). The ecosystem approach as a framework for understanding knowledge utilization. *Environment and Planning. C, Government & Policy*, *32*(2), 301–319. doi:10.1068/c1329j

Henriques, J. C. (2009). Agricultura urbana e resiliência em Lisboa: O papel do governo municipal. *Urban Agriculture Magazine*, *22*, 82–90. Retrieved from http://www.agriculturaurbana.org.br/RAU/AU22/rau22_r_lisboa.pdf

Hope, D., Gries, C., Zhu, W., Fagan, W. F., Redman, C., Grimm, N. B., ... Kinzig, A. (2003). Socioeconomics Drive Urban Plant Diversity. *Proceedings of the National Academy of Sciences of the United States of America*, *100*(15), 8788–8792. doi:10.1073/pnas.1537557100 PMID:12847293

Instituto Brasileiro de Geografia e Estatística (IBGE). (2010). *Sinopse do Censo Demográfico 2010: Minas Gerais*. Retrieved June 13, 2017, from http://www.censo2010.ibge.gov.br/sinopse/index.php?uf=31

Kim, G. G. (2016). The Public Value of Urban Vacant Land: Social Responses and Ecological Value. *Sustainability*, *8*(486), 1–19. doi:10.3390u8050486

Kingsley, J., & Townsend, M. (2006). 'Dig In' to Social Capital: Community Gardens as Mechanisms for Growing Urban Social Connectedness. *Urban Policy and Research*, *24*(4), 525–537. doi:10.1080/08111140601035200

Koschke, L., Fürst, C., Frank, S., & Makeschin, F. (2012). A multi-criteria approach for an integrated land-cover-based assessment of ecosystem services provision to support landscape planning. *Ecological Indicators*, *21*, 54–66. doi:10.1016/j.ecolind.2011.12.010

Kremer, P., Hamstead, Z. A., & MacPhearson, T. (2013). A Social-ecological Assessment of Vacant Lots in New York City. *Landscape and Urban Planning*, *120*, 218–233. doi:10.1016/j.landurbplan.2013.05.003

Krylová, R., & Luz, A. L. (2017). Voids in the cities: Obstacles or Opportunities? The Practice of Urban Agriculture in Lisbon. *Espaços Vividos e Espaços Construídos: Estudos sobre a cidade*, *1*(4), 34-48.

Lima, F. M. (2012, September). *Depopulation and Public Spaces: The case study of Lisbon*. Paper presented at the meeting of the 3rd International Conference on Degrowth for Ecological Sustainability and Social Equity. Università IUAV di Venezia, Italy.

Luz, A., & Pires, I. (2015). Regularização das hortas urbanas na cidade de lisboa - as hortas sociais do Vale de Chelas. In I. C. da Silva, M. Pignatelli, & S. M. Viegas (Eds.), *Livro de Atas do 1º Congresso da Associação Internacional de Ciências Sociais e Humanas em Língua Portuguesa 2015* (pp. 6935–6952). Lisbon, Portugal: Associação Internacional de Ciências Sociais e Humanas em Língua Portuguesa. Retrieved from http://repositorio.ul.pt/handle/10451/20797

Madaleno, I. M., & Armijo, G. (2004). Agricultura urbana en metrópolis iberoamericanas: estudio de casos en Santiago de Chile y Lisboa, Portugal. *Investigaciones Geográficas (Mx), 54*, 36-54. Retrieved from http://www.redalyc.org/articulo.oa?id=56905404

Maes, J., Egoh, B., Willemen, L., Liquete, C., Vihervaara, P., Schagner, J. P., ... Bidoglio, G. (2012). Mapping ecosystem services for policy support and decision making in the European Union. *Ecosystem Services*, *1*(1), 31–39. doi:10.1016/j.ecoser.2012.06.004

Martín-López, B., Gómez-Baggethun, E., Garcia-Llorente, M., & Montes, C. (2014). Trade-offs across value-domains in ecosystem services assessment. *Ecological Indicators, 37*(Part A), 220-228. doi:10.1016/j.ecolind.2013.03.003

Martín-López, B., Iniesta-Arandia, I., Garcia-Llorente, M., Palomo, I., Casado-Arzuaga, I., Del Amo, D. G., ... Montes, C. (2012). Uncovering Ecosystem Service Bundles through Social Preferences. *PLoS One*, *7*(6), e38970. doi:10.1371/journal.pone.0038970 PMID:22720006

Martins, D. C. (2012). *Urban and Peri-Urban Agriculture in Lisbon Metropolitan Area - Contributions to Food Sovereignty* (Master's dissertation). University College London.

Matos, R. S., & Batista, D. S. (2013). Urban Agriculture: The Allotment Gardens as Structures of Urban Sustainability. In M. Ozyavuz (Ed.), *Advances in Landscape Architecture* (pp. 457–512). London: InTech. doi:10.5772/55892

McMichael, A., Scholes, R., Hefny, M., Pereira, E., Palm, C., & Foale, S. (2005). Linking ecosystem services and human well-being. In D. Capistrano, C. Samper, M. J. Lee, & C. Raudsepp-Hearne (Eds.), *Ecosystems and Human Well-Being: Multiscale Assessments: Findings of the Sub-Global Assessments Working Group* (pp. 43–60). Washington, DC: Island Press.

McPhearson, T., Kremer, P., & Hamstead, Z. A. (2013). Mapping ecosystem services in New York City: Applying a social-ecological approach in urban vacant land. *Ecosystem Services*, *5*, 11–26. doi:10.1016/j.ecoser.2013.06.005

Milbourne, P. (2012). Everyday (in)justices and ordinary environmentalisms: Community gardening in disadvantaged urban neighborhoods. *Local Environment*, *17*(9), 943–957. doi:10.1080/13549839.2011.607158

Millenium Ecosystem Assessment (MEA). (2005). *Ecosystems and Human Well-Being: Biodiversity Synthesis*. Washington, DC: World Resources Institute.

Mitlin, D., & Thompson, J. (1994). Editorial: Addressing the gaps or dispelling the myths?: Participatory approaches in low-income urban communities. *RRA Notes, 21*, 3-12. Retrieved from http://pubs.iied.org/pdfs/6090IIED.pdf

Mitlin, D., & Thompson, J. (1995). Participatory approaches in urban areas: Strengthening civil society or reinforcing the status quo? *Environment and Urbanization*, *7*(1), 231–250. doi:10.1177/095624789500700113

Montreal Urban Ecology Centre (MUEC). (2015). *Participatory urban planning - planning the city with and for its citizens*. Montreal, Canada: MUEC. Retrieved from http://www.ecologieurbaine.net/documents/boutique/LR-CEUM_brochure_urba_parti_ang.pdf

Municipal de Lisboa, C. (CML) (2016). *Mapa de parques horticolas de Lisboa*. Retrieved June, 2017, from http://www.cm-lisboa.pt/fileadmin/VIVER/Ambiente/Hortas_Urbanas/Parques_Horticolas_2016.pdf

Pagano, M. A., & Bowman, A. O. (2000). Vacant Land in Cities: An Urban Resource. *The Brookings Institution Survey Series, 1*, 1-9. Retrieved from https://www.brookings.edu/wp-content/uploads/2016/06/paganofinal.pdf

Pickett, S. T. A., Cadenasso, M. L., Grove, J. M., Groffman, P. M., Band, L. E., Boone, C. G., ... Wilson, M. A. (2008). Beyond Urban Legends: An Emerging Framework of Urban Ecology, as Illustrated by the Baltimore Ecosystem Study. *Bioscience*, *58*(2), 139–150. doi:10.1641/B580208

Primmer, E., Jokinen, P., Blicharska, M., Barton, D. N., Bugter, R., & Potschin, M. (2015). Governance of Ecosystem Services: A framework for empirical analysis. *Ecosystem Services*, *16*, 158–166. doi:10.1016/j.ecoser.2015.05.002

Rall, E., & Haase, D. (2011). Creative intervention in a dynamic city: A sustainability assessment of an interim use for brownfields in Leipzig, Germany. *Landscape and Urban Planning*, *100*(3), 189–201. doi:10.1016/j.landurbplan.2010.12.004

Ramos, A. R. (2011). *A Integração de Espaços de Cultivo Agrícola em Contextos Urbanos. Proposta de Intervenção para a Requalificação Urbana do Vale de Chelas, Lisboa* (Master's dissertation). Instituto Superior Técnico, University of Lisbon.

Rosenberg, M., Syrbe, R., Vowinckel, J., & Walz, U. (2014). Scenario Methodology for Modelling of Future Landscape Developments as Basis for Assessing Ecosystem Services. *Landscape Online*, *33*, 1–20. doi:10.3097/LO.201433

Roth, M., Frixen, M., Tobisch, C., & Scholle, T. (2015). Finding Spaces for Urban Food Production – Matching Spatial and Stakeholder Analysis with Urban Agriculture Approaches in the Urban Renewal Area of Dortmund-Hörde, Germany. *Journal on Food Agricultura y Sociedad*, *3*(1), 79–88.

Santos, M. (Ed.). (2015). Biodiversidade na Cidade de Lisboa: uma estratégia para 2020 | Documento técnico (3rd ed.). Lisbon, Portugal: Câmara Municipal de Lisboa.

Slocum, N. (2005). *Participatory methods toolkit. A practitioner's manual*. Brussels, Belgium: King Baudouin Foundation.

Souza, M. L. (2010). *Mudar a cidade: uma introdução crítica ao planejamento e à gestão urbanos*. Rio de Janeiro, Brazil: Bertrand Brasil.

Speak, A. F., Mizgajski, A., & Borysiak, J. (2015). Allotment gardens and parks: Provision of ecosystems services with an emphasis on biodiversity. *Urban Forestry & Urban Greening*, *14*(4), 772–781. doi:10.1016/j.ufug.2015.07.007

Street, P. (1997). Scenario workshops: A participatory approach to sustainable urban living? *Futures*, *29*(2), 139–158. doi:10.1016/S0016-3287(96)00073-0

Strohbach, M., Haase, D., & Kabisch, N. (2009). Birds and the City: Urban Biodiversity, Land Use, and Socioeconomics. *Ecology and Society*, *14*(2), 31. doi:10.5751/ES-03141-140231

Swan, C. M., Pickett, S. T. A., Szlavecz, K., Warren, P., & Willey, K. T. (2011). Biodiversity and Community Composition in Urban Ecosystems: Coupled Human, Spatial, and Metacommunity Processes. In J. Niemelä, J. H. Breuste, T. Elmqvist, G. Guntenspergen, P. James, & N. E. McIntyre (Eds.), Urban Ecology: Patterns, Processes, and Applications (pp. 179-186). Oxford, UK: Oxford University Press. doi:10.1093/acprof:oso/9780199563562.003.0021

Syrbe, R., Rosenberg, M., & Vowinckel, J. (2015). Ascertainment and Assessment of ES – Scenario - Development and Participative Methods. In K. Grunewald & O. Bastian (Eds.), *Ecosystem Services – Concept, Methods and Case Studies* (pp. 104–110). Berlin, Germany: Springer.

Telles, G. R. (1997). *Plano Verde de Lisboa: Componente do Plano Director Municipal de Lisboa*. Lisbon, Portugal: Edições Colibri.

Troy, A., & Wilson, M. A. (2006). Mapping ecosystem services: Practical challenges and opportunities in linking GIS and value transfer. *Ecological Economics*, *60*(2), 435–449. doi:10.1016/j.ecolecon.2006.04.007

Turnpenny, J., Russel, D., & Jordan, A. (2014). The Challenge of Embedding an Ecosystem Services Approach: Patterns of Knowledge Utilization in Public Policy Appraisal. *Environment and Planning. C, Government & Policy*, *32*(2), 247–262. doi:10.1068/c1317j

van Berkel, D. B., Carvalho-Ribeiro, S. M., Verburg, P. H., & Lovett, A. (2011). Identifying Assets and Constraints for Rural Development with Qualitative Research Tools: A Case Study of Castro Laboreiro, Portugal. *Landscape and Urban Planning*, *102*(2), 127–141. doi:10.1016/j.landurbplan.2011.03.016

Vemuri, A. W., Grove, J. M., Wilson, M. A., & Burch, W. R. Jr. (2011). A Tale of Two Scales: Evaluating the Relationship Among Life Satisfaction, Social Capital, Income, and the Natural Environment at Individual and Neighborhood Levels in Metropolitan Baltimore. *Environment and Behavior*, *43*(1), 3–25. doi:10.1177/0013916509338551

Villamor, G. B., Palomo, I., Santiago, C. A. L., Oteros-Rosa, E., & Hill, J. (2014). Assessing stakeholders' perceptions and values towards social-ecological systems using participatory methods. *Ecological Processes*, *3*(1), 1–22. doi:10.118613717-014-0022-9

ADDITIONAL READING

Deakin, M., Mitchell, G., Nijkamp, P., & Vreeker, R. (Eds.). (2007). Sustainable Urban Development: Vol. 2. *The Environmental Assessment Methods*. London, United Kingdom: Routledge.

Elmquist, T., Fragkias, M., Goodness, J., Guneralp, B., Marcotullio, B. P. J., McDonald, R. I., ... Wilkinson, C. (2013). *Urbanization, Biodiversity and Ecosystem Services: Challenges and Opportunities. A Global Assessment*. New York, NY/ London, United Kingdom: Springer; Retrieved from https://link.springer.com/content/pdf/10.1007%2F978-94-007-7088-1.pdf

Grunewald, K., & Bastian, O. (Eds.). (2015). *Ecosystem Services – Concept, Methods and Case Studies*. Berlin, Germany: Springer; doi:10.1007/978-3-662-44143-5

Krämer, R., Banzhaf, E., Cabral, I., Engelmann, R., Keim, J., & Bonn, A. (2016). *Mapping ecosystem services and tree diversity of urban gardens using a hybrid remote sensing approach*. Paper presented at the meeting of the GEO BON Open Science Conference, Leipzig, Germany. doi:10.13140/RG.2.1.5190.0408

Niemela, J., Breuste, J. H., Guntenspergen, G., McIntyre, N. E., Elmqvist, T., & James, P. (Eds.). (2012). *Urban Ecology: Patterns, Processes, and Applications*. Oxford, United Kingdom: Oxford University Press.

Opitz, I., Specht, K., Berges, R., Siebert, R., & Piorr, A. (2016). Toward Sustainability: Novelties, Areas of Learning and Innovation in Urban Agriculture. *Sustainability*, *8*(4), 356. doi:10.3390u8040356

KEY TERMS AND DEFINITIONS

Biophysical Attributes: Characteristics that combine biological and physical features such as vegetation, hydrography, relief, climate, among others.

Ecosystem: An assemblage composed of the interactions among biological components (such as plants, animals, microorganisms) and abiotic components (chemical and physical elements like water, air, soil, minerals).

Land Cover: The biophysical land type that comprehends the features that cover the earth´s surface.

Land Use: The function associated with the features that cover the earth´s surface, through its use by people.

Plenary Sessions: Meetings to be attended by all the members' parties and stakeholders defined, aiming at presentations and panel discussions.

Socioecology: The study of systems composed of the interactions between people and nature, with the human component seen as part of nature.

Urban Sustainability: An expected state of urban circumstances that continues in time, including: evaluations on the resource consumption practice, made at the expense of the loss of ecosystems; sound use of resources allowing a generational equity; conservation of the natural environment; low use of non-renewable resources; economic solidarity and diversity; community autonomy; people wellbeing; basic human needs fulfillment.

This research was previously published in the Handbook of Research on Methods and Tools for Assessing Cultural Landscape Adaptation edited by Isabel de Sousa Rosa, Joana Corte Lopes, Ricardo Ribeiro, and Ana Mendes, pages 183-220, copyright year 2018 by Engineering Science Reference (an imprint of IGI Global).

Related Readings

To continue IGI Global's long-standing tradition of advancing innovation through emerging research, please find below a compiled list of recommended IGI Global book chapters and journal articles in the areas of green cities, environmental management, and sustainable urban development. These related readings will provide additional information and guidance to further enrich your knowledge and assist you with your own research.

Ahuja, K., & Khosla, A. (2019). A Framework to Develop a Zero-Carbon Emission Sustainable Cognitive City. In K. Ahuja & A. Khosla (Eds.), *Driving the Development, Management, and Sustainability of Cognitive Cities* (pp. 1–26). Hershey, PA: IGI Global. doi:10.4018/978-1-5225-8085-0.ch001

Akkucuk, U. (2016). SCOR Model and the Green Supply Chain. In U. Akkucuk (Ed.), *Handbook of Research on Waste Management Techniques for Sustainability* (pp. 108–124). Hershey, PA: IGI Global. doi:10.4018/978-1-4666-9723-2.ch006

Albuquerque, C. M. (2017). Cities Really Smart and Inclusive: Possibilities and Limits for Social Inclusion and Participation. In L. Carvalho (Ed.), *Handbook of Research on Entrepreneurial Development and Innovation Within Smart Cities* (pp. 229–247). Hershey, PA: IGI Global. doi:10.4018/978-1-5225-1978-2.ch011

Angelucci, F., Di Girolamo, C., & Zazzero, E. (2018). New Designing Codes for Urban Infrastructures: A Hypothesis of a Transdisciplinary Approach. In G. Carlone, N. Martinelli, & F. Rotondo (Eds.), *Designing Grid Cities for Optimized Urban Development and Planning* (pp. 209–237). Hershey, PA: IGI Global. doi:10.4018/978-1-5225-3613-0.ch012

Avdimiotis, S., & Tilikidou, I. (2017). Smart Tourism Development: The Case of Halkidiki. In L. Carvalho (Ed.), *Handbook of Research on Entrepreneurial Development and Innovation Within Smart Cities* (pp. 491–513). Hershey, PA: IGI Global. doi:10.4018/978-1-5225-1978-2.ch021

Ayvaz, B., & Görener, A. (2016). Reverse Logistics in the Electronics Waste Industry. In U. Akkucuk (Ed.), *Handbook of Research on Waste Management Techniques for Sustainability* (pp. 155–171). Hershey, PA: IGI Global. doi:10.4018/978-1-4666-9723-2.ch008

Bagnato, V. P. (2018). Technology and Urban Structure: The Grid City Between Technological Innovation and New Public Space System. In G. Carlone, N. Martinelli, & F. Rotondo (Eds.), *Designing Grid Cities for Optimized Urban Development and Planning* (pp. 238–253). Hershey, PA: IGI Global. doi:10.4018/978-1-5225-3613-0.ch013

Bala, I., & Singh, G. (2019). Green Communication for Cognitive Cities. In K. Ahuja & A. Khosla (Eds.), *Driving the Development, Management, and Sustainability of Cognitive Cities* (pp. 87–110). Hershey, PA: IGI Global. doi:10.4018/978-1-5225-8085-0.ch004

Bernardino, S., & Santos, J. F. (2017). Building Smarter Cities through Social Entrepreneurship. In L. Carvalho (Ed.), *Handbook of Research on Entrepreneurial Development and Innovation Within Smart Cities* (pp. 327–362). Hershey, PA: IGI Global. doi:10.4018/978-1-5225-1978-2.ch015

Bernardo, M. D. (2017). Smart City Governance: From E-Government to Smart Governance. In L. Carvalho (Ed.), *Handbook of Research on Entrepreneurial Development and Innovation Within Smart Cities* (pp. 290–326). Hershey, PA: IGI Global. doi:10.4018/978-1-5225-1978-2.ch014

Bhatia, P., & Singh, P. (2019). Technological and Gamified Solutions for Pollution Control in Cognitive Cities. In K. Ahuja & A. Khosla (Eds.), *Driving the Development, Management, and Sustainability of Cognitive Cities* (pp. 234–249). Hershey, PA: IGI Global. doi:10.4018/978-1-5225-8085-0.ch010

Bhatt, R. (2017). Zero Tillage for Mitigating Global Warming Consequences and Improving Livelihoods in South Asia. In W. Ganpat & W. Isaac (Eds.), *Environmental Sustainability and Climate Change Adaptation Strategies* (pp. 126–161). Hershey, PA: IGI Global. doi:10.4018/978-1-5225-1607-1.ch005

Bolívar, M. P. (2018). Governance in Smart Cities: A Comparison of Practitioners' Perceptions and Prior Research. *International Journal of E-Planning Research, 7*(2), 1–19. doi:10.4018/IJEPR.2018040101

Bostanci, S. H., & Albayrak, A. N. (2017). The Role of Eco-Municipalities in Climate Change for a Sustainable Future. In W. Ganpat & W. Isaac (Eds.), *Environmental Sustainability and Climate Change Adaptation Strategies* (pp. 213–231). Hershey, PA: IGI Global. doi:10.4018/978-1-5225-1607-1.ch008

Brkljačić, T., Majetić, F., & Tarabić, B. N. (2017). Smart Environment: Cyber Parks (Connecting Nature and Technology). In L. Carvalho (Ed.), *Handbook of Research on Entrepreneurial Development and Innovation Within Smart Cities* (pp. 150–172). Hershey, PA: IGI Global. doi:10.4018/978-1-5225-1978-2.ch008

Cai, Y. Z., Wu, F., Li, J., Wang, J., & Huang, M. (2019). Application of UAV Technology to Planning Study on Chinese Villages in Guanzhong. In J. Vargas-Hernández & J. Zdunek-Wielgołaska (Eds.), *Bioeconomical Solutions and Investments in Sustainable City Development* (pp. 180–195). Hershey, PA: IGI Global. doi:10.4018/978-1-5225-7958-8.ch008

Calisto, M. D., & Gonçalves, A. (2017). Smart Citizens, Wise Decisions: Sustainability-Driven Tourism Entrepreneurs. In L. Carvalho (Ed.), *Handbook of Research on Entrepreneurial Development and Innovation Within Smart Cities* (pp. 20–43). Hershey, PA: IGI Global. doi:10.4018/978-1-5225-1978-2.ch002

Capelo, C. (2017). Exploring the Dynamics of an Energy Service Venture. In L. Carvalho (Ed.), *Handbook of Research on Entrepreneurial Development and Innovation Within Smart Cities* (pp. 269–289). Hershey, PA: IGI Global. doi:10.4018/978-1-5225-1978-2.ch013

Caridi, G. (2018). Calabria 1783: The Orthogonal Grid as a Physical and Ideological Device of Reconstruction. In G. Carlone, N. Martinelli, & F. Rotondo (Eds.), *Designing Grid Cities for Optimized Urban Development and Planning* (pp. 176–187). Hershey, PA: IGI Global. doi:10.4018/978-1-5225-3613-0.ch010

Carlone, G. (2018). Cities and Extension Plans in the Kingdom of the Two Sicilies: Borgo Murattiano of Bari (1812-1859). In G. Carlone, N. Martinelli, & F. Rotondo (Eds.), *Designing Grid Cities for Optimized Urban Development and Planning* (pp. 1–18). Hershey, PA: IGI Global. doi:10.4018/978-1-5225-3613-0.ch001

Carvalho, L. C. (2017). Entrepreneurial Ecosystems: Lisbon as a Smart Start-Up City. In L. Carvalho (Ed.), *Handbook of Research on Entrepreneurial Development and Innovation Within Smart Cities* (pp. 1–19). Hershey, PA: IGI Global. doi:10.4018/978-1-5225-1978-2.ch001

Castagnolo, V. (2018). Analyzing, Classifying, Safeguarding: Drawing for the Borgo Murattiano Neighbourhood of Bari. In G. Carlone, N. Martinelli, & F. Rotondo (Eds.), *Designing Grid Cities for Optimized Urban Development and Planning* (pp. 93–108). Hershey, PA: IGI Global. doi:10.4018/978-1-5225-3613-0.ch006

Castilla-Polo, F., Gallardo-Vázquez, D., Sánchez-Hernández, M. I., & Ruiz-Rodríguez, M. D. (2017). Cooperatives as Responsible and Innovative Entrepreneurial Ecosystems in Smart Territories: The Olive Oil Industry in the South of Spain. In L. Carvalho (Ed.), *Handbook of Research on Entrepreneurial Development and Innovation Within Smart Cities* (pp. 459–490). Hershey, PA: IGI Global. doi:10.4018/978-1-5225-1978-2.ch020

Charalabidis, Y., & Theocharopoulou, C. (2019). A Participative Method for Prioritizing Smart City Interventions in Medium-Sized Municipalities. *International Journal of Public Administration in the Digital Age*, *6*(1), 41–63. doi:10.4018/IJPADA.2019010103

Chatsiwa, J., Mujere, N., & Maiyana, A. B. (2016). Municipal Solid Waste Management. In U. Akkucuk (Ed.), *Handbook of Research on Waste Management Techniques for Sustainability* (pp. 19–43). Hershey, PA: IGI Global. doi:10.4018/978-1-4666-9723-2.ch002

Chawla, R., Singhal, P., & Garg, A. K. (2019). Impact of Dust for Solar PV in Indian Scenario: Experimental Analysis. In K. Ahuja & A. Khosla (Eds.), *Driving the Development, Management, and Sustainability of Cognitive Cities* (pp. 111–138). Hershey, PA: IGI Global. doi:10.4018/978-1-5225-8085-0.ch005

Cılız, N., Yıldırım, H., & Temizel, Ş. (2016). Structure Development for Effective Medical Waste and Hazardous Waste Management System. In U. Akkucuk (Ed.), *Handbook of Research on Waste Management Techniques for Sustainability* (pp. 303–327). Hershey, PA: IGI Global. doi:10.4018/978-1-4666-9723-2.ch016

Collignon de Alba, C., Haberleithner, J., & López, M. M. (2017). Creative Industries in the Smart City: Overview of a Liability in Emerging Economies. In L. Carvalho (Ed.), *Handbook of Research on Entrepreneurial Development and Innovation Within Smart Cities* (pp. 107–126). Hershey, PA: IGI Global. doi:10.4018/978-1-5225-1978-2.ch006

Corum, A. (2016). Remanufacturing, an Added Value Product Recovery Strategy. In U. Akkucuk (Ed.), *Handbook of Research on Waste Management Techniques for Sustainability* (pp. 347–367). Hershey, PA: IGI Global. doi:10.4018/978-1-4666-9723-2.ch018

Crosas, C. (2018). Latin American Cities: Modern Grids From 1850s. In G. Carlone, N. Martinelli, & F. Rotondo (Eds.), *Designing Grid Cities for Optimized Urban Development and Planning* (pp. 39–51). Hershey, PA: IGI Global. doi:10.4018/978-1-5225-3613-0.ch003

D'Onofrio, S., Habenstein, A., & Portmann, E. (2019). Ontological Design for Cognitive Cities: The New Principle for Future Urban Management. In K. Ahuja & A. Khosla (Eds.), *Driving the Development, Management, and Sustainability of Cognitive Cities* (pp. 183–211). Hershey, PA: IGI Global. doi:10.4018/978-1-5225-8085-0.ch008

Dalzero, S. (2018). The Time of the Finished World Has Begun: A New Map of the World – National Borders Partially or Fully Fenced-Off. In G. Carlone, N. Martinelli, & F. Rotondo (Eds.), *Designing Grid Cities for Optimized Urban Development and Planning* (pp. 254–275). Hershey, PA: IGI Global. doi:10.4018/978-1-5225-3613-0.ch014

Damurski, L. (2016). Recent Progress in Online Communication Tools for Urban Planning: A Comparative Study of Polish and German Municipalities. *International Journal of E-Planning Research*, *5*(1), 39–54. doi:10.4018/IJEPR.2016010103

Damurski, L. (2016). Smart City, Integrated Planning, and Multilevel Governance: A Conceptual Framework for e-Planning in Europe. *International Journal of E-Planning Research*, *5*(4), 41–53. doi:10.4018/IJEPR.2016100103

Das, K. K., & Sharma, N. K. (2016). Post Disaster Housing Management for Sustainable Urban Development: A Review. *International Journal of Geotechnical Earthquake Engineering*, *7*(1), 1–18. doi:10.4018/IJGEE.2016010101

Das, S., & Nayyar, A. (2019). Innovative Ideas to Manage Urban Traffic Congestion in Cognitive Cities. In K. Ahuja & A. Khosla (Eds.), *Driving the Development, Management, and Sustainability of Cognitive Cities* (pp. 139–162). Hershey, PA: IGI Global. doi:10.4018/978-1-5225-8085-0.ch006

Deakin, M., & Reid, A. (2017). The Embedded Intelligence of Smart Cities: Urban Life, Citizenship, and Community. *International Journal of Public Administration in the Digital Age*, *4*(4), 62–74. doi:10.4018/IJPADA.2017100105

Djukic, A., Stupar, A., & Antonic, B. M. (2018). The Orthogonal Urban Matrix of the Towns in Vojvodina, Northern Serbia: Genesis and Transformation. In G. Carlone, N. Martinelli, & F. Rotondo (Eds.), *Designing Grid Cities for Optimized Urban Development and Planning* (pp. 128–156). Hershey, PA: IGI Global. doi:10.4018/978-1-5225-3613-0.ch008

Dolunay, O. (2016). A Paradigm Shift: Empowering Farmers to Eliminate the Waste in the Form of Fresh Water and Energy through the Implementation of 4R+T. In U. Akkucuk (Ed.), *Handbook of Research on Waste Management Techniques for Sustainability* (pp. 368–379). Hershey, PA: IGI Global. doi:10.4018/978-1-4666-9723-2.ch019

Dowling, C. M., Walsh, S. D., Purcell, S. M., Hynes, W. M., & Rhodes, M. L. (2017). Operationalising Sustainability within Smart Cities: Towards an Online Sustainability Indicator Tool. *International Journal of E-Planning Research, 6*(4), 1–17. doi:10.4018/IJEPR.2017100101

Ekman, U. (2018). Smart City Planning: Complexity. *International Journal of E-Planning Research, 7*(3), 1–21. doi:10.4018/IJEPR.2018070101

Erdogan, S. (2016). The Effect of Working Capital Management on Firm's Profitability: Evidence from Istanbul Stock Exchange. In U. Akkucuk (Ed.), *Handbook of Research on Waste Management Techniques for Sustainability* (pp. 244–261). Hershey, PA: IGI Global. doi:10.4018/978-1-4666-9723-2.ch013

Eudoxie, G., & Roopnarine, R. (2017). Climate Change Adaptation and Disaster Risk Management in the Caribbean. In W. Ganpat & W. Isaac (Eds.), *Environmental Sustainability and Climate Change Adaptation Strategies* (pp. 97–125). Hershey, PA: IGI Global. doi:10.4018/978-1-5225-1607-1.ch004

Fasolino, I. (2018). Rules for a New Town After a Disaster: The Gridded Schemes in the Plans. In G. Carlone, N. Martinelli, & F. Rotondo (Eds.), *Designing Grid Cities for Optimized Urban Development and Planning* (pp. 157–175). Hershey, PA: IGI Global. doi:10.4018/978-1-5225-3613-0.ch009

Fernandes, C. M., & Dias de Sousa, I. (2017). Digital Swarms: Social Interaction and Emergent Phenomena in Personal Communications Networks. In L. Carvalho (Ed.), *Handbook of Research on Entrepreneurial Development and Innovation Within Smart Cities* (pp. 44–59). Hershey, PA: IGI Global. doi:10.4018/978-1-5225-1978-2.ch003

Fernandes, V., Moreira, A., & Daniel, A. I. (2017). A Qualitative Analysis of Social Entrepreneurship Involving Social Innovation and Intervention. In L. Carvalho (Ed.), *Handbook of Research on Entrepreneurial Development and Innovation Within Smart Cities* (pp. 417–438). Hershey, PA: IGI Global. doi:10.4018/978-1-5225-1978-2.ch018

Gabriel, B. F., Valente, R. A., Dias-de-Oliveira, J., Neto, V. F., & Andrade-Campos, A. (2017). Methodologies for Engineering Learning and Teaching (MELT): An Overview of Engineering Education in Europe and a Novel Concept for Young Students. In L. Carvalho (Ed.), *Handbook of Research on Entrepreneurial Development and Innovation Within Smart Cities* (pp. 363–391). Hershey, PA: IGI Global. doi:10.4018/978-1-5225-1978-2.ch016

Garrick, T. A., & Liburd, O. E. (2017). Impact of Climate Change on a Key Agricultural Pest: Thrips. In W. Ganpat & W. Isaac (Eds.), *Environmental Sustainability and Climate Change Adaptation Strategies* (pp. 232–254). Hershey, PA: IGI Global. doi:10.4018/978-1-5225-1607-1.ch009

Gencer, Y. G. (2016). Mystery of Recycling: Glass and Aluminum Examples. In U. Akkucuk (Ed.), *Handbook of Research on Waste Management Techniques for Sustainability* (pp. 172–191). Hershey, PA: IGI Global. doi:10.4018/978-1-4666-9723-2.ch009

Gencer, Y. G., & Akkucuk, U. (2016). Reverse Logistics: Automobile Recalls and Other Conditions. In U. Akkucuk (Ed.), *Handbook of Research on Waste Management Techniques for Sustainability* (pp. 125–154). Hershey, PA: IGI Global. doi:10.4018/978-1-4666-9723-2.ch007

Gonçalves, J. M., Martins, T. G., & Vilhena da Cunha, I. B. (2017). Local Creative Ecosystems as a Strategy for the Development of Low-Density Urban Spaces. In L. Carvalho (Ed.), *Handbook of Research on Entrepreneurial Development and Innovation Within Smart Cities* (pp. 127–149). Hershey, PA: IGI Global. doi:10.4018/978-1-5225-1978-2.ch007

Goundar, S., & Appana, S. (2017). Mainstreaming Development Policies for Climate Change in Fiji: A Policy Gap Analysis and the Role of ICTs. In W. Ganpat & W. Isaac (Eds.), *Environmental Sustainability and Climate Change Adaptation Strategies* (pp. 1–31). Hershey, PA: IGI Global. doi:10.4018/978-1-5225-1607-1.ch001

Grochulska-Salak, M. (2019). Urban Farming in Sustainable City Development. In J. Vargas-Hernández & J. Zdunek-Wielgołaska (Eds.), *Bioeconomical Solutions and Investments in Sustainable City Development* (pp. 43–64). Hershey, PA: IGI Global. doi:10.4018/978-1-5225-7958-8.ch003

Gruszecka, K. (2019). Ecological Centre of Warsaw as a Development Path. In J. Vargas-Hernández & J. Zdunek-Wielgołaska (Eds.), *Bioeconomical Solutions and Investments in Sustainable City Development* (pp. 117–150). Hershey, PA: IGI Global. doi:10.4018/978-1-5225-7958-8.ch006

Hesapci-Sanaktekin, O., & Aslanbay, Y. (2016). The Networked Self: Collectivism Redefined in Civic Engagements through Social Media Causes. In U. Akkucuk (Ed.), *Handbook of Research on Waste Management Techniques for Sustainability* (pp. 262–276). Hershey, PA: IGI Global. doi:10.4018/978-1-4666-9723-2.ch014

Ikeda, M. (2019). Developing a Sustainable Eco-City in Pre-Olympic Tokyo: Potential of New Methods and Their Limits in an Urban Era. In J. Vargas-Hernández & J. Zdunek-Wielgołaska (Eds.), *Bioeconomical Solutions and Investments in Sustainable City Development* (pp. 196–223). Hershey, PA: IGI Global. doi:10.4018/978-1-5225-7958-8.ch009

İşcan, E. (2019). Strategies of Sustainable Bioeconomy in the Industry 4.0 Framework for Inclusive and Social Prosperity. In J. Vargas-Hernández & J. Zdunek-Wielgołaska (Eds.), *Bioeconomical Solutions and Investments in Sustainable City Development* (pp. 21–42). Hershey, PA: IGI Global. doi:10.4018/978-1-5225-7958-8.ch002

J., J., Samui, P., & Dixon, B. (2016). Determination of Rate of Medical Waste Generation Using RVM, MARS and MPMR. In U. Akkucuk (Ed.), Handbook of Research on Waste Management Techniques for Sustainability (pp. 1-18). Hershey, PA: IGI Global. doi:10.4018/978-1-4666-9723-2.ch001

Jiménez, R., Rodríguez, P. L., & Fernández, R. R. (2019). Green Spaces of the Metropolitan Area of Guadalajara. In J. Vargas-Hernández & J. Zdunek-Wielgołaska (Eds.), *Bioeconomical Solutions and Investments in Sustainable City Development* (pp. 151–179). Hershey, PA: IGI Global. doi:10.4018/978-1-5225-7958-8.ch007

Kais, S. M. (2017). Climate Change: Vulnerability and Resilience in Commercial Shrimp Aquaculture in Bangladesh. In W. Ganpat & W. Isaac (Eds.), *Environmental Sustainability and Climate Change Adaptation Strategies* (pp. 162–187). Hershey, PA: IGI Global. doi:10.4018/978-1-5225-1607-1.ch006

Khan, B. (2019). Bio-Economy: Visions, Strategies, and Policies. In J. Vargas-Hernández & J. Zdunek-Wielgołaska (Eds.), *Bioeconomical Solutions and Investments in Sustainable City Development* (pp. 1–20). Hershey, PA: IGI Global. doi:10.4018/978-1-5225-7958-8.ch001

Khanna, B. K. (2017). Indian National Strategy for Climate Change Adaptation and Mitigation. In W. Ganpat & W. Isaac (Eds.), *Environmental Sustainability and Climate Change Adaptation Strategies* (pp. 32–63). Hershey, PA: IGI Global. doi:10.4018/978-1-5225-1607-1.ch002

Khanna, B. K. (2017). Vulnerability of the Lakshadweep Coral Islands in India and Strategies for Mitigating Climate Change Impacts. In W. Ganpat & W. Isaac (Eds.), *Environmental Sustainability and Climate Change Adaptation Strategies* (pp. 64–96). Hershey, PA: IGI Global. doi:10.4018/978-1-5225-1607-1.ch003

Kocasoy, G. (2016). Economic Instruments for Sustainable Environmental Management. In U. Akkucuk (Ed.), *Handbook of Research on Waste Management Techniques for Sustainability* (pp. 192–211). Hershey, PA: IGI Global. doi:10.4018/978-1-4666-9723-2.ch010

Lallo, C. H., Smalling, S., Facey, A., & Hughes, M. (2017). The Impact of Climate Change on Small Ruminant Performance in Caribbean Communities. In W. Ganpat & W. Isaac (Eds.), *Environmental Sustainability and Climate Change Adaptation Strategies* (pp. 296–321). Hershey, PA: IGI Global. doi:10.4018/978-1-5225-1607-1.ch011

Li, Z., Wang, Y., & Chen, Q. (2017). Real-Time Monitoring of Intercity Passenger Flows Based on Big Data: A Decision Support Tool for Urban Sustainability. *International Journal of Strategic Decision Sciences*, *8*(4), 120–128. doi:10.4018/IJSDS.2017100106

Loi, N. K., Huyen, N. T., Tu, L. H., Tram, V. N., Liem, N. D., Dat, N. L., ... Minh, D. N. (2017). Sustainable Land Use and Watershed Management in Response to Climate Change Impacts: Case Study in Srepok Watershed, Central Highland of Vietnam. In W. Ganpat & W. Isaac (Eds.), *Environmental Sustainability and Climate Change Adaptation Strategies* (pp. 255–295). Hershey, PA: IGI Global. doi:10.4018/978-1-5225-1607-1.ch010

López-Arranz, M. A. (2017). The Role Corporate Social Responsibility Has in the Smart City Project in Spain. In L. Carvalho (Ed.), *Handbook of Research on Entrepreneurial Development and Innovation Within Smart Cities* (pp. 439–458). Hershey, PA: IGI Global. doi:10.4018/978-1-5225-1978-2.ch019

Lucas, M. R., Rego, C., Vieira, C., & Vieira, I. (2017). Proximity and Cooperation for Innovative Regional Development: The Case of the Science and Technology Park of Alentejo. In L. Carvalho (Ed.), *Handbook of Research on Entrepreneurial Development and Innovation Within Smart Cities* (pp. 199–228). Hershey, PA: IGI Global. doi:10.4018/978-1-5225-1978-2.ch010

Magrinho, A., Neves, J., & Silva, J. R. (2017). The Triple Helix Model: Evidence in the Internationalization of the Health Industry. In L. Carvalho (Ed.), *Handbook of Research on Entrepreneurial Development and Innovation Within Smart Cities* (pp. 60–79). Hershey, PA: IGI Global. doi:10.4018/978-1-5225-1978-2.ch004

Maiorano, A. C. (2018). Urban Fronts in Murattiano Neighbourhood of Bari: A Selective Survey of the Built Environment. In G. Carlone, N. Martinelli, & F. Rotondo (Eds.), *Designing Grid Cities for Optimized Urban Development and Planning* (pp. 78–92). Hershey, PA: IGI Global. doi:10.4018/978-1-5225-3613-0.ch005

Martinelli, N., & Mangialardi, G. (2018). Cities With Grid Layout: Ubiquitousness and Flexibility of an Urban Model. In G. Carlone, N. Martinelli, & F. Rotondo (Eds.), *Designing Grid Cities for Optimized Urban Development and Planning* (pp. 188–208). Hershey, PA: IGI Global. doi:10.4018/978-1-5225-3613-0.ch011

Mizutani, S., Liao, K., & Sasaki, T. G. (2019). Forest-River-Ocean Nexus-Based Education for Community Development: Aiming at Resilient Sustainable Society. In J. Vargas-Hernández & J. Zdunek-Wielgołaska (Eds.), *Bioeconomical Solutions and Investments in Sustainable City Development* (pp. 224–248). Hershey, PA: IGI Global. doi:10.4018/978-1-5225-7958-8.ch010

Moreira, A., & Ferreira, M. A. (2017). Strategic Challenges of the Portuguese Molds Industry: A Sectoral Innovation Perspective. In L. Carvalho (Ed.), *Handbook of Research on Entrepreneurial Development and Innovation Within Smart Cities* (pp. 534–560). Hershey, PA: IGI Global. doi:10.4018/978-1-5225-1978-2.ch023

Mujere, N., & Moyce, W. (2017). Climate Change Impacts on Surface Water Quality. In W. Ganpat & W. Isaac (Eds.), *Environmental Sustainability and Climate Change Adaptation Strategies* (pp. 322–340). Hershey, PA: IGI Global. doi:10.4018/978-1-5225-1607-1.ch012

Mundula, L., & Auci, S. (2017). Smartness, City Efficiency, and Entrepreneurship Milieu. In L. Carvalho (Ed.), *Handbook of Research on Entrepreneurial Development and Innovation Within Smart Cities* (pp. 173–198). Hershey, PA: IGI Global. doi:10.4018/978-1-5225-1978-2.ch009

Nath, R. (2019). Parametric Evaluation of Beam Deflection on Piezoelectric Material Using Implicit and Explicit Method Simulations: A Study in Energy Engineering. In J. Vargas-Hernández & J. Zdunek-Wielgołaska (Eds.), *Bioeconomical Solutions and Investments in Sustainable City Development* (pp. 65–87). Hershey, PA: IGI Global. doi:10.4018/978-1-5225-7958-8.ch004

Nayyar, A., Jain, R., Mahapatra, B., & Singh, A. (2019). Cyber Security Challenges for Smart Cities. In K. Ahuja & A. Khosla (Eds.), *Driving the Development, Management, and Sustainability of Cognitive Cities* (pp. 27–54). Hershey, PA: IGI Global. doi:10.4018/978-1-5225-8085-0.ch002

Odabasi, A., & Tiryaki, C. S. (2016). An Empirical Review of Long Term Electricity Demand Forecasts for Turkey. In U. Akkucuk (Ed.), *Handbook of Research on Waste Management Techniques for Sustainability* (pp. 227–243). Hershey, PA: IGI Global. doi:10.4018/978-1-4666-9723-2.ch012

Okay, E. (2016). Towards Smart Cities in Turkey?: Transitioning from Waste to Creative, Clean and Cheap Eco-Energy. In U. Akkucuk (Ed.), *Handbook of Research on Waste Management Techniques for Sustainability* (pp. 277–302). Hershey, PA: IGI Global. doi:10.4018/978-1-4666-9723-2.ch015

Pawlikowska-Piechotka, A., Łukasik, N., Ostrowska-Tryzno, A., & Sawicka, K. (2017). A Smart City Initiative: Urban Greens and Evaluation Method of the Sport and Recreation Potentials (SEM). In L. Carvalho (Ed.), *Handbook of Research on Entrepreneurial Development and Innovation Within Smart Cities* (pp. 561–583). Hershey, PA: IGI Global. doi:10.4018/978-1-5225-1978-2.ch024

Raisinghani, M. S., & Idemudia, E. C. (2016). Green Information Systems for Sustainability. In U. Akkucuk (Ed.), *Handbook of Research on Waste Management Techniques for Sustainability* (pp. 212–226). Hershey, PA: IGI Global. doi:10.4018/978-1-4666-9723-2.ch011

Rathee, D. S., Ahuja, K., & Hailu, T. (2019). Role of Electronics Devices for E-Health in Smart Cities. In K. Ahuja & A. Khosla (Eds.), *Driving the Development, Management, and Sustainability of Cognitive Cities* (pp. 212–233). Hershey, PA: IGI Global. doi:10.4018/978-1-5225-8085-0.ch009

Rotondo, F. (2018). The Grid Cities: Between Tradition and Innovation. In G. Carlone, N. Martinelli, & F. Rotondo (Eds.), *Designing Grid Cities for Optimized Urban Development and Planning* (pp. 109–127). Hershey, PA: IGI Global. doi:10.4018/978-1-5225-3613-0.ch007

Sánchez-Fernández, M. D., & Cardona, J. R. (2017). The Perception of the Effect of Tourism on the Local Community before the Ibiza Smart Island Project. In L. Carvalho (Ed.), *Handbook of Research on Entrepreneurial Development and Innovation Within Smart Cities* (pp. 392–416). Hershey, PA: IGI Global. doi:10.4018/978-1-5225-1978-2.ch017

Santos, B. (2017). Improving Urban Planning Information, Transparency and Participation in Public Administrations. *International Journal of E-Planning Research*, *6*(4), 58–75. doi:10.4018/IJEPR.2017100104

Sardà, J. F. (2018). Cerdà/Barcelona/Eixample: 1855-2017 … A Work in Progress. In G. Carlone, N. Martinelli, & F. Rotondo (Eds.), *Designing Grid Cities for Optimized Urban Development and Planning* (pp. 19–38). Hershey, PA: IGI Global. doi:10.4018/978-1-5225-3613-0.ch002

Selvi, M. S. (2016). Physical Distribution Problems of Textile Companies in Turkey. In U. Akkucuk (Ed.), *Handbook of Research on Waste Management Techniques for Sustainability* (pp. 328–346). Hershey, PA: IGI Global. doi:10.4018/978-1-4666-9723-2.ch017

Signorile, N. (2018). Pride and Prejudice: The Murattiano-Modernism. In G. Carlone, N. Martinelli, & F. Rotondo (Eds.), *Designing Grid Cities for Optimized Urban Development and Planning* (pp. 52–77). Hershey, PA: IGI Global. doi:10.4018/978-1-5225-3613-0.ch004

Singh, G., Kapoor, R., & Khosla, A. K. (2019). Intelligent Anomaly Detection Video Surveillance Systems for Smart Cities. In K. Ahuja & A. Khosla (Eds.), *Driving the Development, Management, and Sustainability of Cognitive Cities* (pp. 163–182). Hershey, PA: IGI Global. doi:10.4018/978-1-5225-8085-0.ch007

Singh-Ackbarali, D., & Maharaj, R. (2017). Mini Livestock Ranching: Solution to Reducing the Carbon Footprint and Negative Environmental Impacts of Agriculture. In W. Ganpat & W. Isaac (Eds.), *Environmental Sustainability and Climate Change Adaptation Strategies* (pp. 188–212). Hershey, PA: IGI Global. doi:10.4018/978-1-5225-1607-1.ch007

Sood, N., Saini, I., Awasthi, T., Saini, M. K., Bhoriwal, P., & Kaur, T. (2019). Fog Removal Algorithms for Real-Time Video Footage in Smart Cities for Safe Driving. In K. Ahuja & A. Khosla (Eds.), *Driving the Development, Management, and Sustainability of Cognitive Cities* (pp. 55–86). Hershey, PA: IGI Global. doi:10.4018/978-1-5225-8085-0.ch003

Stone, R. J. (2017). Modelling the Frequency of Tropical Cyclones in the Lower Caribbean Region. In W. Ganpat & W. Isaac (Eds.), *Environmental Sustainability and Climate Change Adaptation Strategies* (pp. 341–349). Hershey, PA: IGI Global. doi:10.4018/978-1-5225-1607-1.ch013

Stratigea, A., Leka, A., & Panagiotopoulou, M. (2017). In Search of Indicators for Assessing Smart and Sustainable Cities and Communities' Performance. *International Journal of E-Planning Research*, *6*(1), 43–73. doi:10.4018/IJEPR.2017010103

Touq, A. B., & Ijeh, A. (2018). Information Security and Ecosystems in Smart Cities: The Case of Dubai. *International Journal of Information Systems and Social Change*, *9*(2), 28–43. doi:10.4018/IJISSC.2018040103

Üç, M., & Elitaş, C. (2016). Life Cycle Costing for Sustainability. In U. Akkucuk (Ed.), *Handbook of Research on Waste Management Techniques for Sustainability* (pp. 96–107). Hershey, PA: IGI Global. doi:10.4018/978-1-4666-9723-2.ch005

Ulker-Demirel, E., & Demirel, E. (2016). Green Marketing and Stakeholder Perceptions. In U. Akkucuk (Ed.), *Handbook of Research on Waste Management Techniques for Sustainability* (pp. 75–95). Hershey, PA: IGI Global. doi:10.4018/978-1-4666-9723-2.ch004

Vargas-Hernández, J. G., López, J. J., & Zdunek-Wielgołaska, J. A. (2019). Entrepreneurial and Institutional Analysis of Biodiesel Companies in Mexico. In J. Vargas-Hernández & J. Zdunek-Wielgołaska (Eds.), *Bioeconomical Solutions and Investments in Sustainable City Development* (pp. 89–115). Hershey, PA: IGI Global. doi:10.4018/978-1-5225-7958-8.ch005

Vázquez, D. G., & Gil, M. T. (2017). Sustainability in Smart Cities: The Case of Vitoria-Gasteiz (Spain) – A Commitment to a New Urban Paradigm. In L. Carvalho (Ed.), *Handbook of Research on Entrepreneurial Development and Innovation Within Smart Cities* (pp. 248–268). Hershey, PA: IGI Global. doi:10.4018/978-1-5225-1978-2.ch012

Velozo, R. A., & Montanha, G. K. (2017). Evaluation of a Mobile Software Development Company. In L. Carvalho (Ed.), *Handbook of Research on Entrepreneurial Development and Innovation Within Smart Cities* (pp. 514–533). Hershey, PA: IGI Global. doi:10.4018/978-1-5225-1978-2.ch022

Index

Purchase Print, E-Book, or Print + E-Book

IGI Global's reference books can now be purchased from three unique pricing formats: Print Only, E-Book Only, or Print + E-Book.
Shipping fees may apply.

www.igi-global.com

Recommended Reference Books

ISBN: 978-1-5225-7727-0
© 2019; 335 pp.
List Price: $195

ISBN: 978-1-5225-5577-3
© 2019; 291 pp.
List Price: $195

ISBN: 978-1-5225-5291-8
© 2019; 252 pp.
List Price: $175

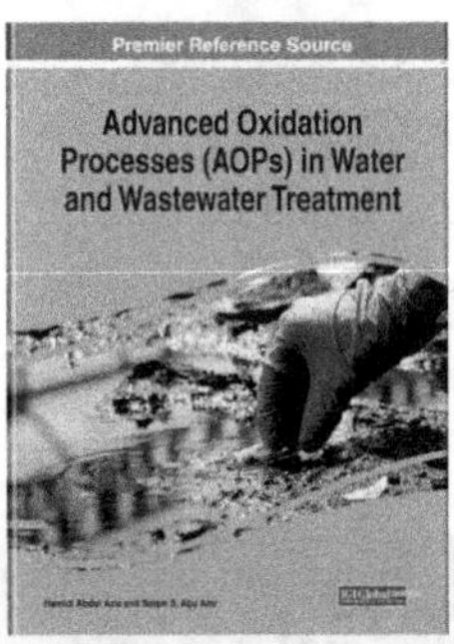

ISBN: 978-1-5225-5766-1
© 2019; 501 pp.
List Price: $195

ISBN: 978-1-5225-5909-2
© 2019; 308 pp.
List Price: $195

ISBN: 978-1-5225-8027-0
© 2019; 337 pp.
List Price: $205

Looking for free content, product updates, news, and special offers?
Join IGI Global's mailing list today and start enjoying exclusive perks sent only to IGI Global members.
Add your name to the list at **www.igi-global.com/newsletters.**

Publisher of Peer-Reviewed, Timely, and Innovative Academic Research

IGI Global
DISSEMINATOR OF KNOWLEDGE

www.igi-global.com Sign up at www.igi-global.com/newsletters facebook.com/igiglobal twitter.com/igiglobal

Ensure Quality Research is Introduced to the Academic Community

Become an IGI Global Reviewer for Authored Book Projects

The overall success of an authored book project is dependent on quality and timely reviews.

In this competitive age of scholarly publishing, constructive and timely feedback significantly expedites the turnaround time of manuscripts from submission to acceptance, allowing the publication and discovery of forward-thinking research at a much more expeditious rate. Several IGI Global authored book projects are currently seeking highly-qualified experts in the field to fill vacancies on their respective editorial review boards:

Applications and Inquiries may be sent to:
development@igi-global.com

Applicants must have a doctorate (or an equivalent degree) as well as publishing and reviewing experience. Reviewers are asked to complete the open-ended evaluation questions with as much detail as possible in a timely, collegial, and constructive manner. All reviewers' tenures run for one-year terms on the editorial review boards and are expected to complete at least three reviews per term. Upon successful completion of this term, reviewers can be considered for an additional term.

If you have a colleague that may be interested in this opportunity, we encourage you to share this information with them.

IGI Global Proudly Partners With eContent Pro International

Receive a 25% Discount on all Editorial Services

Editorial Services

IGI Global expects all final manuscripts submitted for publication to be in their final form. This means they must be reviewed, revised, and professionally copy edited prior to their final submission. Not only does this support with accelerating the publication process, but it also ensures that the highest quality scholarly work can be disseminated.

English Language Copy Editing

Let eContent Pro International's expert copy editors perform edits on your manuscript to resolve spelling, punctuaion, grammar, syntax, flow, formatting issues and more.

Scientific and Scholarly Editing

Allow colleagues in your research area to examine the content of your manuscript and provide you with valuable feedback and suggestions before submission.

Figure, Table, Chart & Equation Conversions

Do you have poor quality figures? Do you need visual elements in your manuscript created or converted? A design expert can help!

Translation

Need your documjent translated into English? eContent Pro International's expert translators are fluent in English and more than 40 different languages.

Hear What Your Colleagues are Saying About Editorial Services Supported by IGI Global

"The service was very fast, very thorough, and very helpful in ensuring our chapter meets the criteria and requirements of the book's editors. I was quite impressed and happy with your service."

– Prof. Tom Brinthaupt,
Middle Tennessee State University, USA

"I found the work actually spectacular. The editing, formatting, and other checks were very thorough. The turnaround time was great as well. I will definitely use eContent Pro in the future."

– Nickanor Amwata, Lecturer,
University of Kurdistan Hawler, Iraq

"I was impressed that it was done timely, and wherever the content was not clear for the reader, the paper was improved with better readability for the audience."

– Prof. James Chilembwe,
Mzuzu University, Malawi

Email: customerservice@econtentpro.com **www.igi-global.com/editorial-service-partners**

IGI Global
DISSEMINATOR OF KNOWLEDGE
www.igi-global.com

Celebrating Over 30 Years of Scholarly Knowledge Creation & Dissemination

InfoSci®-Books

A Database of Over 5,300+ Reference Books Containing Over 100,000+ Chapters Focusing on Emerging Research

InfoSci®-Books Database

The **InfoSci®-Books** database is a collection of over 5,300+ IGI Global single and multi-volume reference books, handbooks of research, and encyclopedias, encompassing groundbreaking research from prominent experts worldwide that span over 350+ topics in 11 core subject areas including business, computer science, education, science and engineering, social sciences and more.

Open Access Fee Waiver (Offset Model) Initiative

For any library that invests in IGI Global's InfoSci-Journals and/or InfoSci-Books databases, IGI Global will match the library's investment with a fund of equal value to go toward **subsidizing the OA article processing charges (APCs) for their students, faculty, and staff** at that institution when their work is submitted and accepted under OA into an IGI Global journal.*

INFOSCI® PLATFORM FEATURES

- No DRM
- No Set-Up or Maintenance Fees
- A Guarantee of No More Than a 5% Annual Increase
- Full-Text HTML and PDF Viewing Options
- Downloadable MARC Records
- Unlimited Simultaneous Access
- COUNTER 5 Compliant Reports
- Formatted Citations With Ability to Export to RefWorks and EasyBib
- No Embargo of Content (Research is Available Months in Advance of the Print Release)

**The fund will be offered on an annual basis and expire at the end of the subscription period. The fund would renew as the subscription is renewed for each year thereafter. The open access fees will be waived after the student, faculty, or staff's paper has been vetted and accepted into an IGI Global journal and the fund can only be used toward publishing OA in an IGI Global journal. Libraries in developing countries will have the match on their investment doubled.*

To Learn More or To Purchase This Database:
www.igi-global.com/infosci-books
eresources@igi-global.com • Toll Free: 1-866-342-6657 ext. 100 • Phone: 717-533-8845 x100

IGI Global
DISSEMINATOR OF KNOWLEDGE
www.igi-global.com

IGI Global
DISSEMINATOR OF KNOWLEDGE
www.igi-global.com

Publisher of Peer-Reviewed, Timely, and
Innovative Academic Research Since 1988

IGI Global's Transformative Open Access (OA) Model:
How to Turn Your University Library's Database Acquisitions Into a Source of OA Funding

In response to the OA movement and well in advance of Plan S, IGI Global, early last year, unveiled their OA Fee Waiver (Offset Model) Initiative.

Under this initiative, librarians who invest in IGI Global's InfoSci-Books (5,300+ reference books) and/or InfoSci-Journals (185+ scholarly journals) databases will be able to subsidize their patron's OA article processing charges (APC) when their work is submitted and accepted (after the peer review process) into an IGI Global journal.*

How Does it Work?

1. When a library subscribes or perpetually purchases IGI Global's InfoSci-Databases including InfoSci-Books (5,300+ e-books), InfoSci-Journals (185+ e-journals), and/or their discipline/subject-focused subsets, IGI Global will match the library's investment with a fund of equal value to go toward subsidizing the OA article processing charges (APCs) for their patrons.

   Researchers: Be sure to recommend the InfoSci-Books and InfoSci-Journals to take advantage of this initiative.
2. When a student, faculty, or staff member submits a paper and it is accepted (following the peer review) into one of IGI Global's 185+ scholarly journals, the author will have the option to have their paper published under a traditional publishing model or as OA.
3. When the author chooses to have their paper published under OA, IGI Global will notify them of the OA Fee Waiver (Offset Model) Initiative. If the author decides they would like to take advantage of this initiative, IGI Global will deduct the US$ 1,500 APC from the created fund.
4. This fund will be offered on an annual basis and will renew as the subscription is renewed for each year thereafter. IGI Global will manage the fund and award the APC waivers unless the librarian has a preference as to how the funds should be managed.

Hear From the Experts on This Initiative:

"I'm very happy to have been able to make one of my recent research contributions, 'Visualizing the Social Media Conversations of a National Information Technology Professional Association' featured in the *International Journal of Human Capital and Information Technology Professionals*, freely available along with having access to the valuable resources found within IGI Global's InfoSci-Journals database."

– **Prof. Stuart Palmer,**
Deakin University, Australia

For More Information, Visit: www.igi-global.com/publish/contributor-resources/open-access or contact IGI Global's Database Team at eresources@igi-global.com

www.ingramcontent.com/pod-product-compliance
Lightning Source LLC
LaVergne TN
LVHW080015210926
848481LV00011B/3026
9781799804420